THE
EXTRATERRESTRIAL
ENCYCLOPEDIA

THE EXTRATERRESTRIAL ENCYCLOPEDIA

Our Search for Life in Outer Space

Revised and Updated

By Joseph A. Angelo, Jr.

Facts On File

New York • Oxford

The Extraterrestrial Encyclopedia: Our Search for Life in Outer Space (Revised and Updated)

Copyright © 1985, 1991 by Facts On File, Inc.

Facts On File, Inc.   Facts On File Limited
460 Park Avenue South   Collins Street
New York NY 10016   Oxford OX4 1XJ
USA   United Kingdom

Library of Congress Cataloging-in-Publication Data
Angelo, Joseph A.
The extraterrestrial encyclopedia / Joseph A. Angelo, Jr—Rev
and updated.
p.   cm.
Includes bibliographical references and index.
ISBN 0-8160-2276-3
1. Life on other planets—Encyclopedias.   2. Interstellar
communication—Encyclopedias.   I. Title.
QB54.A523   1991
574.5999—dc20   90-22192

A British CIP catalogue record for this book is available from the British
Library.

Facts On File books are available at special discounts when purchased in
bulk quantities for businesses, associations, institutions or sales
promotions. Please call our Special Sales Department in New York at
212/683-2244 (dial 800/322-8755 except in NY, AK or HI) or in Oxford at 865/728399.

Jacket design by Ron Monteleone
Composition by The Maple-Vail Book Manufacturing Group
Manufactured by The Maple-Vail Book Manufacturing Group
Printed in the United States of America

10 9 8 7 6 5 4 3 2

This book is printed on acid-free paper.

This book is dedicated to
my wife, Joan,
and to my daughter, Jennifer,
on her sixteenth birthday.

Contents

Acknowledgments

This book could not have been prepared without the generous support and assistance of the National Aeronautics and Space Administration. Special acknowledgment must also be given to those individuals who were particularly helpful in the development of this book (first and/or second edition). These individuals include Mr. Les Gaver (Media Services Branch, NASA Headquarters); Mr. Edward (Larry) Noon (Jet Propulsion Laboratory); Ms. Lisa Vazquez (Media Services Corp., Johnson Space Center); Mr. Terry White (Public Information Office, Johnson Space Center); Ms. Vera Buescher (Ames Research Center); Dr. John Billingham (Ames Research Center); Dr. Bernard Oliver (Ames Research Center); Mrs. Alice Price (Air Force Art Collection); Mr. William A. Rice (Boeing Aerospace Company); Mr. Georg von Tiesenhausen (Marshall Space Flight Center); Mr. Donald Engel (Public Information Office, Eastern Space and Missile Center); Dr. Henry Robitaille (EPCOT–The LAND, Walt Disney World); Mr. David Buden (Idaho National Engineering Laboratory); Dr. Gary L. Bennett (Propulsion, Power and Energy Division, NASA Headquarters); Ms. Jane Mellors (European Space Agency–Washington, D.C. Office); Mr. Mike Gentry (Media Services Corp., Johnson Space Center); Mr. R. Joseph Sovie (Lewis Research Center); Dr. Wendell Mendell (Johnson Space Center); the men and women of the Audiovisual Branch, NASA Headquarters; and the library staff at the Florida Institute of Technology. The late Dr. Krafft Ehricke provided a special spark of imagination and excitement about the entire concept of our extraterrestrial civilization, and his eternally optimistic thoughts form a cornerstone of this book. Finally, I would like to publicly thank my family for patiently tolerating the many weekends consumed in manuscript research and development.

Introduction

Long before the start of the Space Age in 1957, technological visionaries imagined human settlements on the planets and man-made cities floating majestically in space. Although such planetary bases and large-scale space settlements are still part of the future, they came a significant step closer to realization on July 20, 1989 (the 20th anniversary of the *Apollo 11* Moon landing) when President George Bush proclaimed:

> In 1961, it took a crisis—the space race—to speed things up. Today we do not have a crisis. We have an opportunity. To seize this opportunity, I am not proposing a 10-year plan like Apollo. I am proposing a long-range, continuing commitment. First, for the coming decade—for the 1990s— Space Station *Freedom*—our critical next step in all our space endeavors. And next—for the new century—back to the Moon. Back to the future. And this time, back to stay. And then—a journey into tomorrow—a journey to another planet—a manned mission to Mars.

This permanent space station will form the core of many future space activities. In it, astronauts will live, work and even play in low Earth orbit for extended periods of time (e.g., typically six months to a year). When completed and operational, this station and the space transportation infrastructure to get us there and back to Earth will represent a modest, but permanent, human beachhead in the Universe. In response to President Bush's Space Exploration Initiative (SEI), space technologists and planners are also looking beyond low Earth orbit toward the Moon and Mars. Lunar bases, human expeditions to Mars and permanent settlements on the Red Planet are all distinct technological possibilities in the next few decades.

But why should people want to live and work in space in the first place? Perhaps the most compelling reason is because it is part of being human—it is of our very essence to want to explore. People need challenge, variety and adventure. These are the stimuli that help civilizations flourish and grow. Unfortunately, we now live at a time— the first in human history—when there are no great new land or sea frontiers on Earth to explore and conquer. Men and women have explored the depths of the oceans, climbed the peaks of the highest mountains, trekked across the most desolate, sun-scorched deserts, sailed alone around the world and journeyed to the North and South Poles. Through the use of space-based remote sensing technologies, every major region on our planet has been mapped and charted.

But while there is not much left to explore *outward* across our planet, there is an entire Universe to explore *upward*—beyond the boundaries of Earth. Space technology has arrived just in time! By its means, outer space, unreachable in all previous ages, now becomes an exciting new frontier. Space provides the human race with our greatest opportunity of all: an essentially limitless environment in which to grow and develop.

Modern advances in space technology are paving the way for the creation of our extraterrestrial civilization—a major development in the evolution of intelligent life on Earth paralleled in planetary history only by the migration of life from the ancient seas to the land eons ago. Today, we are all witnesses to the start of the migration of intelligent life from the land to the stars. As we watch (and some of us actually participate), human beings will begin to occupy first cislunar space and then heliocentric space on our march to neighboring star systems. The exciting news is that we are no longer creatures confined to a limited *"one-planet"* civilization; through space technology, we are acquiring the tools needed to both intelligently steward the precious biospheric heritage of our home planet and to respond to our overall cosmic destiny among the stars.

When we think in terms of life among the stars, we also ask the obvious question: Are we alone in this vast Universe? Is the human being the best the Universe has been able to do in the evolution of intelligent life over the past 15 to 20 billion years? If we are indeed alone, then it is our destiny to become the first intelligent species to sweep through the Galaxy spreading life, knowledge and love where there is now only emptiness. Perhaps, on the other hand, the Universe is teeming with advanced life-forms. Our emergence as a *Solar System civilization* might provide us with the levels of maturity and technology necessary to participate in any such galactic community. Either alternative is fascinating. In fact, we have now reached a level of technological sophistication and planetary development where it is the Universe or nothing! We can choose to seek our destinies among the stars, at peace with one another and with our advanced technologies, or sadly, we can elect to turn our backs on the stars and perish quietly on one tiny planet in an average star system at the fringes of the Galaxy.

This book has been written for everyone who wonders about things beyond the planet Earth, about things and beings extraterrestrial. In one volume, you will find a discussion of the major space technologies and developments that mark our search for extraterrestrial life and the initiation of our extraterrestrial civilization. One thought has especially stimulated and inspired me during the development of this book: Somewhere out there, possibly reading this very sentence, are the young man and young woman who will be the parents of the first Martians. Regarding the possibility of other intelligent life in the Universe, we can now say (at the very least): "We have met the extraterrestrials, and they are US!"

Dr. Joseph A. Angelo, Jr.
Cape Canaveral

THE
EXTRATERRESTRIAL
ENCYCLOPEDIA

A

abiotic Not involving living things; not produced by living organisms.

abundance of elements (in the Universe) Analyses of solar and stellar spectra have shown that hydrogen and helium are by far the most abundant elements in the Universe. All other elements taken together make up only some 2 percent of the mass of the Universe. The "cosmic abundance" of an element can be expressed as its percentage of the total mass found in the Universe. Using the abundances found in our own Solar System as a "cosmic standard" and incorporating other astrophysical data (such as stellar spectra), the estimated cosmic abundance of the most common materials in the Universe is:

Element	Abundance (percentage of total mass)
Hydrogen (H)	73
Helium (He)	25
Oxygen (O)	0.8
Carbon (C)	0.3
Nitrogen (N)	0.1
Neon (Ne)	<0.1
Silicon (Si)	<0.1
Magnesium (Mg)	<0.1

active galaxies At the center of some galaxies is a small highly luminous region, or active nucleus. In such a galaxy, this active nucleus emits more energy from a region about the size of our Solar System than is emitted from all the stars in that galaxy taken collectively. The emitted energy extends over a wide portion of the electromagnetic spectrum, from radio waves to X-rays. Three major types of these active galactic nuclei (AGN) have been categorized at present: quasars, BL Lac objects and Seyfert galaxies.

Quasars are the brightest type of AGN. A bright quasar can have a luminosity that is a million times that of a typical (normal) galaxy. Because of their great luminosity and distinctive optical spectra, these mysterious objects have been observed at distances in the Universe far greater than any other type of celestial object.

BL Lac objects (named after their prototype, the variable object BL Lac in the constellation Lacerta [the Lizard]) are now thought to be a particular type of quasar in which material is being shot out of the nucleus at near-light speed in a narrow beam. One intriguing characteristic of BL Lac objects is their essentially featureless spectrum, which does not display emission lines. Yet, BL Lac objects display rapid variations in brightness across the radio, infrared and visible portions of the electromagnetic spectrum.

Seyfert galaxies are now thought to be low-luminosity quasars.

Scientists have not yet identified the mechanism that produces the variable, bright luminosities of these AGNs —although many contemporary theories involve the accretion (drawing in) of matter into a massive black hole.

See also: **astrophysics; galaxy**

Advanced X-Ray Astrophysics Facility (AXAF) A planned NASA X-ray astrophysics facility designed to complement visual and radio-frequency observations made from the ground and from space-based observatories such as the Hubble Space Telescope. The basic objectives of the AXAF are to determine the positions of X-ray sources; their physical properties, such as composition and structure; and the processes involved in X-ray production. The Advanced X-ray Astrophysics Facility will study stellar structure and evolution, large-scale galactic phenomena, active galaxies, clusters of galaxies, quasars and cosmology. This facility will be a grazing incidence X-ray telescope with nested pairs of mirrors and a 9- to 12-meter focal length. It will have an overall length of approximately 13 to 16 meters and a mass of between 10,000 and 12,000 kilograms (see fig. 1). The AXAF will be placed in a 28.5-degree inclina-

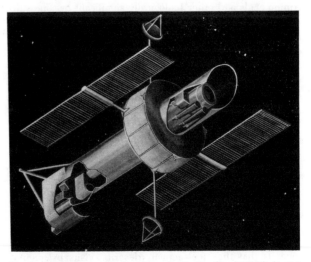

Fig. 1 The Advanced X-Ray Astrophysics Facility (AXAF). (Artist rendering courtesy of NASA).

tion, approximately 500-kilometer altitude orbit by the Space Shuttle, which will revisit it periodically for maintenance, repair or the inclusion of new experiment packages. X-ray observations are now providing basic clues to the evolution of giant systems of galaxies. However, we have only a few tantalizing clues at present, and the Advanced X-Ray Astrophysics Facility, planned for launch in late 1996, should make the picture much clearer.

See also: **astrophysics; X-ray astronomy**

albedo   The fraction of incident (that is, falling or striking) light or electromagnetic radiation that is reflected by a surface or an entire object.

alien   (In the sense used throughout this book) an extraterrestrial; an inhabitant (presumably intelligent) of another world.

alien life form (ALF)   A contemporary phrase used to describe extraterrestrial beings.

See also: **extraterrestrial civilizations**

Alpha Centauri   The star system nearest to our own Sun, approximately 4.3 light-years away. It is actually a triple star system, with two stars orbiting around each other and the third star, called Proxima Centauri, revolving around the pair at some distance. In various celestial configurations, Proxima Centauri becomes the closest known star to our Solar System—approximately 4.2 light-years away.

amino acid   An acid containing the amino (NH_2) group, a group of molecules necessary for life. More than 80 amino acids are presently known, but only some 20 occur naturally in living organisms, where they serve as the building blocks of proteins. Amino acids have been synthesized nonbiologically under conditions that simulate those that may have existed on the primitive Earth, followed by the synthesis of most of the biologically important molecules. Amino acids and other biologically significant organic substances have been found to occur naturally in meteorites and are not considered to have been produced by living organisms. Meteorites, chunks of extraterrestrial matter arriving on Earth, represent a very interesting source of information about the occurrence of prebiotic (prelife) chemistry beyond the Earth. Recent investigations have shown that many of the molecules considered necessary for life are also present in meteorites. As a result of these observations, it appears that the chemistry of life is not unique to the planet Earth itself.

See also: **exobiology; life in the Universe**

ancient astronauts   Was the Earth visited in the past by extraterrestrial explorers, by powerful intelligent beings from another planet or even from the stars?

Some people think so. As evidence, they may point to legends that they feel actually describe visits by ancient astronauts. Many of the world's cultures, for instance, have traditional tales of beings with superhuman powers who helped early humans.

Take the story of how prehistoric people first learned to use fire, which comes to us from Greek mythology. Prometheus, whose name means "forethought," and his brother Epimetheus, whose name means "afterthought," were members of a group of giant gods called Titans. It was Epimetheus's task to give the animals of the Earth the powers they needed to survive. But he gave them so many powers that when it came time to help humans, he had no gift left to give.

Prometheus, seeing the helplessness of humans, took pity on them. A far wiser god than his brother, he stole the gift of fire from the other gods and brought it down from the heavens to give humankind. With it, people could light small fires at the entrance of their caves and protect themselves from dangerous prowling animals. And with it, eventually, they could develop a technology that would make them the supreme animals on Earth.

Zeus, the king of the gods, was greatly angered at Prometheus's foresight. He had him chained to a rock, says the legend, and sent an eagle each day to tear at his liver. Prometheus suffered for thousands of years until Zeus's anger finally subsided. Then Hercules killed the eagle and freed Prometheus from his chains.

On one level, a legend like this seems to offer the tantalizing suggestion that gods such as the Titans were really ancient astronauts who helped early humans. Unfortunately, there is no evidence whatsoever to show that such legends are based on historic fact.

There are other theories about ancient astronauts, too. The major claim is that extraterrestrials affected the biological development of primitive apelike creatures and caused them to become intelligent human beings. How else, the reasoning goes, could primitive humans have produced advanced civilizations such as those that arose in ancient Egypt and Central America? This idea is sometimes offered as a complement to the two traditional theories of human origins. The first, the creationist theory, holds that God created humans and all the other animals as unique and separate species. The second, the evolutionist theory, holds that animals evolved, or developed, from earlier species; in the case of humans, from apelike ancestors.

There are several variations on the idea that human development was affected by extraterrestrials. One version has it that a migratory wave of ancient astronauts visited the Earth, planted the seeds of civilization among the primitive creatures they found on the planet and then continued on to the edge of the Galaxy.

Another version of the theory suggests that extraterrestrial visitors performed some form of genetic engineering on apes or even primitive humans. The result, it suggests, was the evolution of human intelligence and the emergence of modern humans.

A third variation on the theme suggests that humans themselves arose on another planet. Traveling from a distant homeland, some of these early humans landed on

Earth and established a flourishing civilization. Then suddenly this advanced civilization was devastated by some kind of cosmic catastrophe. Right now, the theory goes, we are rebuilding civilization on the ruins of that ancient society.

Such ideas have given rise to many popular books and movies that seek to link the achievements of ancient human civilizations with visits from extraterrestrials. But what is the proof? Where is the scientific evidence? It rests, insist people who believe in the theory, on the existence of enormous and astounding ancient architectural structures that they believe "primitive" humans could not have built alone.

The main examples offered are Egypt's spectacular pyramids, built beginning around 2650 B.C. as tombs for the pharaohs, or kings. The largest, the Great Pyramid, was one of the Seven Wonders of the World. Its base alone covers some 13 acres (5.25 hectares), an area large enough to hold 10 football fields. Rising originally to a height of 481 feet (147 meters), it is built of more than 2 million stone blocks, each cut to fit snugly with the next, and each weighing an average of 5,000 pounds (2,270 kg).

How, indeed, did primitive humans achieve such a colossal feat? Well, in fact, they didn't—simply because people of this period were *not* primitive, at least not in the sense of lacking intelligence in the first place. All archaeological evidence shows that their brains, both in form and function, were exactly the same as our own. Geologists have been able to construct a picture of Earth's history that spans billions of years, and humans' part in that history goes back not the mere 5,000 years to the time of ancient Egypt, but back some 3½ million years or more. Ancient Egyptians were modern people: curious, ingenious and intelligent, just like their 20th-century descendants.

It is true, of course, that their technology was crude and primitive by today's standards; they had no machinery or iron tools, for instance. Nevertheless, they were capable of building the pyramids with the technology available at the time. They were able to cut the stone blocks with copper chisels and saws, drag them from the quarries to the site on rope-bound sledges, then haul them into position up ramps made of earth, wood and sun-dried brick.

Then, too, a pharaoh had enormous manpower at his disposal. Herodotus, the Greek historian, said 400,000 workers labored for 20 years to build the Great Pyramid. They worked on the tomb during the annual flood, when the river Nile overflowed its banks onto the nearby fields and made farming impossible.

The apparently sophisticated practices of early astronomers are also cited as a reason for believing in the intervention of ancient astronauts. But again, early astronomers did have tools. Working with a transit (a pole or tube with which to mark the passage of celestial objects across a meridian), they could have become quite practiced in observing the heavens. The astronomers of ancient Egypt, for example, discovered that the seasonal flooding of the Nile River corresponded with the time when the star Sirius first became visible in the morning sky. As a result, they were able to create a very accurate calendar, consisting of 365.25 days.

Perhaps the most solid proof against the ancient-astronaut theory is the sheer lack of evidence. Almost two centuries of serious archaeological and geological exploration of our planet have failed to yield the slightest shred of credible evidence that ancient astronauts once walked with early humans. Archaeologists have, on the other hand, uncovered thousands of purely terrestrial ancient tombs, cities and campsites.

Where, then, one might ask, are the extraterrestrial visitors' former landing sites, installations and exploratory equipment? Any starfaring civilization would surely have left behind at least one or two samples of their sophisticated technology. Did we of Earth not leave our footprints and exploratory equipment on the Moon? Are not two Viking craft even now silently guarding the Martian landscape?

See also: **unidentified flying object**

Androcell An innovative and bold concept, proposed by Dr. Krafft A. Ehricke, involving a human-made new world that is an independent self-contained human biosphere not located on any naturally existing celestial object. These human-made miniworlds, or planetellas, would use mass far more effectively than the natural worlds of our Solar System, which formed out of the original nebular material. For example, naturally formed celestial bodies are essentially "solid" spherical objects of great mass. Their surface gravity forces result from the attraction of a large quantity of matter. However, except for the first kilometer or so, the interior of these natural worlds is essentially "useless" from a human exploitation and habitation point of view.

Instead of large quantities of matter, the Androcell would use rotation (centrifugal inertia) to provide variable levels of artificial gravity. The unutilizable solid interior of a natural celestial body is now replaced (through human ingenuity) with many useful, inhabitable layers of cylinders. Therefore, Androcell inhabitants would be able to enjoy a truly variable life-style in a multi-gravitational-level world. There would be a maximum gravity level at the outer edges of the Androcell, tapering off to essentially zero gravity in the inner cylinder levels closest to the central hub.

The Androcell will not be tied to the Earth-Moon system but rather—with its giant space-based factories, farms and fleets of merchant spacecraft—will be free to seek political and economic development throughout heliocentric (Sun-centered) space. Its inhabitants might trade with Earth, the Moon, Mars or other Androcells. These giant space settlements of 10,000 to perhaps 100,000 or more inhabitants are most analogous to the city-state of ancient Greece. The multi-gravity-level life-style would encourage migration to and from other worlds—perhaps a terraformed Mars or subdued Venus, or maybe even one of the moons of the giant outer planets. In essence, the Androcell represents the "cellular division" of humanity—since as autonomous extraterrestrial city-states, their inhabitants could choose to pursue culturally diverse life-styles.

Of course, we already have our initial, natural Andro-cell—we call it "Spaceship Earth." In time, inhabitants of our parent world will be able to use their technical skills and human intelligence to fashion a series of such Andro-cells, or large space settlements, throughout the Solar System. As the number of such artificial human habitats grows, a swarm of settlements might eventually encircle the Sun, capturing and using its entire energy output. At that point our extraterrestrial civilization will have created a Dyson sphere, and the next stage of cosmic mytosis, migration to the stars, would occur.

See also: **Dyson sphere; extraterrestrial civilizations; space settlement**

android A term from science fiction describing a robot with near-human form or features.

See also: **cyborg**

Andromeda Galaxy The Great Spiral Galaxy in Andromeda is our neighboring galaxy and the most distant object visible to the naked eye of an observer on Earth. It is as large as our own Milky Way, or larger, and is some 2.2 million light-years (670 kiloparsecs) away. The Andromeda Galaxy is also called M31.

See also: **astrophysics; stars**

androsphere A term, developed by the late Dr. Krafft A. Ehricke, that describes the synthesis of the terrestrial and extraterrestrial environments. It relates to our productive integration of the Earth's biosphere, which contains the major terrestrial environmental regimes, and the material and energy resources of the Solar System, such as the Sun's radiant energy and the Moon's mineral resources.

angstrom [symbol: Å] A unit of length commonly used to measure wavelengths of electromagnetic radiation in the visible, near infrared and near ultraviolet portions of the spectrum.

$$1 \text{ angstrom (Å)} = 10^{-10} \text{ meter} = 0.1 \text{ nanometer}$$

This unit is named after Anders Jonas Ångstrom (1814–74), a Swedish physicist, who quantitatively described the solar spectrum in 1868.

annihilation The conversion of a particle and its corresponding antiparticle into electromagnetic radiation (annihilation radiation) upon collision. This annihilation radiation has a minimum energy equivalent to the resting masses (m_o) of the two colliding particles. For example, when a positron and an electron collide, the minimum annihilation radiation consists of a pair of gamma rays, each of 0.511-million electron volts (MeV) energy. The energy of annihilation is derived from the mass of the disappearing particles according to the famous Einstein mass-energy equivalence formula, $E = m_o c^2$.

See also: **antimatter**

antigalaxy A galaxy composed of antimatter.

See also: **antimatter; antimatter cosmology**

antimatter Matter in which the ordinary nuclear particles (such as neutrons, protons and electrons) are conceived of as being replaced by their corresponding antiparticles—that is, antineutrons, antiprotons, positrons and so on. For example, an "antiparticle hydrogen atom" would consist of a negatively charged antiproton with an orbital positron (positively charged electron). Normal matter and antimatter mutually annihilate each other upon contact and are converted into pure energy, called annihilation radiation. Although individual antiparticles have been discovered and their behavior observed in laboratories, bulk quantities of antimatter have yet to be found in the Universe.

See also: **annihilation**

antimatter cosmology A cosmological model proposed by the Swedish scientists Alfvén and Klein as an alternate to the Big Bang model. In their model the early Universe is assumed to consist of a huge, spherical cloud, called a metagalaxy, containing equal amounts of matter and antimatter. As this cloud collapsed under the influence of gravity, its density increased and a condition was reached in which matter and antimatter collided—producing large quantities of annihilation radiation. The radiation pressure from the annihilation process caused the Universe to stop collapsing and to expand. In time, clouds of matter and antimatter formed into equivalent numbers of galaxies and antigalaxies. (An antigalaxy is a galaxy composed of antimatter.)

There are many technical difficulties with the Alfvén-Klein cosmological model. For example, no observational evidence has yet been obtained of large quantities of antimatter existing in the Universe. If these antigalaxies existed, large quantities of annihilation (gamma-ray) radiation would certainly be emitted at the interface points between the matter and antimatter regions of our Universe.

See also: **antimatter; astrophysics; "Big Bang" theory; cosmology**

antiparticle Every elementary particle has a corresponding real (or hypothetical) antiparticle, which has equal mass but opposite electric charge (or other property, as in the case of the neutron and antineutron). The antiparticle of the electron is the positron; of the proton, the antiproton; and so on. However, the photon is its own antiparticle. When a particle and its corresponding antiparticle collide, they are converted into energy in a process called annihilation.

See also: **annihilation; antimatter**

aperture synthesis The use of a variable-aperture radio interferometer to mimic the "full dish" of a huge equivalent radio telescope.

See also: **radio astronomy**

arc-minute   one/sixtieth of a degree of angle. This unit is associated with precise measurements of motions and positions of celestial objects as occurs in the science of astrometry.

$$1° = 60 \text{ arc-min} = 60'$$

See also: **arc-second; astrometry**

arc-second   one/three thousand six hundredth of a degree of angle. This unit is associated with very precise measurements of stellar motions and positions in the science of astrometry.

$$1' \ (2 \text{arc-min}) = 60 \text{ arc-sec} = 60''$$

See also: **arc-minute; astrometry**

Arecibo Interstellar Message   To help inaugurate the powerful radio/radar telescope of the Arecibo Observatory in the tropical jungles of Puerto Rico, an interstellar message of friendship was beamed to the fringes of the Milky Way Galaxy. On November 16, 1974, this interstellar radio signal was transmitted toward the Great Cluster in Hercules (Messier 13 or M13, for short), which lies about 25,000 light-years away from Earth. The globular cluster M13 contains about 300,000 stars within a radius of approximately 18 light-years.

```
0 0 0 0 0 0 1 0 1 0 1 0 1 0 0 0 0 0 0 0 0 0 0 0 1 0 1 0 0 0 0 0 1 0 1 0
0 0 0 0 0 1 0 0 1 0 0 0 1 0 0 0 1 0 0 0 1 0 0 1 0 1 1 0 0 1 0 1 0 1 0 1
0 1 0 1 0 1 0 1 0 1 0 0 1 0 0 1 0 0 0 0 0 0 0 0 0 0 0 0 0 0 0 0 0 0 0 0
0 0 0 0 0 0 0 0 0 0 0 0 0 0 1 1 0 0 0 0 0 0 0 0 0 0 0 0 0 0 0 0 0 0 0 0
1 1 0 1 0 0 0 0 0 0 0 0 0 0 0 0 0 0 0 0 0 1 1 0 1 0 0 0 0 0 0 0 0 0 0 0
0 0 0 0 0 0 0 0 1 0 1 0 1 0 0 0 0 0 0 0 0 0 0 0 0 0 0 0 1 1 1 1 1 0
0 0 0 0 0 0 0 0 0 0 0 0 0 0 0 0 0 0 0 0 0 0 0 0 0 0 0 1 1 0 0 1 0
1 1 1 0 0 0 1 1 0 0 0 0 1 1 0 0 1 0 0 0 0 0 0 0 0 0 0 0 1 1 0 0 1 0
0 0 0 1 1 0 1 0 0 0 1 1 0 0 0 1 1 0 0 0 0 1 1 0 1 0 1 1 1 1 0 1 1 1 1 1
0 1 1 1 1 0 1 1 1 1 1 0 0 0 0 0 0 0 0 0 0 0 0 0 0 0 0 0 0 0 0 0 0 0 0
0 1 0 0 0 0 0 0 0 0 0 0 1 0 0 0 0 0 0 0 0 0 0 0 0 0 0 0 0 0 0 0 0 0 0 0
0 0 0 0 0 0 0 0 0 0 1 0 0 0 0 0 0 0 0 0 0 0 0 0 0 0 0 1 1 1 1 1 0 0
0 0 0 0 0 0 0 0 0 1 1 1 1 1 0 0 0 0 0 0 0 0 0 0 0 0 0 0 0 0 0 0 0 0
0 0 1 1 0 0 0 0 1 1 0 0 0 0 1 1 1 0 0 0 1 1 0 0 0 1 0 0 0 0 0 0 1 0 0 0
0 0 0 0 0 0 1 0 0 0 0 1 1 0 1 0 0 0 0 0 1 1 0 0 0 1 1 1 0 0 1 1 0 1 0 1 1 1
1 1 0 1 1 1 1 1 0 1 1 1 1 1 0 1 1 1 1 0 0 0 0 0 0 0 0 0 0 0 0 0 0 0
0 0 0 0 0 0 0 0 1 0 0 0 0 0 1 1 0 0 0 0 0 0 0 0 0 0 1 0 0 0 0 0 0 0
0 0 1 1 0 0 0 0 0 0 0 0 0 0 0 0 0 1 0 0 0 0 1 1 0 0 0 0 0 0 0 0 0 0
1 1 1 1 1 0 0 0 0 1 1 0 0 0 0 0 1 1 1 1 0 0 0 0 0 0 0 0 0 0 0 1 1 0
0 0 0 0 0 0 0 0 0 0 1 0 0 0 0 0 0 0 1 0 0 0 0 0 0 0 0 0 1 0 0 0 0 0 1
0 0 0 0 0 1 1 0 0 0 0 0 0 1 0 0 0 0 0 0 0 1 1 0 0 0 0 1 1 0 0 0 0 0
1 0 0 0 0 0 0 0 0 0 1 1 0 0 0 1 0 0 0 1 0 0 0 0 1 0 0 0 0 0 0 0 0 0 0
0 1 1 0 0 1 1 0 0 0 0 0 0 0 0 1 1 0 0 0 1 0 0 0 1 1 0 0 0 0
0 0 0 0 1 1 0 0 0 0 1 1 0 0 0 0 0 1 0 0 0 0 0 0 0 1 0 0 0 0 0 1 0 0 0
0 0 0 0 0 1 0 0 0 0 0 1 0 0 0 0 0 1 1 0 0 0 0 0 0 0 1 0 0 0 1 0 0 1 0 0 0
0 0 0 0 0 1 1 0 0 0 0 0 0 1 0 0 0 1 0 0 0 0 0 0 0 0 1 0 0 0 0 0 0 0
1 0 0 0 0 0 1 0 0 0 0 0 0 0 1 0 0 0 0 0 0 1 0 0 0 0 0 1 0 0 0 0 0 0 0
0 0 0 0 0 0 1 1 0 0 0 0 0 0 0 1 1 0 0 0 0 0 0 1 1 0 0 0 0 0 0 0 0 0
0 1 0 0 0 1 1 1 0 1 0 1 0 1 0 0 0 0 0 0 0 0 1 0 0 0 0 0 0 0 0 1 0 0 0 0
0 0 0 0 0 0 0 0 0 1 0 0 0 0 1 1 1 1 0 0 0 0 0 0 0 0 0 0 0 0 1 0 0 0
0 1 0 1 1 1 0 1 0 0 1 0 1 1 0 1 1 0 0 0 0 0 1 0 0 1 1 1 0 0 1 0 0 1 1 1
1 1 1 1 0 1 1 1 0 0 0 0 1 1 0 0 0 0 1 1 0 1 1 1 0 0 0 0 0 0 0 0 0 1 0
1 0 0 0 0 0 1 1 1 0 1 1 0 0 1 0 0 0 0 0 0 1 0 1 0 0 0 0 0 0 1 1 1 1 1 0 0
1 0 0 0 0 0 0 1 0 1 0 0 0 0 0 1 1 0 0 0 0 0 1 0 0 0 0 0 0 1 1 0 1 1 0 0 0
0 0 0 0 0 0 0 0 0 0 0 0 0 0 0 1 1 1 0 1 0 1 0 0 0 1 0 1 0 1 0 1 0 1
0 0 0 1 0 0 0 0 0 0 0 0 0 0 0 0 0 1 1 1 0 1 0 1 0 0 0 1 0 1 0 1 0 1 0 1
0 1 0 0 1 1 1 0 0 0 0 0 0 0 0 0 1 0 1 0 1 0 0 0 0 0 0 0 0 0 0 0 0 0
0 0 1 0 1 0 0 0 0 0 0 0 0 0 0 0 0 1 1 1 1 0 0 0 0 0 0 0 0 0 0 0 0 0
0 0 0 1 1 1 1 1 1 1 0 0 0 0 0 0 0 0 0 0 0 1 1 1 0 0 0 0 0 0 0 0 1 1
0 0 0 0 0 0 0 1 1 0 0 0 0 0 0 0 0 0 1 1 0 0 0 0 0 0 1 1 0 1 0 0
0 0 0 0 0 0 1 0 1 1 0 0 0 0 1 1 0 0 1 1 0 0 0 0 0 0 0 0 0 1 1 0 1 0 0 0
0 0 1 0 0 0 1 0 1 0 0 0 0 0 0 1 0 1 0 0 1 0 0 0 1 0 0 0 1 0 0 1 0 0 1
0 0 1 0 0 0 1 0 1 0 0 0 0 0 0 0 1 0 0 1 0 1 0 0 0 1 0 0 0 0 0 0 0 0 0
0 1 0 0 0 0 1 0 0 0 0 1 0 0 0 0 0 0 0 1 0 0 0 0 0 0 1 0 0 0 0 0 0 1 0 0
0 0 0 0 0 0 0 0 0 1 0 0 1 0 1 0 0 0 0 0 0 0 0 0 0 0 1 1 1 1 0 0 1 1
1 1 1 0 1 0 0 1 1 1 1 0 0 0
```

Fig. 1   The Arecibo Message of 1974 in binary notation. (Courtesy of Frank D. Drake and the staff of the National Astronomy & Ionosphere Center, which is operated by Cornell University under contract with the National Science Foundation.)

This transmission, often called the Arecibo Interstellar Message, was made at the 2380 megahertz radio frequency with a 10 hertz bandwidth. The average effective radiated power was 3×10^{12} watts (3 terawatts) in the direction of transmission. The signal is considered to be the strongest radio signal yet beamed out into space by our planetary civilization. Perhaps 25,000 years from now a radio telescope operated by members of an intelligent alien civilization somewhere in the M13 cluster will receive and decode this interesting signal. If they do, they will learn that intelligent life has evolved here on Earth!

The Arecibo Interstellar Message of 1974 consisted of 1679 consecutive characters. It was written in a binary format—that is, only two different characters were used. As shown in figure 1, the characters can be denoted as "0" and "1". In the actual transmission, each character was represented by one of two specific radio frequencies and the message was transmitted by shifting the frequency of the Arecibo Observatory's radio transmitter between these two radio frequencies in accordance with the plan of the message.

The message itself was constructed by the staff of the National Astronomy and Ionosphere Center (NAIC). It can be decoded by breaking up the message into 73 consecutive groups of 23 characters each and then arranging these groups in sequence one under the other. The numbers, 73 and 23, are prime numbers. Their use should facilitate the discovery by any alien civilization receiving the message, that the above format is the right way to interpret the message. Figure 2 shows the decoded message: The first character transmitted (or received) is located in the upper right hand corner.

This message describes some of the characteristics of terrestrial life that the scientific staff at the National Astronomy and Ionosphere Center felt would be of particular interest and technical relevance to an extraterrestrial civilization. The NAIC staff interpretation of the interstellar message is as follows.

The Arecibo message begins with a "lesson" that describes the number system being used. This number system is the binary system, where numbers are written in powers of two (2) rather than of ten (10) as in the decimal system used in everyday life. Staff scientists believe that the binary system is one of the simplest number systems. It is also particularly easy to code in a simple message. Written across the top of the message (from right to left) are the numbers one (1) through ten (10) in binary notation. Each number is marked with a "number label"—that is, a single character, which denotes the start of a number.

The next block of information sent in the message occurs just below the numbers. It is recognizable as five numbers. From right to left these numbers are: 1, 6, 7, 8, and 15. This otherwise unlikely sequence of numbers should eventually be interpreted as the atomic numbers of the elements hydrogen, carbon, nitrogen, oxygen and phosphorus.

Next in the message are twelve groups on lines 12 through 30 that are similar groups of five numbers. Each of these groups represents the chemical formula of a mol-

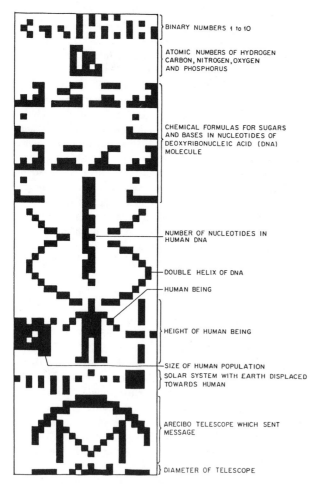

BINARY NUMBERS 1 to 10

ATOMIC NUMBERS OF HYDROGEN CARBON, NITROGEN, OXYGEN AND PHOSPHORUS

CHEMICAL FORMULAS FOR SUGARS AND BASES IN NUCLEOTIDES OF DEOXYRIBONUCLEIC ACID (DNA) MOLECULE

NUMBER OF NUCLEOTIDES IN HUMAN DNA

DOUBLE HELIX OF DNA

HUMAN BEING

HEIGHT OF HUMAN BEING

SIZE OF HUMAN POPULATION
SOLAR SYSTEM WITH EARTH DISPLACED TOWARDS HUMAN

ARECIBO TELESCOPE WHICH SENT MESSAGE

DIAMETER OF TELESCOPE

Fig. 2   Decoded form of the Arecibo Interstellar Message. (Courtesy of Frank D. Drake and the staff of the National Astronomy & Ionosphere Center, which is operated by Cornell University under contract with the National Science Foundation.)

ecule or radical. The numbers from right to left in each case provide the number of atoms of hydrogen, carbon, nitrogen, oxygen, and phosphorus, respectively, that are present in the molecule or radical.

Since the limitations of the message did not permit a description of the physical structure of the radicals and molecules, the simple chemical formulas do not define in all cases the precise identity of the radical or molecule. However, these structures are arranged as they are organized within the macromolecule described in the message. Intelligent alien organic chemists somewhere in the M13 cluster should eventually be able to arrive at a unique solution for the molecular structures being described in the message.

The most specific of these structures, and perhaps the one that should point the way to correctly interpreting the others, is the molecular structure that appears four times on lines 17 through 20 and lines 27 through 30. This is a structure containing one phosphorus atom and four oxygen

atoms, the well-known phosphate group. The outer structures on lines 12 through 15 and lines 22 through 25 give the formula for a sugar molecule, deoxyribose. The two sugar molecules on lines 12 through 15 have between them two structures: the chemical formulas for thymine (left structure) and adenine (right structure). Similarly, the molecules between the sugar molecules on lines 22 through 25 are: guanine (on the left) and cytosine (on the right).

The macromolecule or overall chemical structure is that of deoxyribonucleic acid (DNA). The DNA molecule contains the genetic information that controls the form, living processes, and behavior of all terrestrial life. This structure is actually wound as a double helix, as depicted in lines 32 through 46 of the message. The complexity and degree of development of intelligent life on Earth is described by the number of characters in the genetic code, that is, by the number of adenine-thymine and guanine-cytosine combinations in the DNA molecule. The fact that there are some four billion such pairs in human DNA is illustrated in the message by the number given in the center of the double helix between lines 27 and 43. Please note that the "number label" is used here to establish this portion of the message as a number and to show where the number begins.

The double helix leads to the "head" in a crude sketch of a human being. The scientists who composed the message hoped that this would establish a connection between the DNA molecule, the size of the helix, and the presence of an "intelligent" creature. To the right of the sketch of a human being is a line that extends from the head to the feet of our "message human." This line is accompanied by the number 14. This portion of the message is intended to convey the fact that the "creature" drawn is 14 units of length in size. The only possible unit of length associated with the message is the wavelength of the transmission, namely 12.6 centimeters. This makes the creature in the message 176 centimeters or about five feet nine inches tall. To the left of the human being is a number, approximately four billion. This number represents the current human population on the planet Earth.

Below the sketch of the human being is a representation of our Solar System. The Sun is at the right, followed by nine planets with some coarse representation of relative sizes. The third planet, Earth, is displaced to indicate that there is something special about it. In fact, it is displaced towards the drawing of the human being, who is centered on it. Hopefully, an extraterrestrial scientist in pondering this message will recognize that Earth is the home of the intelligent creatures that sent it.

Below the Solar System and centered on the third planet is an image of a telescope. The concept of "telescope" is described by showing a device that directs rays to a point. The mathematical curve leading to such a diversion of paths is crudely indicated. The telescope is not upside down, but rather "up" with respect to the symbol for the planet Earth.

At the very end of the message (bottom of figure 2), the size of the telescope is indicated. Here, it is both the size of the largest radio telescope on Earth and also the size of the telescope that sent the message (namely, the Arecibo telescope). It is shown as 2,430 wavelengths across, or roughly 305 meters. (No one, of course, expects an alien civilization to have the same unit system we use here on Earth—but physical quantities, such as the wavelength of transmission, provide a common reference frame.)

This interstellar message was transmitted at a rate ten characters per second and it took 169 seconds to transmit the entire information package. It is interesting to realize, that just one minute after completion of transmission, our interstellar greetings passed the orbit of Mars. After 35 minutes, the message passed the orbit of Jupiter; and after 71 minutes it silently crossed the orbit of Saturn. Some five hours and 20 minutes after transmission, the message passed the orbit of Pluto, leaving the Solar System and entering interstellar space. It will be detectable by telescopes anywhere in our Galaxy of approximately the same size and capability as the Arecibo facility that sent it!

If you had to prepare a message to the stars, what type of information would you decide to transmit?

See also: **Arecibo Observatory; interstellar communication**

Arecibo Observatory The Arecibo Observatory, the world's largest radio/radar telescope, is located in the lush tropical jungles of Puerto Rico. (See fig. 1.) It is the main observing instrument of the National Astronomy and Ionosphere Center, a national center for radio and radar astronomy and ionospheric physics operated by Cornell University under contract with the National Science Foundation. The 305-meter (1,000-foot) dish of the observatory's giant telescope fills a spherical bowl that was naturally formed by the collapse of huge limestone caves.

The enormity of the giant concrete, steel and aluminum structure provides an interesting contrast to the surrounding green carpet of lush tropical jungle. Yet, a certain harmony or synthesis is also present. The giant radio telescope actually uses nature to help explore nature. This observatory was located in the tropics because the Moon and the planets pass nearly overhead at the lower latitudes. Astronomers depend on the remote setting and the surrounding hills to reduce radio-wave interference. The enormous telescope itself (three football fields across) sits, as mentioned, in a natural limestone sinkhole, the use of which avoided very expensive excavation costs during construction. Wild tropical vegetation under the bowl prevents erosion of the underlying terrain. The minimal annual temperature variations enhance structural stability.

When it operates as a radio receiver, the large Arecibo telescope listens for signals from celestial objects at the farthest reaches of the Universe. As a radar transmitter/receiver, it assists astronomers and planetary scientists by bouncing signals off the Moon, off nearby planets and their satellites, or even off the layers in the Earth's ionosphere.

The white steel triangle suspended over the giant dish (see fig. 1) is a support structure for the equipment that receives and amplifies the radio waves collected from space.

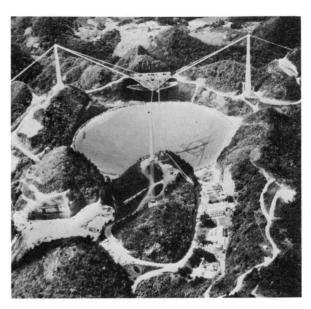

Fig. 1 The Arecibo radio/radar observatory is the largest facility of this kind on the planet Earth. It is located in the tropical jungles of Puerto Rico. (Courtesy of National Astronomy and Ionosphere Center.)

Once amplified, these signals are then sent to a control building on the ground below for data processing and evaluation.

Most radio telescopes have a steerable dish or reflector that collects the incoming radio signals. At the Arecibo Observatory, the giant reflector dish lies immobile in the Earth, while the receiving equipment (all 600 tons of it) hangs suspended some 130 meters (426 ft) above it. This receiving equipment is steered and pointed with the assistance of remote-control devices. In fact, the entire giant observatory has been so designed that a single experimenter can operate it alone.

There is a circular railroad track, winding around the inside of the white steel triangle, for steering the feed arm. The feed arm is the large section that hangs down directly below the steel triangle. (See fig. 1.) Approximately 100 meters long, the feed arm directs the observatory's transmitting and receiving equipment to various part of the sky. The two carriage houses (containing the transmitting and receiving equipment) can be moved to selected positions along the curved underside of the feed arm by means of a second railroad track. In this way, astronomers can "point" the Arecibo telescope through a combination of feed-arm and carriage-house movements.

The long structures that project from the carriage houses are the telescope's antennas. Their shapes are determined by the reflecting dish's spherical configuration; that is, radio signals hitting the dish are reflected into a "focal line" rather than a focal point. The largest of these "line feeds" is 29 meters.

In addition to receiving radio signals collected by the giant reflector dish, this feed can also direct radio energy down on the reflector for transmitting signals into space. The Arecibo Observatory has four radar transmitters: the 3–15-megahertz (600 kilowatt) transmitter for ionospheric investigations; the 50-megahertz (1 kilowatt) transmitter for ionospheric and lunar studies; the 430-megahertz (160 kilowatt) transmitter for lunar and planetary studies; and the 2,380-megahertz (S-band 400 kW) transmitter for planetary studies.

The giant Arecibo telescope can also play a role in our search for extraterrestrial intelligence (SETI). It can not only listen for interstellar radio signals from intelligent alien civilizations; it can also be used as a transmitter to send our own messages to the stars. In fact, the 400,000-watt radar output of the observatory's S-band transmitter, when concentrated into a narrow beam by the giant reflector, has an effective power 100 times greater than the total electric power production of all the world's generating plants. In other words, the Arecibo telescope can transmit the strongest signal now leaving the Earth. This human-made radio-frequency "beacon" is actually powerful enough to be detected by similar radio observatories located anywhere in the Galaxy! Just think, alien radio astronomers may someday be searching our region of the Milky Way when suddenly they detect an Arecibo transmission that "shines" with a (radio-frequency) brilliance 10 billion times stronger than the Sun. In 1974 a very special message was sent to the stars from this facility.

See also: **Arecibo Interstellar Message; interstellar communication; radar astronomy; radio astronomy; Very Large Array**

artificial intelligence (AI)   A term commonly taken to mean the study of thinking and perceiving as general information-processing functions—or the science of machine intelligence (MI). In the past few decades, computer systems have been programmed to diagnose diseases; prove theorems; analyze electronic circuits; play complex games such as chess, poker and backgammon; solve differential equations; assemble mechanical equipment using robotic manipulator arms and end effectors (the "hands" at the end of the manipulator arms); pilot unmanned vehicles across complex terrestrial terrain, as well as through the vast reaches of interplanetary space; analyze the structure of complex organic molecules; understand human speech patterns; and even write other computer programs. All of these computer-accomplished functions require a degree of "intelligence" similar to mental activities performed by the human brain. Someday, a general theory of intelligence may emerge from the current efforts of scientists and engineers who are now engaged in the field of artificial intelligence. This general theory would help guide the design and development of even "smarter" robot spacecraft and exploratory probes, allowing us to more fully explore and use the resources that await us throughout the Solar System.

Artificial intelligence generally includes a number of elements or subdisciplines. Some of these are: planning and problem solving; perception; natural language; expert systems; automation, teleoperation and robotics; distributed data management; and cognition and learning. Each of these AI subdisciplines will now be discussed briefly.

All artificial intelligence involves elements of planning and problem solving. The problem-solving function implies a wide range of tasks, including decision making, optimization, dynamic resource allocation and many other calculations or logical operations.

Perception is the process of obtaining data from one or more sensors and processing or analyzing these data to assist in making some subsequent decision or taking some subsequent action. The basic problem in perception is to extract from a large amount of (remotely) sensed data some feature or characteristic that then permits object identification.

One of the most challenging problems in the evolution of the digital computer has been the communication that must occur between the human operator and the machine. The human operator would like to use an everyday, or natural, language to gain access to the computer system. The process of communication between machines and people is very complex and frequently requires sophisticated computer hardware and software.

An expert system permits the scientific or technical expertise of a particular human being to be stored in a computer for subsequent use by other human beings who have not had the equivalent professional or technical experience. These expert systems have been developed for use in such diverse fields as medical diagnosis, mineral exploration and mathematical problem solving. To create such an expert system, a team of software specialists will collaborate with a scientific expert to construct a computer-based interactive dialogue system that is capable, at least to some extent, of making the expert's professional knowledge and experience available to other individuals. In this case, the computer, or "thinking machine," not only stores the scientific or professional expertise of one human being but also permits ready access to this valuable knowledge base because of its artificial intelligence, which guides other human users.

Automatic devices are those that operate without direct human control. NASA has used many such automated smart machines to explore alien worlds. For example, the two Viking Landers placed on the Martian surface in 1976 represent one of the great triumphs of robotic space exploration. After separation from the Viking Orbiter spacecraft, the lander (protected by an aeroshell) descended into the thin Martian atmosphere at speeds of approximately 16,000 kilometers per hour (10,000 miles per hour). It was slowed down by aerodynamic drag until its aeroshell was discarded. The lander slowed down further by releasing parachutes and finally achieved a gentle landing by automatically firing retro-rockets. This entire sequence was successfully accomplished automatically by both Viking Landers.

Teleoperation implies that a human operator is in remote control of a mechanical system. Control signals can be sent by means of "hardwire" (if the device under control is nearby) or via electromagnetic signals (for example, laser or radio frequency) if the robot system is some distance away. Of course, in dealing with great distances in interplanetary exploration, a situation is eventually reached when even electromagnetic-wave transmission cannot accommodate real-time control. When the device to be controlled on an alien world is light-minutes or even light-hours away, teleoperation must yield to increasing levels of autonomous, machine-intelligence-dependent robotic operation.

Robot devices are computer-controlled mechanical systems that are capable of manipulating or controlling other machine devices, such as end effectors. Robots may be mobile or fixed in place and either fully automatic or teleoperated.

Large quantities of data are frequently involved in the operation of automatic robotic devices. The field of distributed data management is concerned with ways of organizing cooperation among independent but mutually interacting data bases.

In the field of artificial intelligence, the concept of cognition and learning refers to the development of a machine intelligence that can deal with new facts, unexpected events and even contradictory information. Today's smart machines handle new data by means of preprogrammed methods or logical steps. Tomorrow's "smarter" machines will need the ability to learn, possibly even to understand, as they encounter new situations and are forced to change their mode of operation.

Perhaps late in the next century, as the field of artificial intelligence matures, we will send fully automatic robot probes on interstellar voyages. Each very smart interstellar probe must be capable of independently examining a new star system for suitable extrasolar planets and, if successful in locating one, beginning the search for extraterrestrial life. Meanwhile, back on Earth, scientists will wait for its electromagnetic signals to travel light-years through the interstellar void, eventually informing its human builders that the extraterrestrial exploration plan has been successfully accomplished.

See also: **robotics in space; Viking Project**

asteroid A small, solid object orbiting the Sun independently of a planet. The majority of the asteroids—which are also called minor planets—have orbits that lie in the "asteroid belt" between Mars and Jupiter (most travel around the Sun at distances of 2.2 to 3.3 astronomical units), and they possess orbital periods between three and six years. There are more than 500 main-belt asteroids over 50 kilometers in diameter and thousands that are smaller. Approximately 2,000 of these have fairly well known orbits and have been assigned names. Their diameters range from approximately 1,000 kilometers (Ceres) down to 20 kilometers and smaller. Beginning in this century, more than 60 asteroids that cross the Earth's orbit have also been

Table 1 Data for Selected Asteroids

Achilles: This asteroid was the first member of the Trojan group to be discovered.

Diameter	~70 km [a]
Perihelion	4.44 AU [b]
Aphelion	5.98 AU
Orbital period	11.77 yr
Inclination	10.3°
Year discovered	1906

Adonis: A member of the Apollo group of minor planets.

Diameter	0.3 km
Perihelion	0.44 AU
Aphelion	3.30 AU
Orbital period	2.56 yr
Inclination	1.4°
Year discovered	1936

Amor: The major member of the Amor group of asteroids.

Diameter	0.5 km
Perihelion	1.08 AU
Aphelion	2.76 AU
Orbital period	2.77 yr
Inclination	11.9°
Year discovered	1932

Aten: An "Earth-crossing" asteroid with an orbital period of less than one year.

Diameter	~1.1 km
Perihelion	0.79 AU
Aphelion	1.14 AU
Orbital period	0.95 yr
Inclination	17.9°
Year discovered	1976
Features	Type S; probably silicate or metal-rich; albedo = 0.17

Apollo: The asteroid for which the Apollo group of Earth-crossing asteroids is named. This group of asteroids have perihelion values (that is, their distances from the Sun at their point of closest approach) falling inside the orbit of the Earth. This group is named after the asteroid Apollo, which approached within 0.07 astronomical units (AU) of the Earth in 1932—a "near miss" by celestial-encounter standards. Because they cross the Earth's orbit, the Apollo group is of particular interest in asteroid mining concepts.

Diameter	~1 km
Perihelion	0.65 AU
Aphelion	2.29 AU
Orbital period	1.78 yr
Inclination	6.36°
Year discovered	1932

Ceres: In 1801 the Italian astronomer Giuseppe Piazzi discovered it—the first asteroid to be found and the largest.

Diameter	1,002 km
Perihelion	2.55 AU
Aphelion	2.98 AU
Orbital period	4.60 yr
Inclination	10.6°
Year discovered	1801
Features	Type C

Table 1   Data for Selected Asteroids (Continued)

Chiron: An unusual minor planet with an orbit lying almost entirely between Saturn and Uranus, making it the most distant asteroid yet found. It is conceivable that Chiron is one of the brighter members of a distant, as yet to be discovered asteroid belt, or else it may possibly be a dormant comet or even an escaped satellite of Saturn or Uranus. Chiron is known to be in a highly perturbed, "chaotic" orbit that will eventually lead either to collision with a planet or to ejection from the Solar System.

Diameter	~300–400 km (uncertain)
Perihelion	8.5 AU
Aphelion	18.9 AU
Orbital period	50.68 yr
Inclination	6.93°
Year discovered	1977

Eros: The largest member of the Amor group of Earth-crossing asteroids. Under favorable circumstances it passes only 0.13 AU (20 million km) from the Earth, at which time it is one of the brightest asteroids. Scientists believe that Eros's surface is stony or stony-iron in composition. It is highly elongated, with possible dimensions of 18 km × 36 km. Some scientists believe that Eros might actually be two celestial objects orbiting in contact or very close proximity (an asteroid binary).

Diameter	~20 km (irregular shape)
Perihelion	1.13 AU
Aphelion	1.78 AU
Orbital period	1.76 yr
Inclination	10.83°
Year discovered	1898
Features	Type S

Hermes: A tiny member of the Apollo group of asteroids. It was discovered in 1937 when it approached within 0.006 AU (780,000 km) of Earth, a very close call by astronautical standards.

Diameter	0.5 km
Perihelion	0.62 AU
Aphelion	2.66 AU
Orbital period	2.10 yr
Inclination	6.2°
Year discovered	1937

Icarus: A member of the Apollo group. It has the smallest perihelion of any known asteroid.

Diameter	~1.4 km
Perihelion	0.19 AU
Aphelion	1.97 AU
Orbital period	1.12 yr
Inclination	20.2°
Year discovered	1949

Juno: Discovered in 1804 by the German astronomer Karl Harding. This asteroid was the third to be found.

Diameter	249 km
Perihelion	1.98 AU
Aphelion	3.35 AU
Orbital period	4.36 yr
Inclination	13°
Year discovered	1804
Features	Type S

Pallas: The second minor planet to be discovered—a feat accomplished by the German astronomer Heinrich Wilhelm Olbers in 1802.

Diameter	583 km
Perihelion	2.11 AU
Aphelion	3.42 AU
Orbital period	4.61 yr
Inclination	34.8°
Year discovered	1802
Features	Type U

Trojan Group: A group of asteroids that lie near the two Lagrangian points (that is, the two "gravity wells," or points of stable equilibrium) in Jupiter's orbit around the Sun. Achilles was the first asteroid in this group to be identified; many subsequently discovered members of this group have been named in honor of the heroes, both Greek and Trojan, of the Trojan War.

Vesta: The brightest minor planet and the third largest, with an approximate diameter of 555 km. Vesta was discovered in 1807 by the German astronomer Heinrich Wilhelm Olbers. Under favorable conditions it can just be seen with the naked (unaided) eye. Vesta is probably among the best studied of all asteroids. It is apparently unique among the larger minor planets in that it appears to have a surface of basaltic lava, indicating a complex geological history of heating and volcanism. Many space scientists believe that Vesta may be the parent body of a class of meteorites called the eucrite class (indicating a type of stony composition). Although this hypothesis is not yet proven, the theory appears to represent the most probable connection between specific meteorites and an asteroid parent body. Telescopic studies show that Vesta is nearly spherical, with several types of igneous rock on its surface. With its low inclination and 2.36 AU (average) distance from the Sun, Vesta is among the most easily reached of the large main-belt asteroids.

Diameter	555 km
Perihelion	2.15 AU
Aphelion	2.57 AU
Orbital period	3.63 yr
Inclination	7.14°
Year discovered	1807

[a] ~70 km = approximately 70 kilometers.
[b] AU = astronomical unit(s).
SOURCE: Developed by the author from the latest NASA data.

discovered, the majority of these within the last decade. Such asteroids are popularly known as "Earth-crossers," or "Earth-crossing asteroids." Table 1 provides selected information about some of these minor planets.

The surfaces of most asteroids appear dark, suggesting significant carbon content. Chemically bound water also appears to have been detected on some surfaces. The mass of the entire asteroid belt is estimated to be approximately 3×10^{21} kilograms—a value that is about 1/20th the mass of our Moon.

Asteroids are usually classified on the basis of their spectral characteristics—that is, by how much light of different colors they reflect. The C-type (carbonaceous) asteroids appear to be covered with a dark, carbon-based material and are assumed to be similar to carbonaceous chondrite meteorites. They may contain up to 10 percent water, 6 percent carbon, significant amounts of sulfur and useful amounts of nitrogen. The S-type (silicaceous) asteroids are thought to be similar to silicaceous or stony-iron meteorites. They are more common near the inner edge of the main belt. Of particular interest in space industrialization is the fact that S-class Earth-crossing asteroids may contain up to 30 percent free metals—that is, alloys of iron, nickel and cobalt, along with high concentrations of precious metals. The M-type (for metallic) asteroids are thought to be the remaining metallic cores of very small, differentiated planetoids that have been stripped of their crusts by collisions with other asteroids. Three minor asteroid designations—R-type, E-type and U-type—are also encountered. The R-type stands for asteroids that have red features, while the E-type asteroids (estatine achondrites) appear to possess various stony compositions. Finally, U-type asteroids display unusual or unknown characteristics.

To date, there is no general consensus among space scientists on the origin and nature of the various types of asteroids or their assumed relationships to different types of meteorites. The earliest theory about the origin of the asteroids was that they are remnants of the explosion of a "missing planet" between Mars and Jupiter. At present, the tendency is to support other theories, ones suggesting that the material that formed the asteroids gradually accumulated in the same way as the larger planets, with the asteroids eventually suffering many collisions among themselves along with experiencing large orbital disturbances from close approaches to massive celestial objects, such as Jupiter. Space scientists have also hypothesized that some of the dark (that is, low-albedo) asteroids whose perihelia (orbital points nearest to the Sun) fall within the orbit of Mars may actually be extinct cometary nuclei.

Earth-crossing asteroids are of special interest in the development of our extraterrestrial civilization, since they would be the easiest to visit, return samples from and eventually exploit for their material content. By convention, these "Earth-crossers" or inner Solar System asteroids are divided into three groups: Aten, Apollo and Amor. The Aten group of asteroids possess orbits that overlap the Earth's orbit at aphelion (the point at which Earth is farthest from the Sun), while the Apollo group have orbits that overlap the Earth's orbit at perihelion. The Amor group of asteroids, on the other hand, are those with perihelia lying between 1.017 astronomical units (Earth's aphelion) and 1.38 astronomical units (an arbitrarily chosen figure) and whose orbits actually cross the orbit of Mars.

Some scientists have postulated that the impact of an Earth-crossing asteroid may have been responsible for the disappearance of the dinosaurs and many other animal species on Earth some 65 million years ago.

See also: **Comet Rendezvous and Asteroid Flyby Mission; extraterrestrial catastrophe theory; Vesta Missions**

asteroid mining The asteroids, especially Earth-crossing asteroids, can serve as "extraterrestrial warehouses" from which future space workers may extract certain critical materials needed in space-based industries and in the construction of large space habitats. On some asteroids scientists expect to find water (trapped), organic compounds and metals. In time, it may be far more efficient to mine an Earth-crossing asteroid than to lift these very same raw materials from the Earth's surface. Thus, asteroid mining could play a major role in the evolution of our extraterrestrial civilization.

See also: **asteroid**

astro- A prefix meaning "star" or "stars" and (by extension) sometimes used as the equivalent of "celestial," as in *astroengineering.*

astroengineering Incredible feats of engineering and technology involving the energy and material resources of an entire star system or several star systems. The detection of such astroengineering projects would be a positive indication of the presence of a Type II, or even Type III, extraterrestrial civilization in our Galaxy. One example of an astroengineering project would be the creation of a "Dyson sphere"—a cluster of structures and habitats made by an intelligent extraterrestrial species to encircle their native star and effectively intercept all of its radiant energy output.

See also: **Dyson sphere; extraterrestrial civilizations**

astrometry The science that is concerned with the very precise measurement of the motion and position of celestial objects; a subset or branch of astronomy.

astronomical unit (AU) An extraterrestrial unit of distance defined as the "mean" distance between the center of the Earth and the center of the Sun—that is, the semimajor axis of the Earth's orbit. One AU is equal to 149.6×10^6 kilometers (approximately 92.9 million miles), or 499.01 light-seconds.

astrophysics Astronomy addresses fundamental questions that have occupied man since his primitive beginnings. What is the nature of the Universe? How did it begin, how is it evolving and what will be its eventual fate? As important as these questions are, there is another motive for astronomical studies. Since the 17th century when Newton's studies of celestial mechanics helped him formulate the three basic laws of motion and the universal law of gravitation, the sciences of astronomy and physics have become intertwined. "Astrophysics" can be defined as the study of the nature and physics of stars and star systems. It provides the theoretical framework for understanding astronomical observations. At times astrophysics can be used to predict phenomena before they have even been observed by astronomers, such as black holes. The vast laboratory of outer space makes it possible to investigate large-scale physical processes that cannot be duplicated in a terrestrial laboratory. Although the immediate, tangible benefits to mankind from progress in astrophysics cannot easily be measured or predicted, the opportunity to extend our understanding of the workings of the Universe is really an integral part of the rise of our extraterrestrial civilization.

Today, astrophysics has within its reach the ability to bring about one of the greatest scientific achievements ever—a unified understanding of the total evolutionary scheme of the Universe. This remarkable revolution in astrophysics is happening now due to the confluence of two streams of technical development: remote sensing and spaceflight. Through the science of remote sensing, we have acquired sensitive instruments capable of detecting and analyzing radiation across the whole range of the electromagnetic (EM) spectrum. Spaceflight, on the other hand, enables astrophysicists to place these remote sensing instruments outside the Earth's atmosphere. The wavelengths transmitted through the interstellar medium and arriving in the vicinity of near-Earth space are spread over approximately 24 decades of the spectrum. (A decade is a group, series or power of 10.) However, most of this electromagnetic radiation never reaches the surface of the Earth, because the terrestrial atmosphere effectively blocks such radiation across most of the spectrum. It should be remembered that the visible and infrared "atmospheric windows" occupy a spectral slice whose width is roughly one decade. Ground-based radio observatories can detect stellar radiation over a spectral range that adds about five more decades to the range of observable frequencies; but the remaining 18 decades of the spectrum are still blocked and are effectively "invisible" to astrophysicists on the Earth's surface. Consequently, information that can be gathered by observers at the bottom of the Earth's atmosphere represents only a small fraction of the total amount of information available concerning extraterrestrial objects. Sophisticated remote sensing instruments placed above the Earth's atmosphere are now capable of sensing electromagnetic radiation over nearly the entire spectrum—and these instruments are rapidly changing our picture of the cosmos.

For example, we previously thought that the interstellar medium was a fairly uniform collection of gas and dust; but spaceborne ultraviolet telescopes are now showing us that its structure is very inhomogeneous and complex. There are newly discovered components of the interstellar medium, such as extremely hot gas that is probably heated by shock waves from exploding stars. There is a great deal of interstellar pushing and shoving going on. Matter gathers and cools in some places because matter elsewhere is heated and dispersed. Besides discovering the existence of the very hot gas, the orbiting telescopes have discovered two potential sources of the gas: the intense stellar winds that boil off hot stars; and the rarer, but more violent, blasts of matter from exploding supernovae.

In addition, X-ray and gamma ray astronomy have contributed substantially to the discovery that the Universe is not relatively serene and unchanging as previously imagined, but is actually dominated by the routine occurrence of incredibly violent events.

And this series of remarkable new discoveries is just beginning. Future astrophysics missions will provide access to the full range of the electromagnetic spectrum at increased angular resolution. They will support experimentation in key areas of physics, especially relativity and gravitational physics. Out of these exciting discoveries, perhaps, will emerge the scientific pillars for constructing an extraterrestrial civilization based on technologies unimaginable in the framework of contemporary physics.

Virtually all the information we receive about celestial objects comes to us through observation of electromagnetic radiation. Cosmic ray particles are an obvious and important exception, as are extraterrestrial material samples that have been returned to Earth (for example, lunar rocks). Each portion of the electromagnetic spectrum carries unique information about the physical conditions and processes in the Universe. Infrared radiation reveals the presence of thermal emission from relatively cool objects while ultraviolet and extreme ultraviolet radiation may indicate thermal emission from very hot objects. Various types of violent events can lead to the production of X-rays and gamma rays.

Although EM radiation varies over many decades of energy and wavelength, the basic principles of measurement are quite common to all regions of the spectrum. The fundamental techniques used in astrophysics can be classified as imaging, spectrometry, photometry and polarimetry. Imaging provides basic information about the distribution of material in a celestial object, its overall structure and, in some cases, its physical nature. Spectrometry is a measure of radiation intensity as a function of wavelength. It provides information on nuclear, atomic and molecular phenomena occurring in and around the extraterrestrial object under observation. Photometry involves measuring radiation intensity as a function of time. It provides information about the time variations of physical processes within and around celestial objects, as well as their absolute intensities. Finally, polarimetry is a measurement of radiation intensity as a function of polarization angle. It provides information on ionized particles rotating in strong magnetic fields.

High-energy astrophysics encompasses the study of extraterrestrial X-rays, gamma rays and energetic cosmic ray particles. Prior to space-based high-energy astrophysics, scientists believed that violent processes involving high-energy emissions were rare in stellar and galactic evolution. Now, because of studies of extraterrestrial X-rays and gamma rays, we know that such processes are quite common rather than exceptional. The observation of X-ray emissions has been very valuable in the study of high-energy events, such as mass transfer in binary star systems, interaction of supernovae remnants with interstellar gas, and quasars (whose energy source is presently unknown). It is thought that gamma rays might be the missing link in understanding the physics of high-energy objects such as pulsars and black holes. The study of cosmic ray particles provides important information about the physics of nucleosynthesis and about the interactions of particles and strong magnetic fields. High-energy phenomena that are suspected sources of cosmic rays include supernovae, pulsars, radio galaxies and quasars.

X-ray astronomy is the most advanced of the three high-energy astrophysics disciplines. Space-based X-ray observatories increase our understanding in the following areas: (1) stellar structure and evolution, including binary star systems, supernovae remnants, pulsar and plasma effects, and relativity effects in intense gravitational fields; (2) large-scale galactic phenomena, including interstellar media and soft X-ray mapping of local galaxies; (3) the nature of active galaxies, including spectral characteristics and the time variation of X-ray emissions from the nuclear or central regions of such galaxies; and (4) rich clusters of galaxies, including X-ray background radiation and cosmology modeling.

Gamma rays consist of extremely energetic photons (that is, energies greater than 10^5 electron volts) and result from physical processes different than those associated with X-rays. The processes associated with gamma ray emissions in astrophysics include: (1) the decay of radioactive nuclei; (2) cosmic ray interactions; (3) curvature radiation in extremely strong magnetic fields; and (4) matter–antimatter annihilation. Gamma ray astronomy reveals the explosive, high-energy processes associated with such celestial phenomena as supernovae, exploding galaxies and quasars, pulsars and black hole candidates.

Gamma ray astronomy is especially significant because the gamma rays being observed can travel across our entire Galaxy, and even across most of the Universe, without suffering appreciable alteration or absorption. Therefore, these energetic gamma rays reach our Solar System with the same characteristics, including directional and temporal features, as they started with at their sources, possibly many light years distant and deep within regions or celestial objects opaque to other wavelengths. Consequently, gamma ray astronomy provides information on extraterrestrial phenomena not observable at any other wavelength in the electromagnetic spectrum and on spectacularly energetic events that may have occurred far back in the evolutionary history of the Universe.

Cosmic rays are extremely energetic particles that extend in energy from one million (10^6) electron volts to over 10^{20} eV and range in composition from hydrogen (atomic number $Z = 1$) to a predicted atomic number of $Z = 114$. This composition also includes small percentages of electrons, positrons and possibly antiprotons. Cosmic-ray astronomy provides information on the origin of the elements (nucleosynthetic processes) and the physics of particles at ultra-

high-energy levels. Such information addresses astrophysical questions concerning the nature of stellar explosions and the effects of cosmic rays on star formation and galactic structure and stability.

Astronomical work in a number of areas will greatly benefit from large, high-resolution optical systems that operate outside the Earth's atmosphere. Some of these areas include investigation of the interstellar medium, detailed study of quasars and black holes, observation of binary X-ray sources and accretion disks, extragalactic astronomy, and observational cosmology. The Hubble Space Telescope (HST) constitutes the very heart of NASA's space-borne ultraviolet/optical astronomy program through the end of this century. Launched in 1990, and when refurbished in the mid-1990s, its ability to cover a wide range of wavelengths, to provide fine angular resolution and to detect faint sources will make it one of the most powerful astronomical telescopes ever built.

Another interesting area of astrophysics involves the extreme ultraviolet (EUV) region of the electromagnetic spectrum. The interstellar medium is highly absorbent at EUV wavelengths (100 to 1,000 angstroms [Å]). EUV data gathered from space-based instruments will be used to confirm and refine contemporary theories of the late stages of stellar evolution, to analyze the effects of EUV radiation on the interstellar medium and to map the distribution of matter in our "solar neighborhood."

Infrared (IR) astronomy involves studies of the electromagnetic (EM) spectrum from 1 to 100 micrometers wavelength, while radio astronomy involves wavelengths greater than 100 micrometers. (A micrometer is one millionth of a meter.) Infrared radiation is emitted by all classes of "cool" objects (stars, planets, ionized gas and dust regions, and galaxies) and cosmic background radiation. Most emissions from objects with temperatures ranging from 3 to 2,000 degrees Kelvin are in the infrared region of the spectrum. In order of decreasing wavelength, the sources of infrared and microwave (radio) radiation are: (1) galactic synchrotron radiation; (2) galactic thermal bremsstrahlung radiation in regions of ionized hydrogen; (3) the cosmic background radiation; (4) 15 degrees Kelvin cool galactic dust and 100 degrees Kelvin stellar-heated galactic dust; (5) infrared galaxies and primeval galaxies; (6) 300 degrees Kelvin interplanetary dust, and (7) 3,000 degrees Kelvin starlight.

Gravitation is the dominant long-range force in the Universe. It governs the large-scale evolution of the Universe and plays a major role in the violent events associated with star formation and collapse. Outer space provides the low-acceleration and low-noise environment needed for the careful measurement of relativistic gravitational effects. A number of interesting experiments have been identified for a space-based experimental program in relativity and gravitational physics.

The ultimate aim of astrophysics is to understand the origin, nature and evolution of the Universe. It has been said that the Universe is not only stranger than we imagine, it is stranger than we *can* imagine! Through the creative use of modern space technology, we will witness many major discoveries in astrophysics in the exciting decades ahead—each discovery helping us understand a little better the magnificent Universe in which we live and the place in which we will build our extraterrestrial civilization.

See also: **Cosmic Background Explorer; Gamma Ray Observatory; Hubble Space Telescope; stars**

Astropolis A visionary urban extraterrestrial facility in near-Earth space proposed by the late Dr. Krafft A. Ehricke. The modular design of Astropolis was selected to make maximum use of the different levels of gravity available in a large, rotating space facility. Astropolis represents a logical growth step beyond the space station. It would contain several thousand inhabitants who would live and work in an unusual multi-gravity-level world. This proposed facility would contain residential sections, a dynarium, space industrial zones, space agricultural facilities, research laboratories and even other world enclosures (OWEs).

Astropolis would have the ability to completely recycle air, water and waste materials. Energy would be supplied by either nuclear power plants or solar arrays. The research section of Astropolis would be dedicated to the long-term use of the space environment for basic and applied research, as well as for eventual industrial exploitation. The other world enclosures would be located at various distances from the hub of Astropolis. Using these special OWE facilities, exobiologists, space scientists, planetary engineers and interplanetary explorers will be able to simulate the gravitational environment of all major celestial objects in the Solar System of interest from the perspective of human visitation and possible settlement. These include the Moon, Mars, Venus, the asteroids and certain moons of the giant outer planets. Pioneering work in the OWE facilities of Astropolis could pave the way for the opening up of both cislunar and heliocentric space (that is, the space between the Earth and the Moon's orbit, as well as all other space surrounding the Sun) to human occupancy.

Astropolis is envisioned as a 4,000- to 15,000-ton-class space complex that would be rotated very slowly, at about 925 revolutions per Earth day (24 hours). Because of its low angular velocity, Coriolis forces (sideward force felt by an astronaut moving radially in a rotating system, such as a space station) would cause little disturbance and discomfort—even at the greatly reduced artificial gravity levels occurring closer to the hub. Therefore, research and industrial projects conducted on an orbiting facility like Astropolis would be able to enjoy excellent variable-gravity-level simulations with minimum Coriolis-force disturbances—in contrast to smaller space stations, which would be spinning more rapidly.

See also: **Androcell; extraterrestrial civilizations; other world enclosures; space settlement; space station.**

B

Barnard's star A red dwarf star approximately six light-years from the Sun, making it the fourth nearest star to our Solar System. The absolute magnitude of Barnard's star is 13.2, and its spectral class is M5V. It has the largest known proper motion, some 10.3 arc-seconds per year. Discovered in 1916 by E.E. Barnard, this star is of popular interest because recent investigations of its wobbling motion have hinted at the possible existence of one or more Jupiter-sized planets orbiting around it.

See also: **extrasolar planets; stars**

Bernal sphere Long before the Space Age began, the British physicist and writer J. Desmond Bernal predicted that the majority of humanity would someday live in "artificial globes" orbiting around the Sun. In his 1929 work *The World, the Flesh, and the Devil*, Bernal boldly speculated about the colonization of outer space.

Bernal's concept of spherical space habitats has influenced both early space-station designs and very recent space-settlement designs.

See also: **space settlement; space station**

"Big Bang" theory A theory in cosmology concerning the origin of the Universe. According to the Big Bang cosmological model, there was a very large explosion, also called the "initial singularity," that started the space and time of our Universe. The Universe itself has been expanding ever since. It is currently thought that the Big Bang event occurred some 15 billion to 20 billion years age. Astrophysical observations and discoveries lend support to the Big Bang model, especially the discovery of the cosmic microwave background (CMB) radiation in 1964.

Actually, there are two general variations of the Big Bang model. In one, it is assumed that the Universe will expand forever. This is called the "open-Universe model." The other variation, called the "closed-Universe model," assumed that the Universe will eventually recollapse—possibly back into another "singularity," or "Big Crunch," as some cosmologists like to call it.

The Big Bang theory assumes that cosmic expansion was started by a primordial fireball, which emitted radiation and pushed matter apart. A very short time after the Big Bang, the concepts of space and time, as we understand them today, started. The interval between the Big Bang and the start of space and time is called the "Planck time" (T_p), described as,

$$T_P = \sqrt{\frac{Gh}{c^5}} = 1.4 \times 10^{-43} \text{ sec}$$

where G is the gravitational constant, h is Planck's constant and c is the speed of light.

Following the Big Bang, as the matter in the Universe expanded, it cooled, condensing into galaxies, which in turn gave rise to stars and eventually other celestial objects such as planets, asteroids, comets and so on. The radiation released from this ancient explosion has now cooled and can be observed as the cosmic microwave background radiation.

Although there are quite probably very many distant objects far out in space that we haven't been able to detect yet, space scientists have discovered radiation from a source quite literally at the edge of the observable Universe. As we look deep into space, we are really looking back into time—that is, we are viewing the most distant regions of the Universe as they were very long ago. In the Big Bang model, before the galaxies and stars formed, the Universe was filled with hot, glowing gas that was opaque. Then, at some point in time during the first million years or so after the Big Bang event, this expanding gas cooled and became transparent. Today, we can look out into space and back into time only until we reach that very distant region where we are observing the early Universe when it transitioned from an opaque to a transparent gas. Beyond that point, space is opaque, so that light waves simply cannot reach us. Instead, we see the glow of that primordial hot gas as it cooled to about 10,000 degrees Kelvin and then cleared. This glow was originally emitted as ultraviolet radiation but has now been Doppler-shifted to longer wavelengths by the expansion of the Universe and currently resembles emission from a dense gas at a temperature of only 2.7 degrees Kelvin. Today, scientists observe it as a diffuse background of microwave and thermal radiation. In fact, you can think of this microwave radiation as a distant spherical wall that surrounds us and delimits the edges of the observable Universe.

Recent astrophysical data, including observations by the *Cosmic Background Explorer* (COBE) and *Roentgen Satellite* (ROSAT), are now puzzling Big Bang cosmologists. They are being challenged to explain how clumpy structures of galaxies could have evolved from a previously assumed "smooth" (i.e., uniform and homogeneous) Big Bang event. They must also explain the existence of large, organized clusters of quasars.

See also: **astrophysics; closed universe; Cosmic Background Explorer; cosmology; open universe; Roentgen satellite**

binary stars   A pair of stars that orbit about their common center of mass. By convention, the star that is nearest the center of mass in a binary star system is called the "primary," while the other (smaller) star of the system is called the "companion." Binary star systems can be further classified as visual binaries, eclipsing binaries, spectroscopic binaries and astrometric binaries. Visual binaries are those systems that can be resolved into two stars by an optical telescope. Eclipsing binaries occur when each star of the system alternately passes in front of the other, obscuring or eclipsing it and thereby causing their combined bright-

ness to diminish periodically. Spectroscopic binaries are resolved by the Doppler shift of their characteristic spectral lines as the stars approach and then recede from the Earth while revolving about their common center of mass. In an astrometric binary, one star cannot be visually observed, and its existence is inferred from the irregularities in the motion of the visible star of the system.

Binary star systems are more common in our Galaxy than generally realized. Perhaps 50 percent of all stars are contained in binary systems. The typical mean separation distance between members of a binary star system is on the order of 10 to 20 astronomical units.

The binary star systems first identified, revealed their "binary characteristic" to us through variations in their optical emissions. A new class of binary star systems, called X-ray binaries, were discovered during the 1970s by means of space-based X-ray observations. In a typical X-ray binary system, a usually massive "optical" star is accompanied by a compact, X-ray-emitting companion that might be a neutron star or even a black hole.

See also: **astrophysics; black holes; stars; X-ray astronomy**

biogenic elements   Those elements generally considered by scientists to be essential for all living systems. Exobiologists usually place primary emphasis on the elements hydrogen (H), carbon (C), nitrogen (N), oxygen (O), sulfur (S) and phosphorous (P). The chemical compounds of major interest are those normally associated with water (H_2O) and with organic chemistry, in which carbon (C) is bonded to itself or to other biogenic elements. Exobiologists also include several "life-essential" elements associated with inorganic chemistry, such as iron (Fe), magnesium (Mg), calcium (Ca), sodium (Na), potassium (K) and chlorine (Cl) in this overall grouping, but these are often given secondary emphasis in cosmic evolution studies.

See also: **exobiology; life in the Universe.**

biosphere   The life zone of a planetary body; for example, the part of the Earth inhabited by living organisms.

See also: **ecosphere; global change**

black holes   Theorized gravitationally collapsed masses from which nothing—light, matter or any other kind of signal—can escape. Scientists today generally speculate that a black hole is the natural end product when a giant star dies and collapses. If the star has three or more solar masses left after exhausting its nuclear fuels (a solar mass is a unit of measure equivalent to the mass of our own Sun), then it can become a black hole. As with the formation of a white dwarf or a neutron star, the collapsing giant star's density and gravity increase with contraction. However, in this case, because of the large mass involved, the gravity of the collapsing star becomes too strong for even neutrons to resist, and an incredibly dense point mass, or "singularity," is formed. Therefore, a black hole is essentially a singularity surrounded by an event region in which the gravity is so strong that absolutely nothing can escape.

Remember, as the massive star collapses, its gravitational escape velocity (the speed an object needs to reach in order to escape from the star) also increases. When a giant star has collapsed to a dimension called the Schwarzschild radius, its gravitational escape velocity is equal to the speed of light. At this point, not even light itself can escape from the black hole!

The Schwarzschild radius is simply the "event horizon," or boundary of no return, for a black hole. This dimension bears the name of the German astronomer Karl Schwarzschild, who wrote the fundamental equations describing a black hole in 1916. Anything crossing this boundary can never leave the black hole. In fact, the event horizon represents the start of a region disconnected from normal space and time. We cannot see beyond this event horizon into a black hole, and time itself is considered to stop there.

As shown in table 1, the event horizon, or Schwarzschild radius, of a black hole is proportional to the mass of the collapsed star. For a star of 10 solar masses, for example, the Schwarzschild radius is 30 kilometers.

Once a black hole has been formed, it crushes anything crossing its event horizon into its incredibly dense singularity. As the black hole devours matter, its event horizon expands. This expansion is limited only by the availability of mass. Gigantic black holes, which contain the crushed remains of billions of stars, are considered theoretically possible. In fact, some astrophysicists speculate that rotating black holes (called Kerr black holes) containing the remains of millions or billions of dead stars may lie at the centers of galaxies. These enormous rotating black holes may be the powerhouses of quasars and active galaxies. Quasars are believed to be galaxies in an early, violent evolutionary stage, while active galaxies are characterized by their extraordinary energy outputs, which occur mostly from their cores, or "galactic nuclei." Scientists think that "normal" galaxies like our own Milky Way are only "quiet" because the black holes at their centers have no more material upon which to feed.

Today, evidence that super-dense stars like white dwarfs and neutron stars really exist has supported the idea that black holes themselves (representing what may be the ultimate in density) must also exist. But how can we detect an object from which nothing, not even light, can escape?

Astrophysicists think they may have found indirect ways of detecting black holes. Their techniques depend upon certain black holes being members of binary star systems. (A binary star system consists of two stars comparatively near to and revolving about each other. Astronomers would say the two stars are "gravitationally bound" to each other.) Unlike our Sun, many stars in the Galaxy belong to binary systems.

If one of the stars in a particular binary system has become a black hole (although invisible), it would betray its existence by the gravitational effects it produces upon the companion star, which is observable. These gravitational effects would actually be in accordance with Newton's universal law of gravitation; that is, the mutual gravitational attraction of the two celestial objects is directly proportional to their masses and inversely proportional to the square of the distance between them. Once beyond the black hole's event horizon, its gravitational influences are the same as exerted by other objects (of equivalent mass) in the "normal" Universe.

Astrophysicists have also speculated that a substantial part of the energy of matter spiraling into a black hole is converted by collision, compression and heating into X rays and gamma rays, which display certain spectral characteristics. This X and gamma radiation emanates from the material as it is pulled toward the black hole. However, once the captured material has been pulled across the black hole's event horizon, this radiation cannot escape.

Black-hole candidates are celestial phenomena that exhibit such black-hole "capture effects" in a binary star system. Several have now been discovered and studied using space-based astronomical observatories (especially X-ray observatories). One very promising candidate is called Cygnus X-1, an invisible object in the constellation Cygnus the Swan. Cygnus X-1 means that it is the first X-ray source discovered in Cygnus. X rays from the invisible object have characteristics like those expected from materials spiraling toward a black hole. This material is apparently being pulled from the candidate black hole's binary companion— a large star of about 30 solar masses. Based upon the candidate black hole's gravitational effects on its visible companion, the hole's mass has been estimated to be about six solar masses. In time, the giant visible companion might itself collapse into a neutron star or a black hole; or else it might be devoured piece by piece by its black-hole companion. This form of stellar cannibalism would significantly enlarge the existing black hole's event horizon. Two other black hole candidates are LMC X-3 and A0620-00.

Big Bang cosmology states that our Universe began with a violent explosion that sent pieces of matter flying outward in all directions. To date, cosmologists and astronomers have not detected enough mass in the Universe to reverse this expansion process. The possibility remains, however, that the missing mass may be locked up in undetectable black holes that are more prevalent than anyone currently anticipates.

Do enough black holes exist within the Universe to reverse this expansion? If so, what will then happen to the

Table 1   Schwarzschild Radius as a Function of Mass

Mass of Collapsed Star (solar masses)[a]	Schwarzschild Radius (km)
5	15
10	30
20	60
50	150

[a] One solar mass = mass of our Sun.

Universe? Will all of the stars, galaxies and other matter in the Universe eventually collapse inward, just like a massive star that has exhausted its nuclear fuels? Will there ultimately be created one very large black hole, within which the Universe will again collapse back to a singularity?

Extrapolating back more than 10 billion to 15 billion years, some cosmologists trace the present Universe to an initial singularity. Thus, is a singularity both the beginning and the end of our Universe? Is our Universe but a pause or a phase between such singularities?

Other scientists have put forward even more intriguing questions and speculations. If the Universe itself is closed and nothing can escape, perhaps we may already be in a megasize black hole!

As we learn more about our extraterrestrial environment and gather further supporting evidence about the existence and properties of black holes, scientists may someday begin to answer these puzzling questions.

See also: **cosmology; stars (white dwarfs, neutron stars and black holes)**

C

Cassini Mission The Cassini mission is a proposed joint NASA and European Space Agency (ESA) project to Saturn, its major moon Titan, and its complex system of at least 16 other moons. As currently planned, the Cassini mission will be launched in April 1996 by an expendable Titan–Centaur launch vehicle configuration. NASA will provide the launch vehicle and the Cassini orbiter spacecraft. The overall mission and the orbiter spacecraft are named after the Italian–French astronomer, Giovanni Domenico Cassini (1625–1712), who discovered the major Saturnian moons of Iapetus, Rhea, Tethys and Dione, and who also observed in 1675 that Saturn's rings were split into two parts by a narrow gap, now called the Cassini division by modern astronomers. The Titan probe, called Huygens, will be provided by ESA. This probe is named in honor of the Dutch scientist and astronomer, Christiaan Huygens (1629–1695), who discovered Saturn's largest moon, Titan, in 1656.

To reach Saturn, the Cassini spacecraft (orbiter and probe configuration) must first execute flybys of the Earth and Jupiter in order to gain the needed "gravity assist" boost in velocity. The first flyby of Earth will take place 26 months after launch, followed by the Jupiter flyby some 19 months later. The Cassini spacecraft will then arrive at the planet Saturn in December 2002. During the first portion of its voyage (that is, after launch but before its Earth flyby in June 1998), the Cassini spacecraft will navigate through a

portion of the asteroid belt and could encounter the asteroid Maja in March 1997. Maja is believed to be a carbonaceous or "C-type" asteroid about 78 kilometers (50 mi) in diameter. The Cassini spacecraft's final planetary encounter before proceeding on to Saturn will be with Jupiter, which it will fly past in February 2000 at a distance of about 3.6 million kilometers (approximately 2.23 million miles)—a distance some 50 times the radius of Jupiter itself.

Upon reaching Saturn in December 2002, the Cassini spacecraft will swing within 1.8 Saturn radii of the planet to initiate the first of approximately three dozen highly elliptical orbits during the remainder of the mission. Eighty-five days after reaching Saturn, while traveling back in toward the planet as it completes its first orbit, the Cassini spacecraft will release the Huygens probe to begin descent to Titan.

Eleven days later the Huygens probe will enter Titan's dense atmosphere, protected at first by a heat shield. When the heat shield has been jettisoned, the probe will deploy its parachute assembly for final descent to Titan's surface, which may be covered with puddles or even an ocean of liquid ethane (C_2H_6). The probe mission ends on impact. After relaying the Huygens probe data back to scientists on Earth, the Cassini orbiter will encircle Saturn and perform targeted and nontargeted flybys of most of the planet's 17 known moons, with particular observational emphasis given to Titan.

The main objectives of the Cassini mission are: (1) to conduct detailed studies of Saturn's atmosphere, rings and magnetosphere; (2) to perform close-up investigations of the Saturnian moon system; and (3) to characterize Titan's interesting atmosphere and little-known surface. Unlike any other moon yet observed in our Solar System, Titan has a thick atmosphere composed mainly of nitrogen and methane. Within this atmosphere a complex process of chemical evolution has occurred and continues to produce organic compounds. Titan is essentially a large, natural laboratory in which scientists can observe on a planetary scale some of the prebiotic processes in organic chemistry that may have ultimately led to the origin of life on Earth. Titan's thick, murky atmosphere completely hides its surface from view and so there is currently little evidence of what this surface is actually like. Because ethane appears to be the dominant organic material produced in Titan's atmosphere, it may have collected to form a global ocean, perhaps a kilometer or so deep. On the other hand, there may be thick deposits of accumulated aerosols (polymerized organic molecules) scattered over the surface of Titan, encircling seas or pools of liquid ethane.

See also: **Saturn**

Cepheid variable One of a group of very bright, supergiant stars that pulsate periodically in brightness.

See also: **stars** (red giants and supergiants)

CETI An acronym that stands for "communication with extraterrestrial intelligence." Compare with *SETI*.

See also: **extraterrestrial civilizations; interstellar communication; search for extraterrestrial intelligence**

Challenger accident The Orbiter *Challenger* lifted off Pad B, Launch Complex 39 at the Kennedy Space Center (KSC) in Florida at 11:38 A.M. on the morning of January 28, 1986, as part of Shuttle mission 51-L. At just under 74 seconds into the flight, an explosion occurred that caused the loss of the vehicle and its entire crew. (See fig. 1.) The crew members lost in this tragic event were Francis R. (Dick) Scobee (Commander), Michael John Smith (Pilot), Ellison S. Onizuka (Mission Specialist One), Judith Arlene Resnik (Mission Specialist Two), Ronald Erwin McNair (Mission Specialist Three), S. Christa McAuliffe (Payload Specialist One), and Gregory Bruce Jarvis (Payload Specialist Two). (See fig. 2.) Christa McAuliffe was a school teacher from New Hampshire (flying the Shuttle as part of NASA's Teacher-in-Space program) and Gregory Jarvis was a engineer representing the Hughes Aircraft Company. The other five were members of NASA's astronaut corps.

This tragic accident interrupted for more than two and one half years one of the most productive engineering, scientific and exploratory programs in history. It also evoked a wide range of deep and lasting public responses. First, there was grief and sadness for the loss of the seven gallant crew members. But, there was also a firm national resolve that in their lasting memory the Shuttle program itself would be strengthened, making this tragedy an important (though costly) milestone on the way to achieving the full promise that space offers mankind in the next century.

In response to this event, President Reagan appointed an independent commission, called the Presidential Commission on the Space Shuttle Challenger Accident. This commission was composed of people not connected with the 51-L mission and was charged to fully investigate the accident and to report their findings and recommendations back to the president. The following summarizes the findings and conclusions of this commission.

THE ACCIDENT

Just after liftoff at 0.678 second into the flight, photographic data show that a strong puff of gray smoke was spurting from the vicinity of the aft field joint on the right solid rocket booster (SRB). Computer graphic analysis of NASA launch photographs indicated that the initial smoke came from the 270 to 310-degree sector of the circumference of the aft field joint of the right SRB. This area of the SRB faces the Shuttle's external tank (ET). The vaporized material streaming from the joint indicated there was not complete sealing action within the joint.

Eight more distinctive puffs of increasingly blacker smoke were recorded between 0.836 and 2.500 seconds. The smoke appeared to puff upwards from the joint. While each smoke puff was being left behind by the upward flight of the Shuttle vehicle, the next fresh puff could be seen near the level of the joint. The multiple smoke puffs in this

Fig. 1 Seconds after *Challenger* was destroyed in a catastrophic explosion some 74 seconds into the 51-L mission. The Shuttle's main engine exhaust, solid rocket booster plume and expanding ball of gas from the external tank are visible. (Courtesy of NASA.)

sequence occurred at about four times per second, approximating the frequency of the structural load dynamics and resultant joint flexing.

As the Shuttle increased its upward velocity, it flew past the emerging and expanding smoke puffs. The last smoke was seen above the field joint at 2.733 seconds. The black color and dense composition of the smoke puffs suggest that the grease, joint insulation and rubber O-rings in the joint seal were being burned and eroded by the hot propellant gases.

At approximately 37 seconds, *Challenger* encountered the first of several high-altitude wind-shear conditions, which lasted until about 64 seconds into the flight. The wind shear created forces on the vehicle with relatively large fluctuations. These were immediately sensed and countered by the Shuttle's guidance, navigation and control system. The steering system (thrust vector control) of the SRB responded to all commands and wind-shear effects. The wind shear caused the steering system to be more active than on any previous Shuttle flight.

Both the three Space Shuttle main engines (SSMEs) and two solid rocket boosters operated at reduced thrust approaching and passing through the region of maximum dynamic pressure on the vehicle (some 720 pounds-force per square foot, or 34.5 kilopascals). Then, the main en-

Fig. 2   The crew members for the ill-fated 51-L mission (left to right): Payload Specialist Christa McAuliffe, Payload Specialist Gregory Jarvis, Mission Specialist Judith Resnik, Commander Francis R. Scobee, Mission Specialist Ronald E. McNair, Pilot Michael J. Smith and Mission Specialist Ellison Onizuka. (Courtesy of NASA.)

gines had been throttled up to 104 percent thrust and the SRBs were increasing their thrust when the initial flickering flame appeared on the right SRB in the area of the aft joint. This first very small flame was detected on image enhanced film at 58.788 seconds into the flight. It appeared to originate at about 305 degrees around the booster circumference at or near the aft field joint.

One film frame later from the same camera, this flame was visible without image enhancement. It grew into a continuous, well-defined plume at 59.262 seconds into the flight. At about the same time (at 60 seconds), telemetry data indicated a pressure differential between the chamber pressures in the right and left solid boosters. The right booster chamber pressure was lower, confirming the growing leak area of the field joint.

As the flame plume increased in size, it was deflected rearward by the aerodynamic slipstream and circumferentially by the protruding structure of the upper ring that attached the SRB to the external tank. These deflections directed the flame plume onto the surface of the ET. This sequence of flame spreading is confirmed by analysis of the recovered wreckage. The growing flame also impinged on the strut attaching the SRB to the ET.

The first visual indication that swirling flame from the right SRB had breached the ET occurred at 64.660 seconds into the flight, when there was an abrupt change in the shape and color of the plume. This indicated that it was mixing with leaking hydrogen from the ET. Telemetered data showing changes in hydrogen tank pressurization confirmed this leak. Within 45 milliseconds of the breach of the tank, a bright sustained glow developed on the black-tiled underside of the Orbiter *Challenger* between it and the tank.

Beginning at about 72 seconds into the flight, a series of events occurred in extremely rapid succession, causing termination of the flight. Telemetered data indicate a wide variety of flight system actions that support the visual evidence of the flight photographs, as the Shuttle struggled futilely against the forces that were destroying it.

At about 72.20 seconds into the flight the lower strut linking the right SRB and the ET was severed or pulled away from the weakened hydrogen tank, permitting the right SRB to rotate around the upper attachment strut. This rotation is indicated by divergent yaw and pitch rates between the left and right SRBs.

At 73.124 seconds, a circumferential white vapor pattern

was observed blooming from the side of the external tank dome. This was the beginning of the structural failure of the hydrogen tank that culminated in the entire aft dome dropping away. This event released large quantities of liquid hydrogen from the tank and created a sudden forward thrust of about 2.8 million pounds-force (12.5 meganewtons), pushing the hydrogen tank upward into the intertank structure. At about the same time, the rotating right SRB impacted the intertank structure and the lower part of the liquid oxygen tank. These structures failed at 73.137 seconds into the flight, as evidenced by the white vapors appearing in the intertank region.

Within milliseconds there was a massive, almost explosive, burning of the hydrogen streaming from the failed tank bottom and the liquid oxygen breach in the area of the intertank.

At this point in its trajectory, while traveling at a Mach number of 1.92 and at an altitude of 46,000 feet (about 14 km), the *Challenger* was totally enveloped in the explosive burn. The *Challenger's* reaction control system (RCS) ruptured and a hypergolic burn of the RCS propellants occurred as the propellants exited the hydrogen–oxygen inferno. The Orbiter, under severe aerodynamic loads, broke into several large sections, which emerged from the fireball. Separate sections of the disassembled flight vehicle that were identified from flight photographs included the main engine/tail section with the engines still burning, one wing of the Orbiter, and the forward fuselage trailing a mass of umbilical lines pulled loose from the Orbiter's cargo bay.

THE CAUSE OF THE ACCIDENT

The consensus of the presidential commission and participating investigative agencies was that the loss of the Space Shuttle *Challenger* was caused by a failure in the joint between the two lower segments of the right solid rocket motor. The specific failure was the destruction of the seals (O rings) that were intended to prevent hot gases from leaking through the joint during the propellant burn of the SRB. The commission further suggested that this joint failure was due to a faulty design that was unacceptably sensitive to a number of factors. These factors included the effects of temperature, physical dimensions, the character of materials, the effects of reusability, processing and the reaction of the joint to dynamic loading.

The commission also found that the decision to launch the *Challenger* on that particular day was flawed and that this flawed decision represented a contributing cause of the accident. Those who made the decision to launch were unaware of the recent history of problems concerning the O-rings and the joint. They were also unaware of the initial written recommendation of the contractor advising against launch at temperatures below 53 degrees Fahrenheit (11.7 degrees Celsius) and of the continuing opposition of the engineers at Thiokol (the manufacturer of the solid rocket motors) after the management reversed its position. Nor did the decision-makers have a clear understanding of Rockwell's (the main NASA contractor—builder of the Or-

Fig. 3 Ice and frost at launch pad 39-B, Kennedy Space Center, on the morning of January 28, 1986, prior to the launching of the 51-L mission. (Shuttle vehicle appears in background.) (Courtesy of NASA.)

biter) concern that it was not safe to launch because of ice on the launch pad. (See fig. 3.) The commission concluded that if the decision-makers had known all of these facts, it is highly unlikely that they would have decided to launch the 51-L Shuttle mission on January 28, 1986.

THE CONSEQUENCES OF THE ACCIDENT

One of the consequences of the *Challenger* accident was its adverse impact on NASA's overall image. As the details of flawed decision-making and marginal technology became known to the American public, the "model, leading-edge" federal agency that had previously accomplished the impossible by placing men on the Moon, was transformed into an organization just like any other federal bureaucracy: fallible, vulnerable to political pressure and cumbersome in its ability to make key decisions. This accident also displaced the Space Shuttle as the major U.S. launch vehicle and triggered a rapid return to expendable launch vehicles. Many military and scientific payloads had to be redesigned to fly into space onboard these expendable boosters. In addition, numerous scientific programs were delayed indefinitely, including many of the astrophysics and life science missions scheduled for Shuttle/Spacelab flights. And finally, the ability of the Space Shuttle fleet to launch and maintain a permanent space station (at a launch rate of some 10 to 12 Shuttle flights per year) is being questioned.

On the positive side, the *Challenger* accident also stimulated many Shuttle System modifications and safety improvements, including a modified nose wheel steering system, an improved main landing gear, and the incorporation of an emergency egress slide. This emergency egress slide provides crew members with a means for rapid and safe exit through the Orbiter hatch after a normal opening of the side hatch or after jettisoning the side hatch at the nominal end-of-mission landing site or at a remote or emergency landing site. In addition, NASA restructured the overall launch decision process, including the implementation of a fully accountable flight readiness certification procedure. These hardware and management changes have now been incorporated into the Space Shuttle program to reduce the overall risk associated with future Shuttle flights and to avoid an accumulation of undesirable technical and management circumstances similar to those that led to the *Challenger* tragedy.

See also: **hazards to space workers; National Aeronautics and Space Administration; Space Transportation System**

circumsolar space   Around the Sun; heliocentric (Sun-centered) space.

circumstellar   Around a star—as opposed to *interstellar*, between the stars.

cislunar   Generally, in or pertaining to the region of outer space between the Earth and the Moon. By convention, the outer limit of cislunar space is considered to be the outer limit (from the Moon) of its sphere of (gravitational) influence.

Civil Space Technology Initiative (CSTI)   A vital component of NASA's current space research and development program. CSTI's objectives are to develop specific technologies critical to the accomplishment of near-term, high-

Table 1   Primary CSTI Thrusts

TRANSPORTATION
• Earth-to-orbit (ETO) propulsion
• Booster technology
• Aeroassist flight experiment

OPERATIONS
• Robotics
• Autonomous systems
• High-capacity power

SCIENCE
• Science sensor technology
• Data: high rate/capacity
• Control of flexible structures
• Precision segmented reflectors

SOURCE: Developed by the author, based on NASA data.

Fig. 1   An Aeroassist Orbit Transfer Vehicle (AOTV) slows itself with a balloon-like drag device while passing through the Earth's upper atmosphere before swinging out into its parking orbit in low Earth orbit.

priority, national space goals. The main CSTI emphasis is being placed on technology for efficient, reliable, access to and operations in low Earth orbit (LEO), and on support of science missions conducted from Earth orbit. CSTI elements (table 1) include Earth-to-orbit (ETO) propulsion; the aeroassist flight experiment (AFE; see fig. 1); telerobotics; artificial intelligence (AI); high-capacity power, and science sensor technology.

As shown in figure 1, for example, an Aeroassist Orbit Transfer Vehicle (AOTV) slows itself with a balloon-like drag device, while passing through the upper regions of the Earth's atmosphere before it swings out into a low-altitude parking orbit. The key advantage of an aeroassist maneuver versus an all-propellant "braking" equivalent is the large savings of propellant mass that would otherwise be required to perform the necessary braking and/or orbital capture engine firings. In some anticipated space mission scenarios, this propellant mass reduction could account for as much as a doubling of useful mission payload.

See also: **Project Pathfinder; robotics in space; space nuclear power; space launch vehicles**

closed universe   The closed or bounded universe model in cosmology assumes that the total mass of the Universe is sufficiently large so that one day the galaxies will slow down and stop expanding due to their mutual gravitational attraction. At that time, the Universe will have reached its maximum size; then, it will slowly start to contract under the influence of gravity. This contraction will continue until the total mass of the Universe is essentially compressed together. (This compressed mass condition is sometimes called the "Big Crunch.") Some advocates of the closed Universe model also speculate that after the Big Crunch a new expansion of matter (another "Big Bang") will occur, as part of an overall "oscillating Universe" cycle. Compare with **open universe**.

See also: **astrophysics; "Big Bang" theory; cosmology**

close encounter (CE) An interaction with an unidentified flying object (UFO).

See also: **unidentified flying object**

cluster of galaxies An accumulation of galaxies. These galactic clusters can occur with just a few member galaxies (say, 10 to 100)—such as the Local Group, of which our Milky Way is a part—or they can occur in great groupings involving thousands of galaxies. Clusters of galaxies are gravitationally bound systems of galaxies within a few million light-years of each other. When observed optically, these clusters are dominated by the (visible) light emission from individual galaxies, but when viewed in the X-ray region of the electromagnetic spectrum, most of their X-ray emission appears to come from hot gas between the galaxies. In fact, scientists now think that there is about as much of this hot gaseous material between the galaxies in a cluster, as there is matter in the galaxies themselves.

See also: **galaxy, Local Group, X-ray astronomy**

comet A dirty ice "rock" orbiting the Solar System that the Sun causes to vaporize, glow visibly and stream out a long, luminous tail. Comets are generally regarded as samples of primordial material from which the planets were formed billions of years ago. The comet's nucleus is a type of dirty ice ball, consisting of frozen gases and dust. (See fig. 1.) As the comet approaches the Sun from the frigid regions of deep space, the Sun's radiation causes these ices to sublime (vaporize) and the resultant vapors form an atmosphere or "coma" with a diameter that may reach 100,000 kilometers (60,000 miles). However, cometary nuclei have diameters of only a few tens of kilometers. Ions produced in the coma are effected by the charged particles in the solar wind; while dust particles liberated from the comet's nucleus are impelled in a direction away from the Sun by the pressure of the solar wind. The results are the formation of the plasma (Type I) and dust (Type II) cometary tails, which can extend for up to 100 million kilometers (60 million miles). The Type I tail, composed of ionized gas molecules, is straight and extends radially outward from the Sun as far as 100 million kilometers (10^8 km). The Type II tail, consisting of dust particles, is shorter, generally not exceeding 10 million kilometers in length. It curves in the opposite direction to the orbital movement of the comet around the Sun. It appears that an enormous cloud of hydrogen atoms also surrounds the visible coma. This hydrogen was first detected in comets in the 1960s.

No astronomical object, other than perhaps the Sun or the Moon, has attracted more attention or interest. Since ancient times, comets have been characterized as harbingers of momentous human events. William Shakespeare wrote in the play *Julius Caesar:*

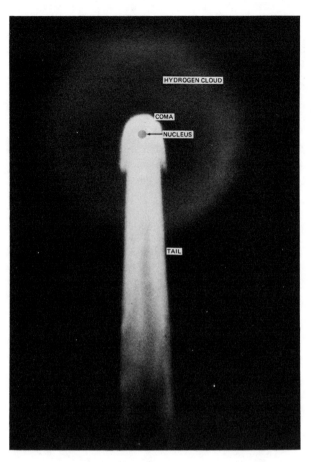

Fig. 1   The major features of a comet. (Courtesy of NASA.)

"When beggars die, there are no comets seen; but the heavens themselves blaze forth the death of princes."

All comets are believed to originate far from the Sun in the Oort Cloud, which is thought to extend out to the limits of the Sun's gravitational attraction, creating a sphere with a radius of between 60,000 and 80,000 astronomical units. This cloud, first described by the Dutch astronomer Jan Hendrik Oort, is thought to contain billions of comets, whose total mass is estimated to be roughly equal to the mass of the Earth. Small numbers of these comets continually enter the planetary regions of our Solar System, possibly through the gravitational perturbations caused by neighboring stars or other "chaotic" phenomena. Once a comet approaches the Solar System, it is also subject to the gravitational influences of the major planets, especially Jupiter, and the comet may eventually achieve a quasi stable orbit within the Solar System.

By convention, comet orbital periods are divided into two classes: long period comets (which have orbital periods in excess of 200 years), and short period comets (which have periods less than 200 years).

Comets may represent interesting sources of extraterrestrial materials. Table 1 lists the atomic and molecular species that have been observed in cometary coma. A

Table 1 Some Observed Atomic and Molecular Compositions of Comets (Coma)

H   (atomic hydrogen)
C, C_2, C_3 (carbon: atomic and molecular)
CH   (carbon-hydrogen free radical)
CN   (cyanogen radical)
HCN   (hydrogen cyanide)
CH_3CN (acetonitrile)
NH   (imine radical)
NH_2 (nitrogen hydride molecule)
NH_3 (ammonia)
O   (atomic oxygen)
OH   (hydroxyl radical)
H_2O (water)
CO   (carbon monoxide)
CO_2 (carbon dioxide)
H_2CO (formaldehyde)
CH_4 (methane)
N_2 (nitrogen)
Na   (sodium)
Fe   (iron)
S   (sulfur)

SOURCE: NASA.

rendezvous with or even an automated sample return mission to a short period comet would answer many of the questions currently puzzling cometary physicists. A highly automated sampling mission, for example, would permit the collection of atomized dust grains and gases directly from the comet's coma. The simplest way to accomplish this exciting space mission would be to use a high-velocity flyby technique, with the automated spacecraft passing as close to the comet's nucleus as possible. This probe would be launched on an Earth-return trajectory. Terrestrial recovery of its cosmic cargo could be accomplished by on-orbit rendezvous or by the ejection of the sample into a small atmospheric entry capsule.

Because they are only very brief visitors to our Solar System, you might wonder what happens to these comets after they blaze a trail across the night sky? Well, the chance of any particular "new" comet being captured in a short-period orbit is quite small. Therefore, most of these cosmic wanderers simply return to the Oort Cloud, presumably to loop back into the Solar System eons later. Or else they are ejected into interstellar space along hyperbolic orbits. Sometimes, however, a comet falls into the Sun. One such event was observed by instruments on a spacecraft in August 1979. Other comets simply break up because of gravitational (tidal) forces or possible outbursts of gases from within. When a comet's volatile materials are exhausted or when its nucleus is totally covered with non-volatile substances, we call the comet "inactive." Some space scientists believe that an inactive comet may become a "dark asteroid," such as those in the Apollo group, or else disintegrate into meteoroids. Finally, on very rare occasions, a comet may even collide with an object in the Solar System, an event with the potential of causing a "cosmic catastrophe."

Perhaps the most well-known comet is Comet Halley. In March 1986, an international armada of five spacecraft encountered this comet as it made its periodic (approximately every 75 years) return to the inner Solar System. These scientific vehicles included: the Soviet *Vega 1* and *Vega 2* spacecraft; the Japanese *Suisei* and *Sakigake* spacecraft; and the European Space Agency (ESA) *Giotto* spacecraft. The ESA *Giotto* spacecraft made and survived the most hazardous encounter as it streaked within 610 kilometers (379 miles) on the sunward-side of the comet's nucleus on March 14 at a relative encounter velocity of 68.4 kilometers per second (42.5 miles per second). Although the United States did not sponsor a special comet-encounter spacecraft mission, American scientists did participate extensively in the overall efforts to explore Comet Halley. For example, the U.S. *Pioneer 12* spacecraft, orbiting between the Earth and Venus, viewed the comet for 12 days as it sped towards the Sun and then was the only spacecraft positioned to observed the comet at perihelion (its closest approach to the Sun—a distance of some 140.8 million kilometers [87.5 million miles]).

As a result of the highly successful international efforts to explore Comet Halley, scientists have been able to confirm their previously postulated dirty ice "rock" model of a comet's nucleus. We learned, for example, that the Halley nucleus is a discrete, single "peanut-shaped" body (some $16 \times 8 \times 7.5$ kilometers [$9.9 \times 5.0 \times 4.7$ miles] in dimension). A very low-albedo (about 0.05), dark dust layer (of yet undetermined thickness) covers the nucleus. The surface temperature of this layer was found to be about 330 Kelvin (57 Celsius); somewhat higher than expected when compared to the temperature of a subliming ice surface, which would be about 190 Kelvin (−83 Celsius). The total gas production rate from the comet's nucleus at the time of the *Giotto* flyby encounter was observed as 6.9×10^{29} molecules per second, of which 5.5×10^{29} molecules per second were water vapor. This observation means that 80 to 90 percent of the comet's nucleus consists of water ice and dust. Other "parent" molecules in the nucleus appear to be carbon dioxide (CO_2), ammonia (NH_3) and methane (CH_4).

Scientists also learned that the abundance ratios of isotopes such as carbon-12 to carbon-13 [C^{12}/C^{13}] and nitrogen-14 to nitrogen-15 [N^{14}/N^{15}] are the same for Comet Halley as they are for the Earth and the rest of the Solar System. Consequently, these observations provided strong physical evidence that comets (as previously postulated) consist of the same primordial material that formed the Sun and the planets. Finally, Halley's nucleus exhibited a rotational period of approximately 2.2 days with an accompanying mutation or wobble phenomenon that occurred every 7.4 days.

If you're planning ahead, Comet Halley's next visit to the inner Solar System will occur in 2061.

See also: **Comet Rendezvous and Asteroid Flyby Mission; Rosetta Mission**

Comet Rendezvous and Asteroid Flyby (CRAF) Mission This mission will extend the international Comet Halley encounter experience (1986) by meeting Comet Kopff near the orbit of Jupiter and traveling along with it for almost three years as the comet loops around the Sun. This sophisticated NASA robot space mission will be the first to use the new Mariner Mark II spacecraft and will also launch a heavily instrumented penetrator/lander probe that will directly sample the comet's nucleus. On the way to encounter Comet Kopff, the CRAF spacecraft will also encounter the asteroid Hamburga.

The target comet, Kopff, is a "short-period" comet whose orbit carries it once around the Sun in 6.46 years. This particular comet travels between a region from just inside the orbit of Jupiter inward to perihelion near the orbit of Mars, some 220 million kilometers (137 million miles) from the Sun.

The main objectives of the comet rendezvous portion of the CRAF mission include (1) characterization of the comet nucleus and coma; (2) a study of the comet tail formation process; and (3) a study of comet tail dynamics, including how a comet tail interacts with the solar wind and solar radiation.

The main objectives of the asteroid flyby portion of the CRAF mission include (1) characterization of the asteroid's structure and geology; (2) a determination of the distribution of minerals, metals and ices (volatiles) on this asteroid; and (3) a measurement of the asteroid's mass and density.

The CRAF spacecraft will be launched by a Titan IV-Centaur expendable launch vehicle configuration in August 1995. NASA's currently planned mission trajectory will carry the CRAF space vehicle out to the asteroid belt, where a propulsive maneuver will then send the spacecraft back toward Earth for a gravity assist boost (July 1997).

In January 1998, while enroute to the Comet Kopff encounter, the CRAF spacecraft will flyby an asteroid called 449 Hamburga. This C-type (carbonaceous) asteroid is about 88 kilometers (55 mi) in diameter. During the asteroid flyby, the CRAF spacecraft will take photographs and make other scientific measurements.

Then, in August 2000, the CRAF spacecraft will arrive at the rendezvous point with Comet Kopff—named after August Adalbert Kopff, who discovered it on August 22, 1906. At this rendezvous point, both the target comet and the CRAF spacecraft will be at the distance of Jupiter's orbit and some 850 days before *perihelion* (that is, before closest approach to the Sun). Sometime in August 2001, the CRAF spacecraft will fire its penetrator/lander probe at the comet's nucleus. It will then continue to fly in an extremely close formation with the comet's nucleus, typically within 10 kilometers (6 mi) range, and will perform close-up exploration before the coma and tail begin to build as the comet approaches the Sun.

Later in the mission, the CRAF spacecraft will move in and out through the comet's coma and down its tail to study their properties, to observe the complex processes brought on by an approach to the Sun and to collect gas and dust samples for detailed automated analyses onboard the robot spacecraft. In December 2002, when the target comet is most active (near perihelion), its nucleus will spew out approximately a ton of gas and dust per second. Therefore, the CRAF mission will gather valuable cometary science data for a total of two and two-thirds years, that is from initial rendezvous (in August 2000) until about 109 days after the robot spacecraft and comet have experienced perihelion and are outward bound again toward aphelion near Jupiter's orbit (late March 2003).

This type of comet rendezvous mission supports the study of cosmic matter that scientists think is the original, relatively unchanged material left behind when a primordial cloud of dust and gas collapsed to form the Solar System some 4.6 billion years ago.

See also: **asteroid, comet, Rosetta Mission, Vesta Missions**

consequences of extraterrestrial contact Just what will happen if we make contact with an extraterrestrial civilization? No one on Earth can really say for sure. However, this contact will very probably be one of the most momentous events in all human history! The contact can be direct or indirect. Direct contact might involve a visit to Earth by a starship from another stellar civilization, or could perhaps take the form of the discovery of an alien probe, artifact or derelict spaceship in the outer regions of our Solar System. Some scientists have speculated, for example, that the hydrogen- and helium-rich giant outer planets might serve or could have served as "fueling stations" for interstellar spaceships from other worlds. Indirect contact, via radio-wave communication, appears to represent the more probable contact pathway (at least from a contemporary terrestrial viewpoint). The consequences of our successful search for extraterrestrial intelligence (SETI) would be nothing short of extraordinary. For example, as part of our own SETI effort, were we to locate and identify but a single extraterrestrial signal, humankind would know immediately one great truth: We are not alone, and it is indeed possible for a civilization to create and maintain an advanced technological society without destroying itself. We might even learn that life, especially intelligent life, is prevalent in the Universe!

The overall impact of this contact will depend on the circumstances surrounding the initial discovery. If it happens by accident or after only a few years of searching, this news, once verified, would surely startle the world. If, however, intelligent alien signals were detected only after an extended effort, lasting generations and involving extensive search facilities, the terrestrial impact of the discovery might be less overwhelming.

The reception and decoding of a radio signal from an extraterrestrial civilization in the depths of space offers the promise of practical and philosophical benefits for all humanity. Response to that signal, however, also involves a potential planetary risk. If we do intercept an alien signal, we can decide (as a planet) to respond—or we may choose not to respond. If we are suspicious of the motives of the alien culture that sent the message, we are under no obligation to respond. There would be no practical way for them to realize that their signal was in fact intercepted, decoded and understood by the intelligent inhabitants of a tiny world called Earth.

Optimistic speculators emphasize the friendly nature of such an extraterrestrial contact and anticipate large technical gains for our planet, including the reception of information and knowledge of extraordinary value. They imagine that there will be numerous scientific and technological benefits from such contacts. However, because of the long round-trip times associated with such radio contacts (perhaps decades or centuries, even with the messages traveling at the speed of light), any information exchange will most likely be in the form of semi-independent transmissions, each containing significant facts about the sending society (such as planetary data, its life-forms, its age, its history, its philosophies and beliefs, and whether it has successfully contacted other alien cultures), rather than an interstellar dialogue with questions asked and answered in rapid succession. Consequently, over the period of a century or more, we Terrans might receive a wealth of information at a gradual enough rate to construct a comprehensive picture of the alien civilization without inducing severe culture shock on Earth.

Some scientists feel that if we are successful in establishing interstellar contact, we would probably not be the first planetary civilization to have accomplished this feat. In fact, they speculate that interstellar communications may have been going on since the first intelligent civilizations evolved in our Galaxy—some four or five billion years ago. One of the most exciting consequences of this type of celestial conversation would be the accumulation by all participants of an enormous body of information and knowledge that has been passed down from alien race to alien race since the beginning of the Galaxy's communicative phase. Included in this vast body of knowledge, something we might call the "galactic heritage," could be the entire natural and social histories of numerous species and planets. Also included, perhaps, would be extensive astrophysical data that extend back countless millennia, providing accurate insights into the origin and destiny of the Universe.

It is felt, however, that these extraterrestrial contacts would lead to far more than merely an exchange of scientific knowledge. Humanity would discover other social forms and structures, probably better capable of self-preservation and genetic evolution. We would also discover new forms of beauty and become aware of different endeavors that promote richer, more rewarding lives. Such contacts might also lead to the development of branches of art and science that simply cannot be undertaken by just one planetary civilization but rather require joint, multiple-civilization participation across interstellar distances. Most significant, perhaps, is the fact that interstellar contact and communication would represent the end of the cultural isolation of the human race. The optimists speculate that we would

be invited to enter a sophisticated "cosmic community" as mature, planetary-civilization "adults" proud of our own human heritage—rather than remaining isolated with a destructive tendency to annihilate ourselves in childish planetary rivalries. Indeed, perhaps the very survival and salvation of the human race depends on finding ourselves cast in a larger cosmic role—a role far greater in significance than any human can now imagine.

If a cosmic community of extraterrestrial civilizations really does exist, it is probably composed of individual cultures and races that have learned to live with themselves and their technologies. Cultural life expectancies might be measured in aeons rather than millennia or even centuries. Identifying with these "super" interstellar civilizations and making contributions to their long-term objectives would definitely provide exciting new dimensions to our own lives and would create an interesting sense of purpose for our planetary civilization.

In considering contact with an extraterrestrial civilization, it is not totally unreasonable to think as well about the possible risks that could accompany exposing our existence to an alien culture—most likely far more advanced and powerful than our own. These risks range from planetary annihilation to humiliation of the human race. For discussion, these risks can be divided into four general categories: (1) invasion, (2) exploitation, (3) subversion and (4) culture shock.

The invasion of the Earth is a very recurrent theme in science fiction. By actively sending out signals into the cosmic void or responding to intelligent signals we've detected and decoded, we would be revealing our existence and announcing the fact that the Earth is a habitable planet. Soon thereafter (this risk scenario speculates) our planet might be invaded by an armada of spaceships carrying vastly superior beings who are set on conquering the Galaxy. After a valiant, but futile, fight, humankind is annihilated or enslaved. While such scenarios make interesting motion pictures and science-fiction novels, a logical review of the overall situation does not appear to support the extraterrestrial-invasion hypothesis or its grim (for Earth) outcome. If, for example, as we currently speculate, direct contact via interstellar travel is enormously expensive and technically very difficult—even for an advanced extraterrestrial civilization—then perhaps only the most extreme crisis would justify mass interstellar travel. Any alien race capable of interstellar travel would most certainly possess the technical skills needed to solve planetary-level population and pollution problems. Hence, the quest for more "living space" as a dominant motive for interstellar migration by an alien civilization does not appear to be a logical premise. It is not altogether inconceivable, of course, that members of an advanced civilization might seek to avoid extinction through mass interstellar migration before their native star leaves the main sequence and threatens to supernova. Again, we can logically conjecture that such a powerful, migrant extraterrestrial race would probably not want to compound the problems of a complex, difficult interstellar journey with the additional problems of interstellar warfare. They would most likely seek habitable, but currently uninhabited, worlds upon which to settle and rebuild their civilization. Such habitable worlds could have been located and identified long in advance, perhaps through the use of sophisticated robot probes.

Of course, interstellar travel might also prove much easier than we now predict. If this is the case, then the Galaxy could be teeming with waves of interstellar expeditions launched by expanding civilizations. Maintaining "radio silence" on a planetary scale is consequently no real protection from such waves of extraterrestrial explorers. We would inevitably be discovered by one of numerous bands of wandering extraterrestrials—without the aid of our electromagnetic-wave "homing beacons." In this situation the real question to be raised is the Fermi paradox: Where are they?

Physical contact between Earth and a benign advanced extraterrestrial civilization might also give rise to a silent, unintentional "invasion." Because of vast differences in biochemistry, alien microorganisms introduced into the terrestrial biosphere during physical contact with an alien civilization could trigger devastating plagues that would annihilate major life-forms on Earth.

Another major contact hazard category is exploitation. Some individuals have speculated that to an advanced alien civilization, human beings might appear to be primitive life-forms—ones that represent interesting experimental animals, unusual pets or even gourmet delicacies. Fortunately, differences in biochemistry might also make us very poisonous to eat. Again, the arguments against the "invasion scenario" apply equally well here. It is very difficult to imagine an advanced civilization expending great resources to cross the interstellar void just to bring home exotic pets, unusual lab animals or—perish the thought—"imported snacks." Perhaps it is more logical to assume that when an alien civilization matures to the level of star travel, such cultural qualities as compassion, empathy and a respect for life (in any form) become dominant.

Another major alien contact hazard category that is frequently voiced is subversion. This appears to be a more plausible and subtle form of contact risk. In this case, an alien race—under the guise of teaching and helping us join a cosmic community—might actually trick us into building devices that allow "Them" to conquer "Us." The alien civilization doesn't even have to make direct contact; the extraterrestrial "Trojan horse" might arrive on radio waves. For example, the alien race could transmit the details of a computer-controlled biochemistry experiment that would then secretly create their own life-forms here. The subversion and conquest of Earth would occur from within! There appears to be no limit to such threats, if we assume terrestrial gullibility and alien treachery. Our only real protection would be to take adequate security precautions during a contact and to maintain a healthy degree of suspicion. A form of extraterrestrial xenophobia may not be totally inappropriate. Perhaps special facilities on the Moon or Mars could serve as extraterrestrial contact and communications "ports of entry" into the Solar System. Any alien attempts at subversion could then be rapidly isolated and, if necessary, terminated long before the terrestrial biosphere itself was endangered.

The fourth major risk category involving extraterrestrial contact is massive culture shock. Some individuals have expressed concerns that even mere contact with a vastly superior extraterrestrial race could prove extremely damaging to human psyches, despite the best intentions of the alien race. Terrestrial cultures, philosophies and religions that now place humans at the very center of creation would have to be "expanded" to compensate for the existence of other, far superior intelligent beings. We would now have to "share the Universe" with someone or something better and more powerful than we are. As the dominant species on this planet, could humans accept this new role? While many scientists currently believe in the existence of intelligent species elsewhere in the Universe, we must keep asking ourselves a more fundamental question: Is humankind in general prepared for the positive identification of such a fact? Will contact with intelligent aliens open up a golden age on Earth or initiate cultural regression?

Historians and sociologists, in studying past contacts between two terrestrial cultures, observe that generally (but not always) the stronger, more advanced culture has dominated the weaker one. This domination, however, has always involved physical contact and usually territorial expansion by the stronger culture. If contact has occurred without aggression, the lesser culture has often survived and even prospered. In the case of extraterrestrial contact by means of interstellar communication, the long delays while messages span the cosmic void at the speed of light should enable our planetary civilization to adapt to the changing cosmic condition. There are no terrestrial examples of cultural domination by radio signals alone, and round-trip exchanges of information would require years or even human generations.

Of course, we cannot assume that contact with an alien civilization is without risk. Four general risk categories have just been discussed. Many individuals now feel that the potential benefits (also previously described) far outweigh any possible concerns. Simply to listen for the signals radiated by other intelligent life does not appear to pose any great danger to our planetary civilization. The real hazard issue occurs if we decide to respond to such signals. Perhaps a planetary consensus will be necessary before we answer an "interstellar phone call." It is also interesting to note here, however, that our ultrahigh-frequency (UHF) television signals are already propagating far out into interstellar space and will possibly be detectable out to some 25 to 50 light-years' distance. Are aliens tonight examining a decades-old episode of "Gunsmoke" as a message from Earth?

It is interesting to recognize that the choice of initiating extraterrestrial contact is no longer really ours. In addition to the radio and television broadcasts that are leaking out into the Galaxy at the speed of light, the powerful radio/radar telescope at the Arecibo Observatory was used to beam an interstellar message of friendship to the fringes of the Milky Way Galaxy on November 16, 1974. We have, therefore, already announced our presence to the Galaxy and should not be too surprised if someone or something answers!

See also: **Arecibo Interstellar Message; extraterrestrial civilizations; Fermi paradox; interstellar communication; search for extraterrestrial intelligence**

Cosmic Background Explorer (COBE)   NASA's *Cosmic Background Explorer* (COBE) spacecraft was successfully launched from Vandenberg Air Force Base, California, by an expendable Delta rocket on November 18, 1989. The 2,268 kg (5,000 lbm) spacecraft was placed in a 900-km (559-mi), 99-degree inclination (polar) orbit, passing from pole to pole along the Earth's terminator (the line between night and day on a planet or moon) to protect its sensitive instruments from the heat and light of the Sun and to prevent the spacecraft's instruments from pointing at the Sun or the Earth.

COBE's one-year space mission was to study some of the most basic questions in astrophysics and cosmology. What was the nature of the primeval explosion (often called the "Big Bang") that started the expanding Universe? What started the formation of galaxies? What caused galaxies to be arranged in giant clusters with vast unbroken voids in between?

Scientists have speculated for decades about the formation of the Universe. Today, the most generally accepted cosmological model is called the Big Bang theory of an expanding Universe. The most important evidence that this gigantic explosion occurred some 15 billion years ago is the uniform diffuse microwave background radiation that reaches the Earth from every direction. This cosmic background radiation was discovered in 1964 as a type of "static from the sky." Its occurrence now appears to be explainable only as a remnant of the Big Bang event.

The COBE spacecraft mapped the sky and measured the radiation emitted by a great variety of objects, in addition to the cosmic background radiation of the Big Bang. COBE's three instruments—the differential microwave radiometer (DMR), the far infrared absolute spectrophotometer (FIRAS) and the diffuse infrared background experiment (DIBRE)—successfully observed the entire sky twice during the one-year mission lifetime. The differential microwave radiometer gathered data that is now helping scientists determine whether the primeval explosion was equally bright in all directions. The far infrared absolute spectrophotometer surveyed the sky to determine whether the cosmic background radiation from the Big Bang has the theoretically predicted spectrum (that is, the anticipated intensity at each wavelength). Finally, the diffuse infrared background experiment searched for the light from the earliest stars and galaxies.

See also: **"Big Bang" theory; cosmology**

cosmic dust   Fine microscopic particles drifting in outer space.

See also: **interplanetary dust**

cosmic rays   Atomic particles (mostly bare atomic nuclei) that have been accelerated to very high velocities and carry great amounts of energy. Cosmic rays move through space at speeds just below the speed of light. They carry an electric charge and spiral along the weak lines of magnetic force that permeate the Galaxy. Although discovered over half a century ago, their origin still remains a mystery. Scientists currently believe they are galactic in origin, but some cosmic rays (especially the most energetic) may actually originate outside the Milky Way Galaxy, making them tiny pieces of extragalactic material.

Galactic cosmic rays represent a unique sample of material from outside the Solar System. Although hydrogen nuclei (that is, protons) make up the highest proportion of the cosmic-ray population (approximately 85%), these particles also range over the entire periodic table of elements, from hydrogen through uranium, and include electrons and positrons. Galactic cosmic rays bring astrophysicists direct evidence of processes like nucleosynthesis and particle acceleration that occur as a result of explosive processes in stars throughout the Galaxy.

Solar cosmic rays consist of protons, alpha particles and other energetic atomic particles ejected from the Sun during solar flare events. Solar cosmic rays are generally lower in energy than galactic cosmic rays.

See also: **astrophysics**

cosmology   May be defined as the study of the origin, evolution and structure of the Universe. Astrophysical discoveries in recent years tend to support the "Big Bang" theory of cosmology—a theory stating that about 15 to 20 billion (10^9) years ago the Universe began in a great explosion (sometimes called the "initial singularity") and has been expanding ever since. For example, the 1964 discovery of the 2.7 degree Kelvin cosmic microwave background radiation with a radio telescope provided the initial observational evidence that there was a very hot phase early in the history of the Universe.

See also: **astrophysics; "Big Bang" theory; closed universe; Cosmic Background Explorer; open universe**

cosmonaut   The name used by the Soviet Union for its space travelers, or "astronauts."

cyborg   A term originating in the science-fiction literature meaning an artificially produced human being. Very sophisticated robots (that is, those robots with near-human qualities) have been called cyborgs, as have human beings who were artificially created.

See also: **android**

D

dark nebula   A cloud of interstellar dust and gas sufficiently dense and thick so that the light from more distant stars and celestial objects is obscured. The Horsehead Nebula (NGC 2024) in Orion is an example of a dark nebula.

See also: **nebula**

Deep Space Network (DSN)   The NASA Deep Space Network (DSN) is a worldwide system for tracking, navigating and communicating with unmanned interplanetary spacecraft, probes and planetary landers. This global network of antennas serves as the radio communications link to distant interplanetary spacecraft and probes, transmitting instructions to them and receiving the data they return to Earth from deep space.

Operated by the Jet Propulsion Laboratory (JPL), the Deep Space Network is a vital element in every space exploration project. It evolved from tracking and data recovery techniques that were developed at JPL during the conduct of missile work for the U.S. Army. For example, in 1958 JPL established a three-station network of receiving stations to gather data from the first U.S. satellite, *Explorer 1*.

The Deep Space Network uses high-sensitivity, large antennas (typically 64-meter [210 feet] and improved 70-meter [230 feet] diameter devices), low-noise receivers and high-power transmitters at locations strategically positioned on three continents (North America, Europe and Australia). The DSN has grown to include a global network of 12 deep-space antenna stations. The three Deep Space Communications Complexes (DSCCs) are located at Goldstone, California (in the Mojave Desert of Southern California); near Madrid, Spain; and near Canberra, Australia. These locations are approximately 120 degrees longitude apart around the Earth, so that the DSN can maintain essentially continuous contact with spacecraft in deep space. Thus, as one antenna dish loses contact due to the Earth's rotation, another one takes over the task of receiving data from the interplanetary spacecraft.

The Deep Space Network has provided tracking and data acquisition support for numerous space exploration projects. These include Ranger (1961–65), Surveyor (1966–68) and Lunar Orbiter (1966–67) to the Moon; the Mariner missions to Venus (1962 and 1967), Mars (1964, 1969 and 1971) and Venus-Mercury (1973); the Pioneer inward and outward heliocentric (Sun-centered) orbiters (1965–68); and the historic *Pioneer 10* and *11* missions to Jupiter and Saturn (1972 and 1973). The DSN has also provided support for the Pioneer Venus orbiter and multiprobe (1978); the joint U.S.–West German Helios spacecraft in orbit around the Sun (1974 and 1976); the Viking orbiter-lander missions to Mars (1975); and the Voyager missions to Jupiter, Saturn, Uranus, Neptune and beyond (1977). (The dates shown are spacecraft launch dates.) The DSN also supported the international spacecraft armada to Comet Halley in March 1986.

After traveling across interplanetary space, the spacecraft signal that reaches a DSN antenna ranges in power from a billionth of a watt (10^{-9}) down to a billionth (10^{-9}) of a trillionth (10^{-12}) of a watt, or some 10^{-21} watt. New technology continually contributes to the network's ability to communicate with ever more distant spacecraft. For example, on June 13, 1983, the DSN successfully recorded data from the *Pioneer 10* spacecraft as it passed the orbit of Neptune to become the first human object to leave the known Solar System. (Pluto's eccentric orbit now has the planet inside of Neptune's orbit until 1999.) That particular communication with the *Pioneer 10* spacecraft covered a distance of more than 4.5 billion kilometers (about 250 light-minutes). It is currently anticipated that the DSN will track and communicate with the *Pioneer 10* and *11* spacecraft and the *Voyager 1* and *2* spacecraft as they leave the Solar System in various directions until at least the year 2000. Furthermore, because of their long-lived nuclear (radioisotope) power supplies, more powerful transmitters, X-band downlink and larger antennas, the *Voyager 1* and *2* spacecraft will most likely be trackable well into the first decade of the next millennium—as they continue to monitor the interstellar medium that surrounds our Solar System.

Although the primary activity of the DSN is to conduct telecommunications with unmanned spacecraft in deep space, the network is also a scientific instrument and has been used for many interesting astronomical activities, including monitoring natural radio sources, such as pulsars and quasars; radar studies of planetary surfaces and the Saturnian ring system; celestial mechanics experiments; lunar gravity experiments; and Earth physics experiments. One of the most interesting (non-spaceflight) experiments for the DSN involves the search for extraterrestrial intelligence (SETI) through the detection and identification of coherent radio messages from alien civilizations.

See also: **images from space; Pioneer 10, 11; search for extraterrestrial intelligence; Voyager**

Doppler shift   The change in (apparent) frequency and wavelength of a source due to the relative motion of the source and an observer. If the source is approaching the observer, the observed frequency is higher and the observed wavelength is shorter. This change to shorter wavelengths is often called the "blueshift." If, on the other hand, the source is moving away from the observer, the observed frequency will be lower and the wavelength will be longer. This change to longer wavelengths is called the "redshift," since for a visible light source this would mean a shift to the longer-wavelength, or red, portion of the visible spectrum.

Drake equation Just where do we look among the billions of stars in our Galaxy for possible interstellar radio messages or signals from extraterrestrial civilizations? That was one of the main questions addressed by the attendees of the Green Bank Conference on Extraterrestrial Intelligent Life held in November 1961 at the National Radio Astronomy Observatory (NRAO), Green Bank, West Virginia. One of the most significant and widely used results from this conference is the Drake equation (named after Dr. Frank Drake), which represents the first attempt to quantify the search for extraterrestrial intelligence (SETI). This "equation" has also been called the Sagan–Drake equation and the Green Bank equation in the SETI literature.

While more nearly a subjective statement of probabilities than a true scientific equality, the Drake equation attempts to express the number (N) of advanced intelligent civilizations that might be communicating across interstellar distances at this time. A basic assumption inherent in this formulation is the principle of mediocrity—namely, that things in our Solar System (and especially on Earth) are nothing particularly special and represent common conditions found elsewhere in the Galaxy. The Drake equation is generally expressed as:

$$N = R^* f_p n_e f_l f_i f_c L \qquad (1)$$

where N is the number of intelligent communicating civilizations in the Galaxy at present

R^* is the average rate of star formation in our Galaxy (stars/year)

f_p is the fraction of stars that have planetary companions

n_e is the number of planets per planet-bearing star that have suitable ecospheres (that is, the environmental conditions necessary to support the chemical evolution of life)

f_l is the fraction of planets with suitable ecospheres on which life actually starts

f_i is the fraction of planetary life starts that eventually evolve to intelligent life-forms

f_c is the fraction of intelligent civilizations that attempt interstellar communication

and L is the lifetime (in years) of technically advanced civilizations.

An inspection of the Drake equation quickly reveals that the major terms cover many disciplines and vary in technical content from numbers that are somewhat quantifiable (such as R^*) to those that are completely subjective (such as L).

For example, astrophysics can provide us a reasonably approximate value for R^*. Namely, if we define R^* as the average rate of star formation over the lifetime of the Galaxy, we obtain

$$R^* = \frac{\text{number of stars in the Galaxy}}{\text{age of the Galaxy}} \qquad (2)$$

We can then insert some typically accepted numbers for our Galaxy to arrive at R^*. Namely,

$$R^* = \frac{100 \text{ billion stars}}{10 \text{ billion years}}$$

$$R^* = 10 \text{ stars/year (approximately)}$$

Generally, the estimate for R^* used in SETI discussions is taken to fall between 1 and 20.

The rate of planet formation in conjunction with stellar evolution is currently the subject of much discussion in modern astrophysics. Do most stars have planets? If so, then the term f_p would have a value approaching unity. On the other hand, if planet formation is rare, than f_p approaches zero. Astronomers and astrophysicists currently think that planets should be a common occurrence in stellar-evolution processes. Therefore, f_p is frequently taken to fall in the range between 0.4 to 1.0 in SETI discussions. The value $f_p = 0.4$ represents a more pessimistic view, while $f_p = 1.0$ is taken as very optimistic.

Similarly, if planet formation is a normal part of stellar evolution, we must next ask how many of these planets are actually suitable for the evolution and maintenance of life? By taking $n_e = 1.0$, we are suggesting that for each planet-bearing star system, there is at least one planet located in a suitable habitable zone, or ecosphere. This is, of course, what we see here in our own Solar System.

We must then ask, given conditions suitable for life, how frequently does it start? One major assumption usually made (again based on the principle of mediocrity) is that wherever life can start, it will. If we invoke this assumption, then f_l equals unity. Similarly, we can also assume that once life starts, it always strives toward the evolution of intelligence, making f_i equal to 1.

This brings us to an even more challenging question: What fraction of intelligent extraterrestrial civilizations want to communicate with other alien civilizations? All we can do here is make a very subjective guess, based on human history. The pessimists take f_c to be 0.1 or less, while the optimists insist that all advanced civilizations desire to communicate and make $f_c = 1.0$.

Finally, we must also speculate on how long an advanced-technology civilization lasts. If we use the Earth as a model here, all we can say is that (at a minimum) L is somewhere between 50 and 100 years. The tools of high technology have emerged on this planet only in the 20th century. Space travel, nuclear energy, electronic communications and so on have arisen on a planet that daily oscillates between the prospects of total destruction and a "golden age" of maturity. Do most other evolving extraterrestrial civilizations follow a similar perilous pattern in which cultural maturity has to desperately race against new technologies that always threaten oblivion if they are unwisely used? Does the development of a technology base necessary for interstellar communication or even interstellar travel also stimulate a self-destructive impulse in advanced planetary civilizations, such that few (if any) survive? Or have

Table 1   Drake Equation Calculations

	R^*	f_p	n_e	f_l	f_i	f_c	L	N	Conclusion
THE BASIC EQUATION: $N = R^* f_p \, n_e \, f_l \, f_i \, f_c \, L$									
Very optimistic values	20	1.0	1.0	1.0	1.0	0.5	10^6	$\sim 10^7$	The Galaxy is full of intelligent life!
Your own values									
Very pessimistic values	1	0.2	1.0	1.0	0.5	0.1	100	~ 1	We are alone!

SOURCE: Developed by the author.

most extraterrestrial civilizations learned to live with their evolving technologies, and do they now enjoy peaceful and prosperous "golden ages" that last for millennia to millions of years? In dealing with the Drake equation, the pessimists place very low values on L (perhaps a hundred or so years), while the optimists insist that L is several thousand to several million years. The ultimate limit of a civilization's lifetime for a sunlike star is established by the main-sequence lifetime of the star itself—namely, several billion years. At that point, even a "super interplanetary" society must develop interstellar travel or perish when its sunlike star leaves the main sequence.

Let's go back now to the Drake equation and put in some "representative" numbers. If we take

$$R^* = 10 \quad \text{stars/year}$$
$$f_p = 0.5 \quad \text{(thereby excluding multiple-star systems)}$$
$$n_e = 1 \quad \text{(based on our Solar System as a common model)}$$
$$f_l = 1 \quad \text{(invoking principle of mediocrity)}$$
$$f_i = 1 \quad \text{(invoking principle of mediocrity)}$$
$$\text{and} \quad f_c = 0.2 \quad \text{(assuming that most advanced civilizations are introverts or have no desire for space travel)}$$

then the Drake equation yields

$$N \approx L$$

This particular result implies that the number of communicative extraterrestrial civilizations in the Galaxy at present is approximately equal to the lifetime (in Earth years) of such alien civilizations.

Let's now take these "results" one step further. If N is about 10 million (a very optimistic Drake equation output), then the average distance between intelligent, communicating civilizations in our Galaxy is approximately 100 light-years. If N is 100,000, then these extraterrestrial civilizations on the average would be about 1,000 light-years apart. But if there were only 1,000 such civilizations existing today, then they would typically be some 10,000 light-years apart. Consequently, even if the Galaxy does contain a few such civilizations, they may be just too far apart to achieve communication within the lifetimes of their respective civilizations. For example, at a distance of 10,000 light-years,

it would take 20,000 years just to start an interstellar dialogue!

By now you might like to try your own hand at estimating the number of intelligent alien civilizations that could be trying to signal us today. If so table 1 has been set up just for you. Simply select (and justify to yourself) typical numbers to be used in the equation, multiply all these terms together and obtain a value for N. Very optimistic and very pessimistic values that have been used in other SETI discussions are included in table 1 to help guide your own SETI efforts.

See also: **extraterrestrial civilizations; Fermi paradox; principle of mediocrity; search for extraterrestrial intelligence**

dwarf galaxy   A small often elliptical galaxy containing a million (10^6) to perhaps a billion (10^9) stars. The Magellanic Clouds, our nearest galactic neighbors, are examples of dwarf galaxies.

See also: **galaxy; Magellanic Clouds**

Dyson sphere   The Dyson sphere is a huge, artificial biosphere created around a star by an intelligent species as part of its technological growth and expansion within a solar system. This giant structure would most likely be formed by a swarm of artificial habitats and mini-planets capable of intercepting essentially all the radiant energy from the parent star. The captured radiant energy would be converted for use through a variety of techniques such as living plants, direct thermal-to-electric conversion devices, photovoltaic cells and perhaps other (as yet undiscovered) energy conversion techniques. In response to the Second Law of Thermodynamics, waste heat and unusable radiant energy would be rejected from the "cold" side of the Dyson sphere to outer space. From our present knowledge of engineering heat transfer, the heat rejection surfaces of the Dyson sphere might be at temperatures of 200 to 300 Kelvin.

This astroengineering project is an idea of the theoretical physicist, Freeman Dyson. In essence, what Dyson has proposed is that advanced extraterrestrial societies, responding to Malthusian pressures, would eventually expand into their local solar system, ultimately harnessing the full extent of its energy and materials resources. Just how much growth does this type of expansion represent?

Well, we must invoke the principle of mediocrity (i.e., things are pretty much the same throughout the Universe) and use our own Solar System as a model. The energy output from our Sun—a G-spectral class star—is approximately 4×10^{26} joules per second. For all practical purposes, our Sun can be treated as a blackbody radiator at approximately 5,800 Kelvin temperature. The vast majority of its energy output occurs as electromagnetic radiation, predominantly in the wavelength range 0.3 to 0.7 micrometers. The available mass in the Solar System for such astroengineering construction projects may be taken as the mass of the planet Jupiter, some 2×10^{27} kilograms. Contemporary energy consumption now amounts to about 10^{13} joules per second, which is about 10 terawatts. Let's now project just a one percent growth in terrestrial energy consumption per year. Within a mere three millennia, mankind's energy consumption needs would reach the energy output of the Sun itself! Today, several billion human beings live in a single biosphere, the planet Earth—with a total mass of some 5×10^{24} kilograms. A few thousand years from now, our Sun could be surrounded by a swarm of habitats, containing trillions of human beings.

The Dyson sphere may therefore be taken as representing an upper limit for growth within our Solar System. It is basically "the best we can do" from an energy and materials point of view in our particular corner of the Universe. The vast majority of these human-made habitats would most probably be located in the "ecosphere" around our Sun—that is, about a one astronomical unit (AU) distance from our parent star. This does not preclude the possibility that other habitats, powered by nuclear fusion energy, might also be found scattered throughout the outer regions of a somewhat dismantled Solar System. (These fusion-powered habitats might also become interstellar space arks.)

Therefore, if we use our own Solar System and planetary civilization as a model, we can anticipate that within a few millennia after the start of industrial development, an intelligent species might rise from the level of planetary civilization (Kardashev TYPE I Civilization) and eventually occupy a swarm of artificial habitats that completely surround their parent star, creating a Kardashev TYPE II civilization. Of course, these intelligent creatures might also elect to pursue interstellar travel and galactic migration, as opposed to completing the Dyson sphere within their home star system (initiating a Kardashev TYPE III civilization).

It was further postulated by Freeman Dyson that such advanced civilizations could be detected by the presence of thermal infrared emission (typically 8.0 to 14.0 micrometer wavelength) from objects in space that had dimensions of one to two astronomical units in diameter.

The Dyson sphere is certainly a grand, far-reaching concept. It is also quite interesting for us to realize that the initial permanent space stations and space bases we construct at the close of the 20th century are, in a sense, the first habitats in the swarm of artificial structures that mankind could eventually build as part of our extraterrestrial civilization. No other generation in human history has had the unique opportunity of constructing the first artificial habitat in our own Dyson sphere!

See also: **extraterrestrial civilizations; principle of mediocrity; space settlement; space station**

Earth The Earth is the third planet from the Sun and the fifth largest in the Solar System. Our planet circles its parent star at an average distance of 149.6 million kilometers (93 million miles). Earth is the only celestial body in the Solar System presently known to support life. Some of the physical and dynamic properties of the Earth as a planet in the Solar System are presented in table 1.

From space, our planet is characterized by its blue waters and white clouds, which cover a major portion of it. The Earth is surrounded by an ocean of air, consisting of 78 percent nitrogen, 21 percent oxygen and the remainder argon, neon and other gases. The standard atmospheric pressure at sea level is 101,325 newtons per square meter (14.7 pounds per square inch). Surface temperatures range from a maximum of about 60 degrees Celsius (140 degrees Fahrenheit) in desert regions along the equator to a minimum of minus 90 degrees Celsius (minus 130 degrees Fahrenheit) in the frigid polar regions. In between, however, surface temperatures are generally much more benign.

The Earth's rapid spin and molten nickel-iron core (see fig. 1) give rise to an extensive magnetic field. This magnetic field, together with the atmosphere, shields us from

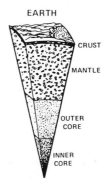

Fig. 1 The interior of the Earth. (Drawing courtesy of NASA.)

Table 1 Dynamic and Physical Properties of the Planet Earth

Radius	
Equatorial	6,378 km
Polar	6,357 km
Mass	5.98×10^{24} kg
Density (average)	5.52 g/cm^3
Surface area	5.1×10^{14} m^2
Volume	1.08×10^{21} m^3
Distance from the Sun (average)	1.496×10^8 km (1 AU)
Eccentricity	0.01673
Orbital period (sidereal)	365.256 days
Period of rotation (sidereal)	23.934 hours
Inclination of equator	23.45 degrees
Mean orbital velocity	29.78 km/sec
Acceleration of gravity, g (sea level)	9.807 m/sec^2
Solar flux at the Earth (above atmosphere)	$1,371 \pm 5$ watts/m^2
Planetary energy fluxes (approximate)	
Solar	10^{17} watts
Geothermal	2.5×10^{13} watts
Tidal friction	3.4×10^{12} watts
Human-made	
Coal-burning	3.06×10^{12} watts
Natural gas-burning	2.00×10^{12} watts
Oil-burning	3.82×10^{12} watts
Nuclear power	0.49×10^{12} watts
Hydroelectric	0.69×10^{12} watts
TOTAL HUMAN-MADE	10.06×10^{12} watts
Number of natural satellites	1 (the Moon)

nearly all of the harmful changed-particle and ultraviolet radiation coming from the Sun and other stars. Furthermore, most meteors burn up in the Earth's protective atmosphere before they can strike the surface.

In the Space Age manned and unmanned spacecraft have enjoyed a unique vantage point from which to observe the Earth, survey its bountiful resources and monitor its delicate biosphere. The Earth's nearest celestial neighbor, the Moon, is also its only natural satellite.

See also: **Earth's trapped radiation belts; global change; Mission to Planet Earth; Moon**

Earthlike planet An extrasolar planet that is located in an ecosphere and has planetary environmental conditions that resemble the terrestrial biosphere—especially a suitable atmosphere, a temperature range that permits the retention of large quantities of liquid water on the planet's surface and a sufficient quantity of energy striking the planet's surface from the parent star. These suitable environmental conditions could then permit the chemical evo-

lution and development of carbon-based life as we know it on Earth. The planet should also have a mass somewhat greater than 0.4 Earth masses (to permit the production and retention of a breathable atmosphere) but less than about 2.4 Earth masses (to avoid excessive surface gravity conditions—that is, to avoid having a gravitational force [g] greater than 1.5).

See also: **ecosphere**

Earth Observing System (EOS) To fully understand global change and the potential influence of human activities, it is very important that we carefully document and scientifically comprehend how the Earth works as a system. Today, the international scientific community is organizing new research efforts to advance our knowledge of both natural and human-induced global change phenomena. Mission to Planet Earth is the central NASA planned contribution to the current U.S. Global Change Research program. There are two new space initiatives included in NASA's Mission to Planet Earth effort: the Earth Observing System (EOS) and Earth Probes.

The Earth Observing System consists of a space-based observing system (called EOSSMS, an acronym for EOS Space Measurement System); a data and information system (called EOSDIS, for EOS Data and Information System); and a scientific research program. EOS represents the start of a truly comprehensive, global observing system with broad and high-resolution spectral and spatial, as well as long-term temporal, coverage of the Earth. The EOS space-based observing system will consist of two series of polar-orbiting platforms, called EOS-A and EOS-B, now planned for launch in the late 1990s. As currently planned, the U.S. EOS space-based observing systems will be complemented by companion polar orbiting platforms from the European Space Agency (ESA) and the National Space Development Agency of Japan (NASDA), as well as the continuing operational meteorological and environmental satellites. The EOS system will continue and integrate the measurements now being taken by short-term research missions. It will also provide the first coordinated simultaneous measurements of the interactions of the atmosphere, oceans, solid Earth, and hydrologic and biogeochemical cycles.

NASA began conceptual studies for the EOS space-based observing system in 1982, and coordination with ESA, Japan, and Canada was initiated in 1986. The planned 15-year observational period will be achieved using three identical EOS satellites per series, each with a five-year design lifetime. The EOS-A series is now focused on atmospheric sounding and surface imaging; the EOS-B series of space platforms will include sensors capable of extending the observations made by NASA's Upper Atmosphere Research Satellite (UARS) and the TOPEX/Poseidon mission. (The UARS mission deals with atmospheric chemistry and the TOPEX/Poseidon mission focuses on oceanic circulation.)

The platform size for the EOS space-based observing

system is being selected to accommodate both atmospheric sounding and surface imaging sensors. This unique grouping of instruments will allow scientists to make and correlate simultaneous measurements of phenomena on the Earth's surface and the condition and composition of the intervening atmosphere. Sometimes gases in the Earth's atmosphere will absorb some of the radiant energy signal being emitted by a surface phenomenon on Earth, making the signal reaching the spacecraft sensor actually lower than the "true" value. If scientists know the composition of the intervening atmosphere at the time of surface phenomenon observation, they can easily correct for this atmospheric absorption and then determine the proper value for the surface event being observed from space.

An EOS platform will be able to accommodate a payload of up to 3,500 kilograms and supply up to 3.2 kilowatts of electric power to the payload. The EOS satellites will be launched into polar orbit from Vandenberg Air Force Base, California, onboard Titan-IV expendable launch vehicles. To provide global coverage every one to three days, the EOS spacecraft will be placed in a Sun-synchronous 98.2-degree inclination orbit at an altitude of 705 kilometers (providing a quasi-two-day coverage repeat). An EOS satellite will have a 1:30 P.M. local time equatorial-crossing.

In support of U.S. Global Change Research program objectives, the EOS platform instruments are being designed to measure the following important environmental variables: (1) cloud properties; (2) the energy exchange between the Earth and space; (3) surface temperature; (4) the structure, composition and dynamics of the atmosphere, winds, lightning and precipitation (rainfall); (5) the accumulation and ablation (melting) of snow; (6) biological activity on land and in near-surface waters; (7) circulation patterns of the oceans; (8) the exchange of energy, momentum and gases between the Earth's surface and its atmosphere; (9) the structure and motion of sea ice; the growth, melting and flow rates of glaciers; (10) the mineral composition of exposed soils and rocks; (11) the changes in stress and surface elevation around global faults; and (12) the input of radiant energy and energetic particles to the Earth from the Sun.

To complement the U.S. EOS platforms, the European Space Agency is planning two series of polar platforms with a climatological and terrestrial focus, and Japan is planning one polar platform. The United States (through NASA and NOAA [National Oceanic and Atmospheric Administration]) will provide sensors to be attached as payloads on the European and Japanese polar-orbiting platforms. In addition, NASA also plans to provide other global change monitoring sensors to be flown as attached payloads on the proposed Space Station *Freedom*.

The EOS program is also placing a major emphasis on its Data and Information System (EOSDIS), which is now planned to accomplish the following objectives: (1) to acquire a comprehensive, global 15-year duration environmental data set; (2) to maximize the utility of these data for scientific purposes; and (3) to facilitate easy access to these important data sets by the global change research community.

NASA's Earth Probes will provide a focus on observing specific Earth processes where smaller platforms and/or orbits different from the EOS system orbits are needed. These Earth Probe missions are now scheduled to start flying in the early 1990s, with a satellite launch approximately every two or three years. The first such Earth Probe mission is the Total Ozone Mapping Spectrometer (TOMS), which will measure atmospheric ozone. The habitability of the planet Earth depends on maintaining the integrity of its ozone layer, which absorbs the harmful ultraviolet radiation emitted by the Sun. One of the next Earth Probe missions will be the Sea-Viewing Wide-Field Sensor (SeaWiFS), an ocean color scanner. Biological processes in the ocean are critical to biogeochemical cycles. SeaWiFS will be able to measure chlorophyll in the open ocean where the signal is low, distinguish between phytoplankton pigments and sediment in coastal waters, and make the necessary atmospheric corrections. The SeaWiFS data will provide the first estimates of the mean and variable rates of primary production of phytoplankton in the world oceans.

The total cost of the EOS system is now estimated to be approximately 17 billion dollars through the year 2000. In the past, NASA had traditionally allocated about 70 percent of a program's funding for spacecraft hardware and 30 percent for ground-based activities, including data analyses. For the EOS program, in recognition of the very large amount of global change data that will be collected and the pressing need for a comprehensive set of Earth system data, NASA plans to allocate about 40 percent of the funding for EOS spacecraft hardware, and will use the remaining 60 percent of the projected program funding for ground-based activities, including development of EOSDIS and support of global change science.

See also: **global change; Mission to Planet Earth; remote sensing**

earthshine A spacecraft or space vehicle in orbit around the Earth is illuminated by sunlight and "earthshine." Earthshine consists of sunlight (0.4- to 0.7-micrometer wavelength radiation, or visible light) reflected by the Earth and thermal radiation (typically 10.6-micrometer wavelength infrared radiation) emitted by the Earth's surface and atmosphere.

Earth's trapped radiation belts The magnetosphere is a region around the Earth through which the solar wind cannot penetrate because of the terrestrial magnetic field. Inside the magnetosphere are two belts or zones of very energetic atomic particles (mainly electrons and protons) that are trapped in the Earth's magnetic field hundreds of kilometers above the atmosphere. These belts were discovered by Professor James Van Allen of the University of Iowa and his colleagues in 1958. Van Allen made the discovery using simple atomic radiation detectors placed onboard *Explorer 1*, the first American satellite. In his honor, these radiation belts are also called the "Van Allen Belts." Their discovery represents one of the first major discoveries about our extraterrestrial environment to occur in the Space Age. The existence of these radiation belts was an unexpected finding, since knowledge about the Earth's magnetosphere was very limited up to the late 1950s.

The two major radiation belts form a doughnut-shaped region around the Earth from about 320 to 32,400 kilometers (200 to 20,000 miles) above the equator (dependent upon solar activity). Energetic protons and electrons are trapped in these belts. The inner Van Allen belt contains both energetic protons (major constituent) and electrons that were captured from the solar wind or were created in nuclear collision reactions between energetic cosmic ray particles and atoms in the Earth's upper atmosphere. The outer Van Allen belt contains mostly energetic electrons that have been captured from the solar wind.

Spacecraft and space stations operating in the Earth's trapped radiation belts are subject to the damaging effects of ionizing radiation from charged atomic particles. These particles include protons, electrons, alpha particles (helium nuclei) and heavier atomic nuclei. Their damaging effects include degradation of material properties and component performance, often resulting in reduced capabilities or even failure of spacecraft systems and experiments. For example, solar cells used to provide electric power for spacecraft are often severely damaged by passage through the Van Allen belts. The Earth's trapped radiation belts also represent a very hazardous environment for human beings traveling in space.

Radiation damage from the Earth's trapped radiation belts can be reduced significantly if the spacecraft or space station is designed with proper radiation shielding. Frequently, crew compartments and sensitive equipment can be located in regions shielded by other spacecraft equipment that is less sensitive to the influence of ionizing radiation. Radiation damage can also be limited by selecting mission orbits and trajectories that avoid long periods of operation where the radiation belts have their highest charged particle populations. For example, for a space station or satellite in low Earth orbit, this would mean avoiding the South Atlantic Anomaly and, of course, the Van Allen Belts themselves.

See also: **cosmic rays; hazards to space workers; magnetosphere; solar wind; South Atlantic Anomaly**

eccentricity (e)   A measure of the ovalness of an orbit. When $e = 0$, the orbit is a circle; when $e = 0.9$, the orbit is a long, thin ellipse.

See also: **orbits of objects in space**

ecliptic   The circle formed by the apparent yearly path of the Sun through the heavens; it is inclined by approximately 23.5 degrees to the celestial equator.

ecosphere   That habitable zone or region around a main-sequence star of a particular luminosity in which a planet can maintain the conditions necessary for the evolution and continued existence of life. For life to occur as we know it on Earth (that is, chemical evolution of carbon-based living organisms), global temperature and atmospheric pressure conditions must permit the retention of a significant amount of liquid water on the planet's surface. Conditions that would prevent a habitable Earthlike planet include circumstances in which all the surface water has been completely evaporated (the runaway greenhouse effect) or in which the liquid water on the planet's surface has become completely frozen or glaciated (the ice catastrophe).

For a star like the Sun, an effective ecosphere would typically extend from about 0.7 to about 1.3 astronomical units (AU). In our Solar System, for example, the inner edge of an ecosphere suitable for human life would reach the orbit of Venus, while the outer edge reaches approximately halfway to the orbit of Mars. Because ecospheres appear to be extremely narrow, a planetary system around an alien star will most likely have only one, or perhaps two, planets that are located in a region suitable for the chemical evolution of carbon-based life.

See also: **global change**

electromagnetic spectrum   When sunlight passes through a prism, it throws a rainbowlike array of colors onto a surface. This display of colors is called the visible spectrum. It represents an arrangement in order of wavelength of the narrow band of electromagnetic (EM) radiation to which the human eye is sensitive.

The electromagnetic spectrum comprises the entire range of wavelengths of electromagnetic radiation, from the shortest-wavelength gamma rays to the longest-wavelength radio waves (see fig. 1). The entire EM spectrum includes much more than meets the eye!

As shown in figure 1, the names applied to the various regions of the EM spectrum are (going from shortest to longest wavelength) gamma ray, X ray, ultraviolet (UV), visible, infrared (IR) and radio. EM radiation travels at the speed of light (that is, about 300,000 kilometers per second) and is the basic mechanism for energy transfer through the vacuum of outer space.

One of the most striking discoveries of 20th-century physics is the dual nature of electromagnetic radiation. Under some conditions electromagnetic radiation behaves like a wave, while under other conditions it behaves like a stream of particles, called photons. The tiny amount of energy carried by a photon is called a quantum of energy

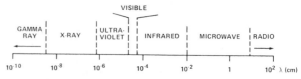

Fig. 1   The electromagnetic spectrum. (Drawing courtesy of NASA.)

(plural: quanta). The word *quantum* comes to us from the Latin and means "little bundle."

The shorter the wavelength, the more energy is carried by a particular form of EM radiation. All things in the Universe emit, reflect and absorb electromagnetic radiation in their own distinctive ways. The way an object does this provides scientists with special characteristics, or a signature, that can be detected by remote sensing instruments. For example, the spectrogram shows bright lines for emission or reflection and dark lines for absorption at selected EM wavelengths. Analyses of the positions and line patterns found in a spectrogram can provide information about the object's composition, surface temperature, density, age, motion and distance.

For centuries, astronomers have used spectral analyses to learn about distant extraterrestrial phenomena. But up until the Space Age, they were limited in their view of the Universe by the Earth's atmosphere, which filters out most of the EM radiation from the rest of the cosmos. In fact, ground-based astronomers are limited to just the visible portion of the EM spectrum and tiny portions of the infrared, radio and ultraviolet regions. Space-based observatories now allow us to examine the Universe in all portions of the EM spectrum. In the Space Age we have also examined the cosmos in the infrared, ultraviolet, X-ray and gamma ray portions of the EM spectrum and have made startling discoveries. We have also developed sophisticated remote sensing instruments to look back on our own planet in many regions of the EM spectrum. Data from these environmental-monitoring and resource-detection spacecraft are providing the tools for a more careful management of our own Spaceship Earth.

See also: **astrophysics, Earth Observing System; global change; Hubble Space Telescope; remote sensing**

electron volt (eV)   A unit of energy equivalent to the energy gained by an electron when it passes through a potential difference of one volt. Larger multiple units of the electron volt are frequently used—as, for example; *keV* for thousand, or kilo, electron volts (10^3eV); *MeV* for million, or mega, electron volts (10^6eV); and *GeV* for billion, or giga, electron volts (10^9eV).

$$1 \text{ electron volt} = 1.602 \times 10^{-19} \text{ joules}$$

escape velocity (V_e)   The minimum velocity needed by a object to climb out of the gravity well (overcome the gravitational attraction) of a celestial body. From classical Newtonian mechanics, the escape velocity from the surface of a celestial body of mass (M) and radius (R) is given by the equation:

$$V_e = \sqrt{(2\,G\,M)\,/R}$$

where G is the universal constant of gravitation (6.672×10^{-11} N $-$ m^2/kg^2)

Table 1   Escape Velocity for Various Objects In The Solar System

Celestial Body	Escape Velocity (V_e) (km/sec)
Earth	11.2
Moon	2.4
Mercury	4.3
Venus	10.4
Mars	5.0
Jupiter	~ 61
Saturn	~ 36
Uranus	~ 21
Neptune	~ 24
Pluto	~ 1
Sun	~618

SOURCE: Developed by the author based on NASA and other astrophysical data.

M is the mass of the celestial object (kg)

N is Newton (unit of force in the *SI*, system)

and R is the radius of the celestial object (m).

Table 1 presents the escape velocity for various objects in the Solar System (or an estimated equivalent V_e for those celestial bodies that do not possess a readily identifiable solid surface, such as the giant outer planets and the Sun).

ETI   An acronym for *extra*terrestrial *i*ntelligence.
   See also: **extraterrestrial civilizations; search for extraterrestrial intelligence**

European Space Agency (ESA)   The joint European space program first got underway in the 1960s. In these early, pioneering days, there were only two basic activities, both conducted by organizations founded in 1964: a scientific research program (conducted by the European Space Research Organization or ESRO) and the development of a space launch vehicle capability (conducted by the European Launcher Development Organization or ELDO). During the following decades, these joint European space activities have expanded enormously through the actions of the European Space Agency (ESA), set up in 1975, which was formed out of and has taken over the activities of both ESRO and ELDO. Today, ESA activities include all fields of peaceful space research and technology application, including space science, microgravity research, communications, Earth observation, launch vehicles and even manned space flight. The driving force behind the decision of European governments to coordinate and pool their efforts in joint space endeavors was quite simply economics. No individual European country could afford to independently sponsor a complete range of space projects and all the necessary infrastructure—although some countries, like France, had and still maintain thriving national space programs in addition to participation in ESA. On the other hand, the nations of Western Europe (taken as an entity) felt that they could not be left out of the "space race" and from an economic perspective, the space program has provided European industry with a tremendous technical stimulus.

The ESA's purpose, as described in its Convention (which entered into force on October 30, 1980) is to "provide for and promote, *for exclusively peaceful purposes,* cooperation among European States in space research and technology and their space applications with a view to their being used for scientific purposes and for space application systems."

The founders of ESA were Belgium, Denmark, the Federal Republic of Germany, France, Ireland, Italy, the Netherlands, Spain, Sweden, Switzerland and the United Kingdom. These nations were joined in full membership on January 1, 1987, by Austria and Norway, and by Finland, which became an associate member state on the same date. In addition, Canada has an agreement for close cooperation with ESA and participates in some of its programs. The *ESA* headquarters are in Paris.

In order to ensure that Europe remains at the forefront of progress, one of ESA's main tasks is to prepare for its member states, at regular intervals, long-term plans proposing the direction European space research should take. ESA also plays a coordination role by following closely the national programs of its member states and, whenever appropriate and feasible, integrating these national plans into ESA programs. A recent example of this type of coordination is the planned spaceplane Hermes. The initial studies for Hermes were conducted by the French Space Agency (CNES), with ESA following its progress. ESA member state ministers expressed their interest in the French spaceplane program at their meeting in January 1985, and the Hermes preparatory program became an ESA responsibility in mid-1986. Then, in November 1987, a decision was made at the Ministers Conference in The Hague to proceed with the Hermes program.

ESA is essentially a research and development organization and does not develop or manufacture its own spacecraft. The definition stages of a space project are the responsibility of ESA's engineers and scientists, who work closely with space experts from member states. However, the actual development work for the space project is carried out by European industry under the supervision and monitoring of ESA staff personnel. This arrangement ensures that each member state receives money back and has a direct share of the technological advances stimulated by space activities, in return for its investment in ESA. All ESA budgets and financial operations are calculated in accounting units equivalent to the European currency unit (ECU). Exchange rates for this Accounting Unit vary each year and are based on the average exchange rates of the different national currencies.

ESA activities are either mandatory or optional. Mandatory ESA space activities are carried out under the general budget and the science program budget. All member states contribute to these mandatory activities on the basis of their average national income. Optional ESA activities are those in which the contributions of member states

reflect each country's interest in a particular field of space science or space application. For example, France has been specifically interested in the development of the European space launch vehicle and has always been the main contributor to the different Ariane programs, while the Federal Republic of Germany has exhibited strong interest in manned space activities, as for example through the ESA Spacelab program.

ESA's main policy-making body is its council, which generally meets once a quarter. Consisting of representatives of ESA member states, this council makes decisions on the overall policy to be followed by the agency and on technical, scientific and administrative matters. For mandatory ESA activities, each member state has one vote, while only participating states may vote on matters pertaining to optional ESA programs. The ESA council is assisted by a number of specialized boards that oversee the management of specific ESA programs. The council is also assisted by the science program committee, the administrative and finance committee, the international relations advisory committee and the industrial policy committee.

The ESA director general is appointed by the council for a four-year term and serves as the agency's chief executive and legal representative. The director general is assisted by eight directors responsible for the following: earth observation and microgravity, science, space station and platforms, space transportation systems, telecommunications, administration, operations and the European Space Research and Technology Center (ESTEC). The ESA director general also receives assistance from his head of cabinet, the inspector general, the associate director for policy coordination, a coordination and monitoring office and the head of ESRIN (European Space Research Institute).

ESA headquarters is located in Paris, France; ESTEC in Noordwijk, the Netherlands; the European Space Operations Center (ESOC) in Darmstadt, Germany; and ESRIN in Frascati, Italy. ESA's Ariane launch complex is located in Kourou, French Guiana, and ESA maintains a liaison office in Washington, D.C. In addition, ESA also has a worldwide network of ground stations (including its ES-TRACK station at Redu in Belgium and the observatory facilities at Villafranca in Spain) for satellite operations, as well as a number of down-range stations to support Ariane launches.

FACILITIES

The European Space Research and Technology Center (ESTEC), located at Noordwijk in the Netherlands, is responsible for the study, development, control and testing of ESA spacecraft, which are built by European industry. Another major responsibility is the ESTEC space technology program, which supports future space missions. The basic technology program at ESTEC generates the new space technology needed for future missions both in the medium and long term; the supporting technology program continues this process and demonstrates the flight-worthiness of new systems; while the technology demonstration program provides for on-orbit demonstration opportunities.

The European Space Operations Center (ESOC) at Darmstadt in Germany is responsible for ESA satellite orbital operations. This responsibility incorporates a wide variety of tasks that must be carried out both before and after the launch of a spacecraft. The hub of activities, once a spacecraft has separated from its launch vehicle, is the Operations Control Center (OCC). The OCC is permanently linked through the ESOC communications network with the entire network of ESA ground stations (known as ESTRACK), which are located around the world. Each ESTRACK station in this network is equipped both to receive spacecraft telemetry and to transmit commands to the spacecraft. Since the late 1970s ESOC has also been responsible for operating a number of telecommunications and meteorology satellites. ESOC has recently been given the task of providing a continuous service into the 1990s on behalf of EUMETSAT, the 16-member-state European Organization for the Exploitation of Meteorological Satellites.

The European Space Research Institute (ESRIN), located at Frascati, Italy (near Rome), is responsible for EARTHNET activities and the ESA Information Retrieval Service (ESA-IRS). The EARTHNET program office at ESRIN provides Europe with a center for the acquisition, preprocessing, archiving and distribution of remote-sensing data. ESA-IRS is one of Europe's leading scientific and technical information sources, with over 130 databases (in English and French) and databanks that are accessible through the ESRIN computer complex. Although the European Space Research Institute that gave ESRIN its name was disbanded in the early 1970s, the establishment's original title still persists today in this ESA information and remote-sensing data facility.

SCIENCE PROGRAM

ESAs science program is the main element of the agency's mandatory activities in which all ESA member states participate. The overall objective of the current ESA long-term plan, Space Science: Horizon 2000, is to continue to keep Europe in the forefront of scientific progress. The corresponding science program objectives are to: (1) contribute to the advancement of fundamental scientific knowledge; (2) establish Europe as a major participant in the worldwide development of space science; (3) offer a balanced distribution of opportunities for "leading-edge" research to the European scientific community; and (4) provide major technological challenges for innovative industrial development.

The four cornerstones of ESA's program, Space Science: Horizon 2000, are:

(1) *Solar–terrestrial science program (STSP)*, a cooperative effort by ESA and NASA, presently consisting of two medium-sized missions, the Solar and Heliospheric Observatory (SOHO) and Cluster (a four-spacecraft space plasma physics mission). The major objective of this cornerstone mission is to address outstanding scientific problems in solar, heliospheric and space plasma physics, through a unified and coordinated approach.

(2) *Planetary science mission,* which would consist of bringing back to Earth pristine samples of material from either an asteroid or a comet.

(3) *X-ray astronomy mission* for the study of the large- and small-scale structures of the hot components of matter (e.g., supernovae) in the Universe.

(4) *Submillimeter wavelength astronomy mission* involving a very-high-resolution spectroscopy instrument.

EARTH OBSERVATION PROGRAM

Over the past 15 years Europe has made a significant contribution in the field of Earth observation from space. Since 1977, for example, the Meteosat satellites have provided data for both weather prediction and scientific research. The success of this program has led to the establishment of a new organization, European Meteorological Services (EUMETSAT). In June 1986, EUMETSAT (with 16 European member states) formally came into being for the purpose of establishing, maintaining and operating the Meteosat operational satellite system. ESA maintains its involvement in the 1990s by procuring the satellites and providing the associated ground control center.

ESA's current earth observation program involves two main fields: (1) the observation of the seas, oceans and land masses of planet Earth and (2) the observation of the terrestrial atmosphere. The first ESA Remote Sensing Satellite (ERS-1) mission will concentrate primarily on global ocean monitoring, although it will also carry a synthetic aperture radar that will provide all-weather, high-resolution imagery over coastal zones, land and ice caps.

The EARTHNET Program also plays an important role in ESA's Earth observation activities and is included within ESA's mandatory programs. EARTHNET's main role is the acquisition (through a network of ground stations located in Europe from the north of Scandinavia to as far south as the Canary Islands) and distribution of satellite Earth observation images. Currently, these images come from U.S. satellites (such as the LANDSAT and TIROS series), the French satellite SPOT-1, and the Japanese satellite MOS-1. These data will be supplemented in the 1990s by data from ERS-1, now scheduled for launch after May 1991.

SPACE TRANSPORTATION SYSTEMS

Since July 1973, when the decision was made to entrust ESA with the development of the European space launch vehicle, the pace of technological development has been such that Europe is now looking beyond the launch of satellites to manned spaceflight systems. In 1973, when ESA member states approved development of a European launch vehicle, they did so for two major reasons. First and perhaps most importantly, this development would give the European member states an independent launch capability for their own spacecraft; and second, it would enable Europe to share in the growing international market for satellite launches.

The development of the Ariane launch vehicle family is an excellent example of European cooperation at an industrial level: In all, some 40 aerospace firms from all over Europe are involved in the development and manufacture of Ariane launch vehicles. Ariane 4, which made its first demonstration flight on June 15, 1988, when it successfully launched three satellites, will serve as Europe's workhorse launch vehicle system for the next decade. Some 70 launch vehicle units will be produced and flown between now and 1998, the planned end of Ariane 4's operational life. There are six versions of the Ariane 4 launch vehicle. These various Ariane 4 configurations can place a variety of payloads, ranging from 2,000 kg (4,410 1bm) to 4,200 kg (9,260 1bm), into a geostationary transfer orbit.

The newest member of the Ariane family, Ariane 5, will give Europe the capability to launch people into space, as well as providing an increased lift capability for future satellites in the 1990s. Ariane 5 will be a three-stage launch vehicle, expected to be operational in the mid-1990s. For manned ESA missions, the Hermes spaceplane will become the Ariane 5 upper stage.

The Ariane vehicles are launched from the Guiana Space Center, near Kourou, French Guiana—a location just 5 degrees north of the equator on the east coast of South America. The two currently operational Ariane launch sites are called ELA 1 and ELA 2 (from the French *Ensemble de Lancement Ariane*), with ELA-2 serving as the prime Ariane 4 launch complex. A third launch complex, ELA-3, is planned for use by the Ariane 5 launch vehicle, including the manned launches with the Hermes spaceplane.

MAN-IN-SPACE PROGRAM

Europe is now preparing to enter what will undoubtedly be one of its major endeavors of the 21st century: the direct exploration and use of space by man himself—an activity limited up to now to the two space "superpowers," the United States and the Soviet Union (who have already flown guest astronauts and cosmonauts from many nations). The decision to acquire its own capabilities for manned spaceflight was formulated by the ESA council at the ministerial level in Rome in January 1985, and then was approved by the European Ministers for Space meeting in The Hague in November 1987. The achievement of an independent on-orbit infrastructure made up of space station elements and a manned space transportation system will constitute a key objective for European space development in the 1990s.

ESA's man-in-space commitment consists of two main parts: (1) the Columbus Space Station program, which includes the Columbus Attached Laboratory as the ESA part of the U.S. Space Station *Freedom*, the Columbus Free-Flying Laboratory, and an unmanned Polar Orbiting Columbus Platform; and (2) an independent space transportation system, which includes the Ariane 5 launch vehicle and the Hermes space vehicle. The Ariane 5 launch

vehicle will place the Columbus Free-Flying Laboratory and the Columbus Polar Platform into orbit and will also launch the Hermes space vehicle. The Hermes space vehicle, in turn, will provide priority service to the Columbus Free-Flying Laboratory.

COLUMBUS PROGRAM

ESA will contribute to the U.S. Space Station *Freedom* through its Columbus development program as approved at an ESA council meeting in November 1987. The Columbus development program involves the development, manufacture and delivery to orbit of three space elements: (1) a Columbus Laboratory Module, to be attached to the core of Space Station *Freedom*; (2) a human-serviced, Columbus Free-Flying Laboratory; and (3) an unmanned Columbus Polar Platform. This program also includes the build-up of the appropriate ground support infrastructure.

The primary objectives of the Columbus program (which started in January, 1988) are as follows: (1) to provide an on-orbit and ground support infrastructure, compatible with the needs of European and international users from the mid-1990s onward; (2) to continue to expand Europe's capabilities in manned spaceflight already initiated with the Spacelab program (Spacelab is the sophisticated, reusable space laboratory developed for ESA by European industry that is flown inside the payload bay of NASA's Space Shuttle; (3) to cooperate with the United States, Japan and Canada in the Space Station *Freedom* program, in which the Columbus Attached Laboratory will represent Europe's major contribution; (4) to ensure the development of Europe's capability to achieve (in the longer term) manned spaceflight autonomy, in particular through cooperation in the *Freedom* program; and (5) to ensure the development in Europe of the key technologies needed for manned spaceflight and for a wide spectrum of on-orbit operations both manned and automated (robotic).

The Columbus Attached Laboratory is a pressurized cylindrical module that will be permanently attached to the core of Space Station *Freedom*. It will be 12.4 meters (40.7 ft) long and 4 meters (13.1 ft) in diameter. This laboratory will be used primarily for payloads and experiments in materials science, fluid physics and life sciences requiring a permanent human presence. This module will be launched and serviced by NASA's Space Shuttle.

The Columbus Free-Flying Laboratory will consist of a two-segment pressurized module that can accommodate payloads and an unpressurized resource module that provides the necessary power, data handling and life support systems. It will accommodate automated and remote-controlled payloads primarily from the materials science, fluid physics, life sciences and technology disciplines that require a long, undisturbed microgravity environment. This free-flying laboratory, together with its initial payload, will be launched by the Ariane 5 vehicle and routinely serviced on-orbit by the Hermes space vehicle at approximately six-month intervals. Initially, these servicing activities will be conducted at Space Station *Freedom*, which the Columbus

Free-Flying Laboratory will also visit every three to four years for major external maintenance.

The Columbus Platform is an unmanned platform designed to operate over a period of several years in a highly inclined sun-synchronous polar orbit. It will be used primarily for Earth observation missions and will be operated in conjunction with one or more platforms provided by international partners. This unmanned platform will be large enough to accommodate European and internationally provided, automated payloads.

EURECA PROGRAM

As a means of bridging the gap between Spacelab and the Space Station *Freedom*, ESA member states decided in 1984 to have the agency develop an unmanned European retrievable carrier, called *Eureca*. *Eureca* will be launched by the Space Shuttle into an orbit that is about 300 km (186 mi) above the Earth.

An on-board propulsion system will then inject the *Eureca* space platform into a higher orbit, at just over 500 km (311 mi) altitude, where it will remain for the duration of its mission. Then, after about six months or so, *Eureca* will descend using its on-board propulsion system to rendezvous with the Space Shuttle for capture and return to Earth.

Not only does *Eureca* represent a bridge between Spacelab and the Space Station *Freedom*, it will also be a very useful tool for both scientific and commercial purposes. It will be used for research in the microgravity sciences and for inflight demonstrations of new technologies throughout the 1990s.

HERMES PROGRAM

The Hermes Spaceplane is intended to carry out a wide variety of missions to satisfy Europe's needs for on-orbit human presence. Hermes' primary mission will be to service the Columbus Free-Flying Laboratory, which will be capable of performing automatic missions independent of a permanent human presence. It is anticipated that Hermes will service this laboratory every six months and that each Hermes mission will last up to 12 days, including 7 days docked to the laboratory. Hermes will dock with the laboratory's pressurized module, using a docking system that will enable the Hermes crew to move freely between the spaceplane's pressurized volume and that of the laboratory, transferring equipment and materials without exposure to the space environment. Hermes will be capable of transporting a three-ton payload of technical, scientific and operational hardware. During these visits, the Hermes crew will also change, configure and initiate experiments and perform maintenance on the free-flying laboratory. As needed, the crew can also perform extravehicular activities (using spacesuits) and can use the Hermes robotic arm (HERA).

The other primary mission of Hermes will be to visit the Space Station *Freedom*, including the ESA Columbus Attached Laboratory Module.

The Hermes spaceplane is scheduled to be operational in the late 1990s and will be capable of carrying out missions for at least 15 years. The Ariane 5 vehicle will launch the Hermes spaceplane from complex ELA-3 at the Guiana Space Center. This Ariane 5 launcher will also be used in an "unmanned mission mode" to place heavy payloads into low Earth orbit or in a geostationary orbit transfer trajectory.

The Hermes space vehicle configuration is being designed to meet both manned mission requirements and the requirements imposed by the Ariane 5 launch vehicle. The Hermes space vehicle will consist of (1) the Hermes spaceplane itself (2) the Hermes resource module (which is adaptable to various mission needs and is jettisoned by the spaceplane prior to atmospheric reentry, and (3) the Hermes propulsion module (which is the propulsion stage associated with the Ariane 5 launch vehicle needed to insert the Hermes spaceplane into its operational orbit). (See fig. 1.)

The Hermes spaceplane is a delta-wing, tailless vehicle with a lift-to-drag ratio sufficient to cope with a lateral range of 2,000 km (1,240 mi) during its return flight. It has a three-crew-member cabin and a pressurized volume divided into a payload section and an area for crew living and mission specific activities. This latter area features facilities for on-board hygiene, food preparation, and sleep. It also provides direct access to the pressurized volume of the resource module which is usable both as a payload section and as an airlock toward space stations or to open space for extravehicular activities. This section is extended by a docking port located at the rear of the resource module and is designed to accommodate two spacesuits (see fig. 2). Table 1 summarizes the various characteristics of the Hermes space vehicle.

Crew safety is a major factor in the overall definition and design of the Hermes space vehicle. The Hermes crew members will be subjected, statistically speaking, to those risks normally associated with the test pilot profession. An early analysis of the risks incurred during the initial launch

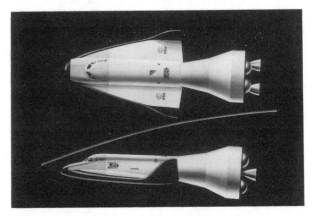

Fig. 1   The Hermes space vehicle, including the Hermes spaceplane, resource module and propulsion module. (Artist rendering courtesy of ESA.)

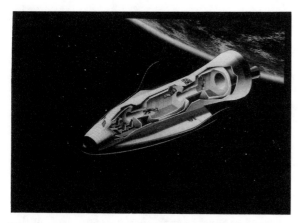

Fig. 2 A cutaway view of the Hermes spaceplane and resource module. (Artist rendering courtesy of ESA.)

Table 1 Hermes Space Vehicle Characteristics and Projected Performance Data

	Spaceplane	Resource Module	Propulsion Module
Length	13 m	6 m	1 m
Wingspan	9 m		
Maximum laden weight in transfer orbit	15,000 kg	8,000 kg	
Maximum laden weight at landing	15,000 kg		
Crew	3 persons		
Maximum payload on orbit	1,500 kg	1,500 kg	
Maximum payload at landing	1,500 kg		
Pressurized volume	33 m³	28 m³	

SOURCE: ESA.

phase of a Hermes mission (corresponding to the ignition and burn of the strapped-on solid rocket boosters) has led to the design decision that the spaceplanes's cockpit will be equipped with an ejection system to displace the crew far enough away from a distressed Ariane 5/Hermes composite vehicle to permit safe recovery on land or sea.

The Ariane 5/Hermes space vehicle will be launched from the ELA-3 complex at Kourou, French Guiana. Landing facilities for the spaceplane include the main runways in French Guiana as well as runways in Europe. There will also be sea recovery facilities and emergency runways to support these manned ESA missions.

> See also: **comet** (Giotto mission); **French Space Agency; National Aeronautics and Space Administration; Rosetta Mission; space station; Space Transportation System; Vesta Missions**

exobiology The multidisciplinary field that involves the study of extraterrestrial environments for living organisms, the recognition of evidence of the possible existence of life in these environments and the study of any nonterrestrial life-forms that may be encountered. The challenges of exobiology are being approached from several different directions. First, material samples from alien worlds in our Solar System can be obtained for study on Earth—as was accomplished during the Apollo lunar expeditions (1969–72); or such samples can be studied on the spot by robot explorers—as was accomplished by the Viking Landers (1976). Lunar rock and soil samples have not revealed any traces of life, while the biological results of the Viking Lander experiments involving Martian soils are still unclear. In particular, the Viking Landers have given us some chemical information about the Martian soil, but we still do not know enough about its nature to predict what reactions will occur when water and nutrients are added to it, as was done in the Viking biological laboratories. Even if the Martian soil is completely sterile and devoid of all life, it is possible that some of the reactions with the added water and nutrients that were observed were just imitations of biological activity. Because of these uncertainties, exobiologists must remain cautious in their final interpretations of the data arising from the Viking Lander biological experiments.

The question about life on Mars still remains open. It is still remotely possible that native Martian life (most likely microscopic in form) now exists in some crevice of the Red Planet. Or perhaps as is more likely, life began there eons ago but then died out. Scientists today are puzzled about how the Martian climatic conditions could have changed so rapidly. It now appears that there were once great rivers and floods of water raging over the plains of the Red Planet. This water has now mysteriously vanished, leaving behind a dry, barren and apparently lifeless desert.

A second major approach in exobiology involves conducting experiments in terrestrial laboratories or in laboratories in space that attempt either to simulate the primeval conditions that led to the formation of life on Earth and extrapolate these results to other planetary environments or to study the response of terrestrial organisms under environmental conditions found on alien worlds.

In 1938 Oparin, a Russian biochemist, proposed a theory of chemical evolution that suggested that organic compounds could be produced from simple inorganic molecules and that life on Earth probably originated by this process. Then, in 1953, at the University of Chicago, American Noble laureate Harold C. Urey and his former student Stanley L. Miller performed what can be considered the first modern experiments in exobiology. Investigating the chemical origin of life, Urey and Miller demonstrated that organic molecules could indeed be produced by irradiating a mixture of inorganic molecules. The historic Urey–Miller experiment simulated the Earth's assumed primitive atmosphere by using a gaseous mixture of methane (CH_4), ammonia (NH_3), water vapor (H_2O) and hydrogen (H_2) in a glass flask. A pool of water was kept gently boiling to promote circulation within the mixture, and an electrical discharge (simulating lightning) provided the energy needed to promote chemical reactions. Within days, the mixture changed colors, indicating that more complex, organic molecules had been synthesized out of this primordial "soup" of inorganic materials.

Other exobiologists and chemists have repeated the overall techniques employed by Urey and Miller. They have experimented with many different forms of energy thought to be present in the Earth's early history, including ultraviolet radiation, high-energy particles and meteorite crashes. These experimenters have subsequently produced in mixtures of inorganic materials many significant organic molecules—organic molecules and compounds found in the complex biochemical structures of terrestrial organisms.

The first compounds synthesized in the laboratory in the classic Urey–Miller experiment were amino acids, the building blocks of proteins. Later experiments have produced sugar molecules, including ribose and dexyribose—essential components of the deoxyribonucleic acid (DNA) and ribonucleic acid (RNA) molecules. The DNA and RNA molecules carry the genetic code of all terrestrial life.

A third general approach in exobiology involves an attempt to communicate with, or at least listen for signals from, other intelligent life-forms within our Galaxy. This effort is often called the search for extraterrestrial intelligence (or SETI, for short). At present, the principal aim of SETI activities throughout the world is to listen for evidence of extraterrestrial radio signals generated by intelligent alien civilizations.

Current theories concerning stellar-formation processes now lead many scientists to believe that planets are normal and frequent companions of most stars. If we consider that the Milky Way Galaxy contains some 100 billion to 200 billion stars, present theories on the chemical evolution of life indicate that it is probably not unique to Earth but may in fact be widespread throughout the Galaxy. Some scientists also speculate that life elsewhere may have evolved to intelligence, curiosity and the ability to build the technical tools required for interstellar transmission and reception of intelligent signals.

Within NASA today, exobiology is an interdisciplinary program of scientific research conducted by the Life Sciences Division. As its overall goal, the NASA Exobiology program seeks to understand the origin, evolution and distribution of life in the Universe. Through both ground-based and space-base research NASA's Exobiology program seeks answers to these fundamental questions: How did the development of the Solar System lead to the formation and persistence of habitable planetary environments? How did life on Earth originate? What factors operating on Earth or at large in the Solar System influenced the course of biological evolution from microbes to intelligence? Where else may life be found in the Universe?

The present understanding of biology and the natural history of life on Earth leads scientists to the conclusion that life originates and evolves on planets and that biological evolution is subject to the numerous factors and circumstances that determine planetary and solar system evolution. For example, the nearness of a supernova explosion or the nearby passage of a rogue star could greatly influence the evolution of a solar system and its potentially life-bearing planets.

Recent results of research supported by NASA's Exobiology program show that water and the prebiotic organic compounds believed to have been required as the building blocks of the chemical precursors to living systems are widespread in our own Solar System and beyond. The somewhat "universal" presence of these compounds allows exobiologists to now hypothesize that the origin of life appears inevitable throughout the Universe wherever these ingredients occur and are accompanied by suitable planetary conditions. Considering the enormous size of the observable Universe, an exciting prediction based on this exobiology hypothesis is that extraterrestrial life is widespread.

Testing the theory that life is a natural consequence of the physical and chemical processes created by the overall evolution of the cosmos requires a broad-based, scientifically rigorous exobiology research program. For example, studies of Comet Halley have revealed the presence of a variety of simple organic compounds and a very interesting, but poorly characterized, complex mixture of higher molecular weight particles—composed only of various combinations of the elements carbon (C), hydrogen (H), oxygen (O) and nitrogen (N). These simple compounds (including formaldehyde and hydrogen cyanide) are among the most abundant that have been observed in interstellar clouds, thereby strongly suggesting to exobiologists that comets may contain components of interstellar origin. Advanced comet encounter and sampling missions, such as the NASA Comet Rendezvous and Asteroid Flyby (CRAF) mission and the European Space Agency Rosetta mission will help exobiologists further explore this intriguing issue.

Data obtained from the Viking missions to Mars are now widely interpreted by most exobiologists to signify the absence of extant life. However, planetary scientists in examining Viking Orbiter images of Mars have detected fluvial features and apparently layered sedimentary deposits that have now been attributed to the action of liquid water on the surface of Mars during the first billion or so years of its history. These observations, indicating that Mars was more "Earth-like" in its early history, raises the intriguing question: Did life indeed arise on Mars and then become extinct due to unfavorable climate changes? Soil samples returned from Mars, through NASA's Mars Sample Return mission (MSRM), would allow exobiologists a more detailed opportunity to determine the possible origin of life on Mars. The discovery and analysis of fossil organisms found in any of these returned Martian rocks would certainly add immeasurably to our understanding of the evolution of extraterrestrial life.

Exobiologists have postulated the existence of a highly chemically reduced atmosphere dominated by methane (CH_4) and nitrogen (N_2) for the primitive Earth; an atmosphere similar to the one found today on Saturn's moon, Titan. Although the Voyager flyby missions revealed traces

Table 1 Proposed Exobiology Experiments to be Performed on Earth-Orbiting Spacecraft (1990–2010)

Observational Exobiology

1. Search for extrasolar planetary systems
2. Study star-forming regions in the galaxy—analogs for the solar nebula
3. Study comets, asteroids, Titan and the giant planets in our own solar system
4. Study the organic chemistry of interstellar molecular clouds

Cosmic Dust Collection

1. Develop and implement capture techniques that preserve biogenic material
2. Develop and implement techniques to determine the orbits of dust particles
3. Refine laboratory methods for the analysis of small particles

In Situ *Experiments*

1. Study the formation, condensation, aggregation and surface chemistry of suspended dust grains
2. Create, release and monitor an artificial comet in space
3. Determine the viability of microorganisms in space

SOURCE: NASA (Ames Research Center).

of many organic compounds in Titan's atmosphere, the degrees of molecular complexity attained in the Titanian atmosphere and the physical processes responsible for their synthesis are still unclear to scientists. The deployment of an instrumented probe into Titan's atmosphere during the Cassini mission should help exobiologists resolve these questions.

Finally, NASA scientists recently investigated a variety of concepts for experiments of exobiological interest that might be conducted on Earth-orbiting spacecraft (including Space Station *Freedom*) over the next few decades. These proposed exobiology experiments are divided into three general groups: observational exobiology, cosmic dust collection, and in-situ experiments. These are summarized in table 1.

See also: **biogenic elements; Cassini Mission; Comet Rendezvous and Asteroid Flyby Mission; extraterrestrial contamination; life in the Universe; Mars; Rosetta Mission; search for extraterrestrial intelligence; Viking Project**

extragalactic Occurring, located or originating beyond our Galaxy (the Milky Way); typically, farther than 100,000 light-years distant.

extrasolar Occurring, located or originating outside of our Solar System; as, for example, extrasolar planets.

extrasolar planets Planets that belong to a star other than the Sun. There are two general methods that can be used to detect extrasolar planets: direct (involving a search for telltale signs of planet radiation) and indirect (involving precise observation of the motion of the parent star on periodic variations in its spectral properties.)

Why should we be interested in the discovery of extrasolar planets? Well, evidence of planets around other stars would help astronomers validate their current hypothesis that planet formation is a normal part of stellar evolution. Evidence of extrasolar planets, especially if we can also determine their frequency of occurrence as a function of the type of star, would also greatly assist scientists in their estimations of the cosmic prevalence of life. If life originates on "suitable" planets whenever it can (as many exobiologists currently hold), then knowing how abundant such suitable planets are in our Galaxy would allow us to make more credible guesses about where to search for extraterrestrial intelligence and what our chances are of finding intelligent life beyond our own Solar System.

See also: **Barnard's star; Drake equation; Project Orion; search for extraterrestrial intelligence**

extraterrestrial Refers to something that occurs, is located or originates outside of the planet Earth and its atmosphere.

extraterrestrial careers The first few decades of the next millennium promise to be some of the most exciting times ever experienced in human history. Building on space technologies developed in the 1960s through 1990s, people will start living in permanent space stations, in lunar bases and settlements and in semi-permanent bases on the planet Mars! Extraterrestrial careers in the next century could include such interesting job titles as space station superintendent; chief engineer, on-orbit manufacturing facility; head exobiology research module (space station); exobiologist-in-charge, Mars expedition; lunar miner; Martian settler; space farmer (lunar base agricultural facility); extraterrestrial geologist; spaceship captain (Earth-to-Moon run); space construction worker; and maybe even "governor" of the lunar settlement. Graduate students interested in advanced degrees in space technology or planetary science might perform their doctoral research on the space station or as part of the scientific and engineering team at the lunar base. We might even witness the start of a fully accredited "University of Space"—with the lunar base serving as our planet's first extraterrestrial campus. Perhaps some of the most exciting breakthroughs in the materials and life sciences will be made by dedicated young investigators exploring the secrets of nature in the microgravity world of the space station. It is not even too unreasonable to speculate that many of tomorrow's Noble prize-winning concepts in physics, chemistry and medicine will be born out of research conducted in our extraterrestrial environment.

With these exciting ideas in mind, you may ask yourself "How do I get started in an extraterrestrial career?" Of course, no one today can predict precisely what career pathways and job titles will emerge as we start to build our extraterrestrial civilization. However, it is a pretty safe bet to assume that until the mid-1990s the traditional "way into space" will be through a career as an astronaut. However, even the concept of what an astronaut is and does, has changed dramatically since the early manned space flight programs. In the Shuttle era, for example, a much larger variety of people have flown into space—broadening the concept of "astronaut" well beyond that of a daring test pilot with nerves of steel who had "the right stuff." This broadening of the meaning of astronaut will continue even further as the space station becomes operational in the mid-1990s and an even wider segment of the technical population experiences the Universe face to face. To help accommodate this expansion of the meaning of astronaut, we will define an astronaut here (for career discussion purposes) as any person who travels above the Earth's sensible atmosphere (150 kilometers). Institutions, such as NASA, may wish to retain the title "astronaut" for selected members of a particularly trained group, such as a Shuttle commander, pilot or mission specialist. However, this specialized distinction will eventually give way to a more general 21st-century concept of a space traveler or worker who engages in activities in the extraterrestrial environment. In fact, with the full-scale operation of a permanent space station and lunar base, we can anticipate the creation of a host of exciting and romantic new job titles.

Unfortunately, not all of us will be able to fly in space and view the Earth gliding silently below while we earn a living meeting the Universe face to face. But don't despair! There will be many other interesting extraterrestrial careers in addition to that of astronaut. These include aerospace craftsperson, aerospace engineer, aerospace engineering technician, astronomer, electronics engineer, geographer, mathematician, meteorologist, physicist, science writer or systems analyst, just to name a few. These career areas support astronauts and their activities in space.

extraterrestrial catastrophe theory   For millions of years giant, thundering reptiles roamed the lands, dominated the skies and swam in the oceans of a prehistoric Earth. Dinosaurs reigned supreme. Then, quite suddenly, some 65 million years ago, they vanished. What happened to these giant creatures and to thousands of other ancient animal species?

From archaeological and geological records, we do know that some tremendous catastrophe occurred about 65 million years ago on this planet. It affected life more extensively than any war, famine or plague in human history. For in that cataclysm approximately 90 percent of all animal species were annihilated—including, of course, the dinosaurs. It is most interesting to observe, however, that as long as those enormous reptiles roamed and dominated the Earth, mammals, including humans themselves, would have had little chance of evolving.

Several years ago the scientists Luis and Walter Alvarez and their colleagues at the University of California at Berkeley discovered that a pronounced increase in the amount of the element iridium in the Earth's surface had occurred at precisely the time of the disappearance of the dinosaurs. Since iridium is quite rare in the Earth's crust and much

more abundant in the rest of the Solar System, they postulated that a large asteroid (or possibly a comet) might have struck the ancient Earth. This cosmic collision would have promoted an environmental catastrophe throughout the planet. The scientists reasoned that such an asteroid would largely vaporize while passing through the Earth's atmosphere, spreading a dense cloud of dust particles—including quantities of extraterrestrial iridium atoms—uniformly around the globe. Many recent studies have observed a worldwide level of enhanced iridium in a thin layer of the Earth's lithosphere (crust) lying between the final geologic formations of the Cretaceous period (which are dinosaur fossil-rich) and the formations of the early Tertiary period (whose rocks are notably lacking in dinosaur fossils). The Alvarez hypothesis further speculated that following this asteroid impact, a dense cloud of dust covered the Earth for many years, obscuring the Sun, blocking photosynthesis and destroying the very food chains upon which many ancient life-forms depended.

There are, of course, many other scientific opinions as to why the dinosaurs vanished. A popular one is that there was a gradual but relentless change in the Earth's climate to which these giant reptiles and many other prehistoric animals simply could not adapt.

Nevertheless, the possibility of an asteroid's or comet's striking the Earth and triggering a catastrophe is quite real. Just look at a recent photograph of Mars or the Moon and ask yourself how those large impact craters were formed. Fortunately, the probability of a large asteroid's striking the Earth is quite low. For example, space scientists now estimate that the Earth will experience one collision with an "Earth-crossing" asteroid (of one kilometer diameter size or greater) every 300,000 years.

Yet, on May 22, 1989, a small "Earth-crossing" asteroid, called 1989FC, passed within 690,000 kilometers (only 0.0046 astronomical unit) of our planet. This cosmic "near-miss" occurred at less than twice the distance to the Moon. Cosmic impact specialists have estimated that if this small asteroid, presumed to be about 200 to 400 meters in diameter, had experienced a straight-in collision with the Earth at a relative velocity of some 16 kilometers per second, it would have impacted with an explosive force of some 400 to 2,000 megatons (MT). (A megaton is the energy of an explosion that is the equivalent to one million tons of TNT.) If this small asteroid had hit a terrestrial landmass, it would have formed a crater some 4 to 7 kilometers across.

See also: **asteroid, comet, Nemesis**

extraterrestrial civilizations   How might we categorize levels or types of extraterrestrial civilizations? According to some scientists, intelligent life in the Universe might be thought of as experiencing three basic levels of civilization, when considered on an astronomical scale. The Soviet astronomer N.S. Kardashev, in examining the issue of information transmission by extraterrestrial civilizations in 1964, first postulated three types of technologically developed civilizations on the basis of their energy use. A Type I civilization would represent a planetary civilization similar to the technology level on Earth today. It would command the use of somewhere between 10^{12} and 10^{16} watts of energy—the upper limit being the amount of solar energy being intercepted by a "suitable" planet in its orbit about the parent star. For example, the solar energy flux at the Earth (the solar constant outside the atmosphere) is 1,371 ± 5 watts per meter squared, and the spectral distribution of sunlight is that of an approximate blackbody radiator at a temperature of 5,760 degrees Kelvin, which peaks in the visible part of the electromagnetic (EM) spectrum—namely, a wavelength of 0.4 to 0,7 micrometer.

A Type II extraterrestrial civilization would engage in feats of planetary engineering, emerging from its native planet through advances in space technology and extending its resource base throughout the local star system. The eventual upper limit of a Type II civilization could be taken as the creation of a "Dyson sphere." A Dyson sphere is a shell or cluster of habitats and structures placed entirely around a star by an advanced civilization to intercept and use basically all the radiant energy from that parent star. What the physicist Freeman J. Dyson proposed in 1960 was that an advanced extraterrestrial civilization would eventually develop the space technologies necessary to rearrange the raw materials of all the planets in its solar system, creating a more efficient composite ecosphere around the parent star. Dyson further postulated that such advanced civilizations might be detected by the presence of thermal infrared emissions from such an "enclosed star system" in contrast to the normally anticipated visible radiation. Once this level of extraterrestrial civilization is achieved, the search for additional resources and the pressures of continued growth could encourage interstellar migrations. This would mark the start of a Type III extraterrestrial civilization.

At maturity a Type III civilization would be capable of harnessing the material and energy resources of an entire galaxy (typically containing some 10^{11} to 10^{12} stars). Energy levels of 10^{37} to 10^{38} watts would be involved!

Command of energy resources might therefore represent a key factor in the evolution of extraterrestrial civilizations. It should be noted that a Type II civilization controls about 10^{12} times the energy resources of a Type I civilization; and a Type III civilization approximately 10^{12} times as much energy as a Type II civilization.

What can we speculate about such civilizations? Well, starting with the Earth as a model (our one and only "scientific data point"), we can presently postulate that a Type I civilization could exhibit the following characteristics: (1) an understanding of the laws of physics; (2) a planetary society (for example, global communication network, interwoven food and materials resource networks); (3) intentional or unintentional emission of electromagnetic radiations (especially radio frequency); (4) the development of space technology and spaceflight—the tools necessary to leave the home planet; (5) (possibly) the development of

nuclear energy technology, both power supplies and weapons; and (6) (possibly) a desire to search for and communicate with other intelligent life-forms in the Universe. Many uncertainties, of course, are present. Given the development of the technology for spaceflight, will the planetary civilization opt to create a solar-system civilization? Do the planet's inhabitants develop a long-range planning perspective that supports the eventual creation of artificial habitats and structures throughout their star system? Or do the majority of Type I civilizations unfortunately destroy themselves with their own advanced technologies before they can emerge from a planetary civilization into a more stable Type II civilization? Does the exploration imperative encourage such creatures to go out from their comfortable, planetary niche into an initially hostile, but resource-rich star system? If this "cosmic birthing" does not occur frequently, perhaps our Galaxy is indeed populated with intelligent life, but at a level of stagnant planetary (Type I) civilizations that have neither the technology nor the motivation to create an extraterrestrial civilization or even to try to communicate with any other intelligent life-forms across interstellar distances.

Assuming that an extraterrestrial civilization does, however, emerge from its native planet and create an interplanetary society, several additional characteristics would become evident. The construction of space habitats and structures, leading ultimately to a Dyson sphere around the native star would reflect feats of planetary engineering and could possibly be detected by thermal infrared emissions as incident starlight in the visible spectrum was intercepted, converted to other more useful forms of energy and the residual energy (determined by the universal laws of thermodynamics) rejected to space as heat at perhaps 300 degrees Kelvin. Type II civilizations might also decide to search in earnest for other forms of intelligent life beyond their star system. They would probably use portions of the electromagnetic spectrum (radio frequency and perhaps X rays or gamma rays) as information carriers between the stars. Remembering that Type II civilizations would control 10^{12} times as much energy as Type I civilizations, such techniques as electromagnetic beacons or feats of astroengineering that yield characteristic X-ray or gamma-ray signatures may lie well within their technical capabilities. Assuming their understanding of the physical Universe is far more sophisticated than ours, Type II civilizations might also use gravity waves or other physical phenomena (perhaps unknown to us now but being sent through our Solar System at this very instant) in their effort to communicate across vast interstellar distances. Type II civilizations could also decide to make initial attempts at interstellar matter transfer. Fully automated robotic explorers would be sent forth on one-way scouting missions to nearby stars. Even if the mode of propulsion involved devices that achieved only a small fraction of the speed of light, Type II societies should have developed the much longer-term planning perspective and thinking horizon necessary to support such sophisticated, expensive and lengthy missions. The Type

II civilization might also utilize a form of panspermia (the diffusion of spores or molecular precursors through space) or even ship microscopically encoded viruses through the interstellar void, hoping that if such "seeds of life" found a suitable ecosphere in some neighboring or distant star system, they would initiate the chain of life—perhaps leading ultimately to the replication (suitably tempered by local ecological conditions) of intelligent life itself. Finally, as the Dyson sphere was eventually completed, some of the inhabitants of this Type II civilization might respond to a cosmic wanderlust and initiate the first "peopled" interstellar missions. Complex space habitats would become "space arks" and carry portions of this civilization to neighboring star systems. Again, however, we must ask what is the lifetime of a Type II civilization? It would appear from an extrapolation of contemporary terrestrial engineering practices that perhaps a minimum of 500 to 1,000 years would be required for even an advanced interplanetary civilization to complete a Dyson sphere.

Throughout the entire Galaxy, however, if just one Type II civilization embarks on a successful interstellar migration program, then—at least in principle—it would eventually (in perhaps 10^8 to 10^9 years) sweep through the Galaxy in a "leapfrogging" wave of colonization, establishing a Type III civilization in its wake. This Type III civilization would eventually control the energy and material resources of up to 10^{12} stars—or the entire Galaxy! Communication or matter transfer would be accomplished by techniques that can now only politely be called "exotic." Perhaps directed beams of neutrinos or even (hypothesized) faster-than-light particles (tachyons) would serve as information carriers for this galactic society. Or, they might use tunneling through black holes as their transportation network. Perhaps again, they would have developed some kind of thought-transference telepathy that could form a basic communication network over the vast regions of interstellar space. In any event, a Type III civilization should be readily evident, since it would be galactic in extent and easily recognizable by its incredible feats of astroengineering.

In all likelihood, our Galaxy at present does not contain a Type III civilization. Or else the Solar System is being ignored—intentionally kept isolated—perhaps as a game

Table 1 Characteristics of Extraterrestrial Civilizations

Level	Characteristics	Energy Consumption	Manifestations
TYPE I	• Planetary society • Developed technology – understanding of the laws of physics – space technology – nuclear technology – electromagnetic communications • Initiation of spaceflight, interplanetary travel; settlement of space • Early attempts at interstellar communication	10^{12}–10^{16} watts • Starting to push planetary resource limits	• Intentional or unintentional electromagnetic emissions (especially radio wave)
TYPE II	• Solar-system society • Construction of space habitats • "Dyson sphere" as an ultimate limit • Search for intelligent life in space • Possible interstellar communication between Type II civilizations • Reasonably long societal lifetimes (10^3 to 10^5 years) • Far-term planning perspectives • Initiation of interstellar travel/colonization	10^{26}–10^{27} watts • Ultimately all radiant energy output of native star is utilized	• Electromagnetic – radio waves – X-rays – gamma rays • Gravity waves • Mass transfer – probes – panspermia – stellar ark
TYPE III	• Galactic civilization • Interstellar communication/travel • Fantastic feats of astroengineering • Very long societal lifetimes (10^8–10^9 years) • Effectively "the immortals," for planning purposes	10^{37}–10^{38} watts • Energy resources of entire Galaxy (10^{11}–10^{12} stars) are commanded	• Feats of astroengineering • Exotic *Communication* – neutrinos – tachyons *Travel* – tunneling through black holes – telepathy – ? ? ? ?

preserve or "zoo," as some have speculated; or maybe it is one of the very last regions to be "filled in."

The other perspective is that if we are indeed alone or the most advanced civilization in our Galaxy, we now stand at the technological threshold of creating the first Type II civilization in the Galaxy; and if successful, we then have the potential of becoming the first interstellar civilization to sweep across the Galaxy, establishing a Type III civilization in the Milky Way!

Table 1 summarizes these speculations on the levels of extraterrestrial civilizations and their potential characteristics.

See also: **consequences of extraterrestrial contact; Drake Equation; Dyson sphere; Fermi paradox; interstellar communication; search for extraterrestrial intelligence; starship; Zoo hypothesis**

extraterrestrial contamination In general, the contamination of one world by life-forms, especially microorganisms, from another world. Using the Earth and its biosphere as a reference, this planetary-contamination process is called *forward contamination* if an extraterrestrial sample or the alien world itself is contaminated by contact with terrestrial organisms, and *back contamination* if alien organisms are released into the Earth's biosphere.

An alien species will usually not survive when introduced into a new ecological system, because it is unable to compete with native species that are better adapted to the environment. Once in a while, however, alien species actually thrive, because the new environment is very suitable and indigenous life-forms are unable to successfully defend themselves against these alien invaders. When this "war of biological worlds" occurs, the result might very well be a permanent disruption of the host ecosphere, with severe biological, environmental and possibly economic consequences.

Of course, the introduction of an alien species into an ecosystem is not always undesirable. Many European and Asian vegetables and fruits, for example, have been successfully and profitably introduced into the North American environment. However, any time a new organism is released in an existing ecosystem, a finite amount of risk is also taken.

Frequently, alien organisms that destroy resident species are microbiological life-forms. Such microorganisms may have been nonfatal in their native habitat, but once released in the new ecosystem, they become unrelenting killers of native life-forms that are not resistant to them. In past centuries on Earth, entire human societies fell victim to alien organisms against which they were defenseless; as for example, the rapid spread of diseases that were transmitted to native Polynesians and American Indians by European explorers.

But an alien organism does not have to directly infect humans to be devastating. Who can easily ignore the consequences of the potato blight fungus that swept through Europe and the British Isles in the 19th century, causing a million people to starve to death in Ireland alone?

In the Space Age it is obviously of extreme importance to recognize the potential hazard of extraterrestrial contamination (forward or back). Before any species is intentionally introduced into another planet's environment, we must carefully determine not only whether the organism is pathogenic (disease-causing) to any indigenous species but also whether the new organism will be able to force out native species—with destructive impact on the original ecosystem. The introduction of rabbits into the Australian continent is a classic terrestrial example of a nonpathogenic life-form creating immense problems when introduced into a new ecosystem. The rabbit population in Australia simply exploded in size because of their high reproduction rate, which was essentially unchecked by native predators.

At the start of the Space Age, scientists were already aware of the potential extraterrestrial-contamination problem—in either direction. Quarantine protocols (procedures) were established to avoid the forward contamination of alien worlds by outbound unmanned spacecraft, as well as the back contamination of the terrestrial biosphere when lunar samples were returned to Earth as part of the Apollo program.

A quarantine is basically a forced isolation to prevent the movement or spread of a contagious disease. Historically, quarantine was the period during which ships suspected of carrying persons or cargo (for example, produce or livestock) with contagious diseases were detained at their port of arrival. The length of the quarantine, generally 40 days, was considered sufficient to cover the incubation period of most highly infectious terrestrial diseases. If no symptoms appeared at the end of the quarantine, then the travelers were permitted to disembark. In modern times, the term *quarantine* has obtained a new meaning: namely, that of holding a suspect organism or infected person in strict isolation until it is no longer capable of transmitting the disease. With the Apollo program and the advent of the lunar quarantine, the term now has elements of both meanings. Of special interest in future space missions to the planets and their moons is how we avoid the potential hazard of back contamination of the Earth's environment when robot spacecraft and human explorers bring back samples for more detailed examination in laboratories on Earth.

A Planetary Quarantine program was started by NASA in the late 1950s at the beginning of the U.S. space program. This quarantine program, conducted with international cooperation, was intended to prevent, or at least minimize, the possibility of contamination of alien worlds by early space probes. At that time, scientists were concerned with forward contamination. In this type of extraterrestrial contamination, terrestrial microorganisms, "hitchhiking" on initial planetary probes and landers, would spread throughout another world, destroying any native life-forms, life precursors or even remnants of past life-forms. If forward contamination occurred, it would compromise future attempts to search for and identify extraterrestrial life-forms that had arisen independently of the Earth's biosphere.

A planetary quarantine protocol was therefore established. This protocol required that outbound unmanned planetary missions be designed and configured to minimize the probability of alien-world contamination by terrestrial life-forms. As a design goal, these spacecraft and probes had a probability of 1 in 1,000 (1×10^{-3}) or less that they could contaminate the target celestial body with terrestrial microorganisms. Decontamination, physical isolation (for example, prelaunch quarantine) and spacecraft design techniques have all been employed to support adherence to this protocol.

One simplified formula for describing the probability of planetary contamination is:

$$P(c) = m \times P(r) \times P(g)$$

where $P(c)$ is the probability of contamination of the target celestial body by terrestrial microorganisms

m is the microorganism burden

$P(r)$ is the probability of release of the terrestrial microorganisms from the spacecraft hardware

and $P(g)$ is the probability of microorganism growth after release on a particular planet or celestial object.

As previously stated, $P(c)$ had a design goal value of less than or equal to 1 in 1,000. A value for the microorganism burden, m, was established by sampling an assembled spacecraft or probe. Then, through laboratory experiments, scientists determined how much this microorganism burden was reduced by subsequent sterilization and decontamination treatments. A value for $P(r)$ was obtained by placing duplicate spacecraft components in simulated planetary environments. Unfortunately, establishing a numerical value for $P(g)$ was a bit more tricky. The technical intuition of knowledgeable exobiologists and some educated "guessing" were blended together to create an estimate for how well terrestrial microorganisms might thrive on alien worlds that had not yet been visited. Of course, today, as we keep learning more about the environments on other worlds of our Solar System, we can keep refining our estimates for $P(g)$. Just how well terrestrial life-forms grow on Mars, Venus, Titan and a variety of other interesting celestial bodies will be the subject of on-site laboratory experiments performed by 21st-century exobiologists.

As a point of history, the early U.S. Mars flyby missions (for example, *Mariner 4*, launched on November 28, 1964, and *Mariner 6*, launched on February 24, 1969) had $P(c)$ values ranging from 4.5×10^{-5} to 3.0×10^{-5}. These missions achieved successful flybys of the Red Planet on July 14, 1965, and July 31, 1969, respectively. Postflight calculations indicated that there was no probability of planetary contamination as a result of these successful precursor missions.

The manned U.S. Apollo missions to the Moon (1969–72) also stimulated a great deal of debate about forward and back contamination. Early in the 1960s, scientists began speculating in earnest, "Is there life on the Moon?" Some of the most bitter technical exchanges during the Apollo program took place over this particular question. If there was life, no matter how primitive or microscopic, we would want to examine it carefully and compare it with life-forms of terrestrial origin. This careful search for microscopic lunar life would, however, be very difficult and expensive because of the forward-contamination problem. For example, all equipment and materials landed on the Moon would need rigorous sterilization and decontamination procedures. There was also the glaring uncertainty about back contamination. If microscopic life did indeed exist on the Moon, did it represent a serious hazard to the terrestrial biosphere? Because of the potential extraterrestrial-contamination problem, time-consuming and expensive quarantine procedures were urged by some members of the scientific community.

On the other side of this early 1960s contamination argument were those exobiologists who emphasized the suspected extremely harsh lunar conditions: virtually no atmosphere; probably no water; extremes of temperature ranging from 120 degrees Celsius at lunar noon to minus 150 degrees Celsius during the frigid lunar night; and unrelenting exposure to lethal doses of ultraviolet, charged-particle and X-ray radiations from the Sun. No life-form, it was argued, could possibly exist under such extremely hostile conditions.

This line of reasoning was countered by other exobiologists who hypothesized that trapped water and moderate temperatures below the lunar surface could sustain very primitive life-forms. And so the great extraterrestrial-contamination debate raged back and forth, until finally the *Apollo 11* expedition departed on the first lunar-landing mission. As a compromise, the *Apollo 11* mission flew to the Moon with careful precautions against back contamination but with only a very limited effort to protect the Moon from forward contamination by terrestrial organisms.

The Lunary Receiving Laboratory (LRL) at the Johnson Space Center in Houston provided quarantine facilities for two years after the first lunar landing. What we learned during its operation serves as a useful starting point for planning new quarantine facilities, Earth-based or space-based. In the future, these quarantine facilities will be needed to accept, handle and test extraterrestrial materials from Mars and other Solar-System bodies of interest in our search for alien life-forms (present or past).

During the Apollo program, no evidence was discovered that native alien life was then present or had ever existed on the Moon. A careful search for carbon was performed by scientists at the Lunar Receiving Laboratory, since terrestrial life is carbon-based. One hundred to 200 parts per million of carbon were found in the lunar samples. Of this amount, only a few tens of parts per million are considered indigenous to the lunar material, while the bulk amount of carbon has been deposited by the solar wind.

Exobiologists and lunar scientists have concluded that none of this carbon appears derived from biological activity. In fact, after the first few Apollo expeditions to the Moon, even back-contamination quarantine procedures were dropped.

There are three fundamental approaches toward handling extraterrestrial samples to avoid back contamination. First, we could sterilize a sample while it is en route to Earth from its native world. Second, we could place it in quarantine in a remotely located, maximum-confinement facility on Earth while scientists examine it closely. Finally, we could also perform a preliminary hazard analysis (called the extraterrestrial protocol tests) on the alien sample in an orbiting quarantine facility before we allow the sample to enter the terrestrial biosphere. To be adequate, a quarantine facility must be capable of (1) containing all alien organisms present in a sample of extraterrestrial material, (2) detecting these alien organisms during protocol testing and (3) controlling these organisms after detection until scientists could dispose of them in a safe manner.

One way to bring back an extraterrestrial sample that is free of potentially harmful alien microorganisms is to sterilize the material during its flight to Earth. However, the sterilization treatment used must be intense enough to guarantee that no life-forms as we currently know them could survive. An important concern here is also the impact the sterilization treatment might have on the scientific value of the alien sample. For example, use of chemical sterilants would most likely result in contamination of the sample, preventing the measurement of certain soil properties. Heat could trigger violent chemical reactions within the soil sample, resulting in significant changes and the loss of important exogeological data. Finally, sterilization would also greatly reduce the biochemical information content of the sample. It is even questionable as to whether any significant exobiology data can be obtained by analyzing a heat-sterilized alien material sample. To put it simply—in their search for extraterrestrial life-forms, exobiologists want "virgin alien samples."

If we do not sterilize the alien samples en route to Earth, we have only two general ways of avoiding possible back-contamination problems. We can place the unsterilized sample of alien material in a maximum quarantine facility on Earth and then conduct detailed scientific investigations, or we can intercept and inspect the sample at an orbiting quarantine facility before allowing the material to enter the Earth's biosphere.

The technology and procedures for hazardous-material containment have been employed on Earth in the development of highly toxic chemical- and germ-warfare agents and in conducting research involving highly infectious diseases. A critical question for any quarantine system is whether the containment measures are adequate to hold known or suspected pathogens while experimentation is in progress. Since the characteristics of potential alien organisms are not presently known, we must assume that the hazard they could represent is at least equal to that of

terrestrial Class IV pathogens. (A terrestrial Class IV pathogen is an organism capable of being spread very rapidly among humans; no vaccine exists to check its spread; no cure has been developed for it; and the organism produces high mortality rates in infected persons.) Judging from the large uncertainties associated with potential extraterrestrial life-forms, it is not obvious that any terrestrial quarantine facility will gain very wide acceptance by the scientific community or the general public. For example, locating such a facility and all its workers in an isolated area on Earth actually provides only a small additional measure of protection. Consider, if you will, the planetary environmental impact controversies that could rage as individuals speculated about possible ecocatastrophes. What would happen to life on Earth if alien organisms did escape and went on a deadly rampage throughout the Earth's biosphere? The alternative to this potentially explosive controversy is quite obvious: locate the quarantine facility in outer space. A space-based facility provides several distinct advantages: (1) it eliminates the possibility of a sample-return spacecraft's crashing and accidentally releasing its deadly cargo of alien microorganisms; (2) it guarantees that any alien organisms that might escape from confinement facilities within the orbiting complex cannot immediately enter the Earth's biosphere; and (3) it ensures that all quarantine workers remain in total isolation during protocol testing (that is, during the testing procedure). (See fig. 1.)

As we expand the human sphere of influence into heliocentric (Sun-centered) space, we must also remain conscious of the potential hazards of extraterrestrial contamination. Scientists, space explorers and extraterrestrial entrepreneurs must be aware of the ecocatastrophes that might occur when "alien worlds collide"—especially on the microorganism level.

With a properly designed and operated orbiting quarantine facility, alien-world materials can be tested for potential hazards. Three hypothetical results of such protocol testing are: (1) no replicating alien organisms are discov-

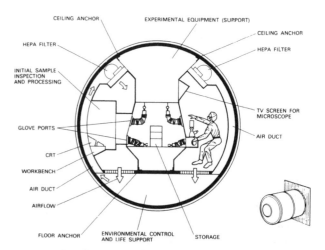

Fig. 1 The laboratory module for an extraterrestrial quarantine facility. (Drawings courtesy of NASA.)

ered; (2) replicating alien organisms are discovered, but they are also found not to be a threat to terrestrial life-forms; or (3) hazardous replicating alien life-forms are discovered. If potentially harmful replicating alien organisms were discovered during these protocol tests, then quarantine workers would either render the sample harmless (for example, through heat- and chemical-sterilization procedures); retain it under very carefully controlled conditions in the orbiting complex and perform more detailed analyses on the alien life-forms; or properly dispose of the sample before the alien life-forms could enter the Earth's biosphere and infect terrestrial life-forms.

See also: **Mars Sample Return Mission**

extraterrestrial intelligence Intelligent life that exists elsewhere in the Universe other than on the planet Earth.

See also: **extraterrestrial civilizations; search for extraterrestrial intelligence**

extraterrestrial life Life-forms that have evolved independent of and now exist outside of the terrestrial biosphere.

See also: **extraterrestrial civilizations; life in the Universe**

extraterrestrial life chauvinisms What characteristics and properties should extraterrestrial life-forms exhibit? Do they resemble terrestrial organisms, or are they entirely different from any living system found on Earth? What properties of terrestrial life are really basic to life found elsewhere in the Universe, and what characteristics of living systems are purely accidents of evolution? These are some of the puzzling questions facing modern scientists and exobiologists as they seek to unravel the mystery of life in the Universe. As you can see, it is not easy to describe what alien life-forms will be like—should they in fact exist elsewhere in the vast reaches of our Galaxy and beyond. At present, we have only one data point (source of information) on the emergence of life in a planetary environment—our own planet Earth. Scientists currently believe that all carbon-based terrestrial organisms have descended from a common, single occurrence of the origin of life in the primeval "chemical soup" of an ancient Earth. How can we project this singular fact to the billions of unvisited worlds in the cosmos? We can only do so with great technical caution, realizing full well that our models of extraterrestrial life-forms and our estimates concerning the cosmic prevalence of life can easily become prejudiced, or chauvinistic, in their findings.

Chauvinism can be defined as a strongly prejudiced belief in the superiority of one's group. Applied to speculations about extraterrestrial life, this word can take on several distinctive meanings, each heavily influencing any subsequent thought on the subject. Some of the more common forms of extraterrestrial life chauvinisms are: G-star chauvinism, planetary chauvinism, terrestrial chauvinism, chemical chauvinism, oxygen chauvinism and carbon chau-

vinism. While such heavily steeped thinking may not actually be wrong, it is important to realize that it also sets limits, intentionally or unintentionally, on contemporary speculations about life in the Universe.

G-star, or Solar-System, chauvinism implies that life can only originate in a star system like our own—namely, a system containing a single G-spectral-class star. Planetary chauvinism assumes that extraterrestrial life has to develop independently on a particular planet, while terrestrial chauvinism stipulates that only "life as we know it on Earth" can originate elsewhere in the Universe. Chemical chauvinism demands that extraterrestrial life be based on chemical processes, while oxygen chauvinism states that alien worlds must be considered uninhabitable if their atmospheres do not contain oxygen. Finally, carbon chauvinism asserts that extraterrestrial life-forms must be based on carbon chemistry.

These chauvinisms, singularly and collectively, impose tight restrictions on the type of planetary system that might support the rise of living systems, possibly to the level of intelligence, elsewhere in the Universe. If they are indeed correct, then our search for extraterrestrial intelligence is now being properly focused on Earthlike worlds around Sunlike stars. If, on the other hand, life is actually quite prevalent and capable of arising in a variety of independent biological scenarios (for example, silicon-based or sulfur-based chemistry), then our contemporary efforts in modeling the cosmic prevalence of life and in trying to describe what "little green men" really look like is somewhat analogous to using the atomic theory of Democritus, the ancient Greek, to help describe the inner workings of a modern nuclear-fission reactor. As we continue to explore our own Solar System—especially Mars and the moons of Jupiter and Saturn—we will be able to better assess how valid these chauvinisms really are.

extraterrestrial resources When people think about outer space, visions of vast emptiness, devoid of anything useful, frequently come to their minds. However, when forward-thinking space technologists gaze into the extraterrestrial environment, they see a new frontier rich with resources, including unlimited solar energy, a full range of raw materials and an environment that is both special (for example, high vacuum, microgravity, physical isolation from the terrestrial biosphere) and reasonably predictable.

Since the start of the Space Age, investigations of the Moon, Mars, the asteroids and meteorites have provided tantalizing hints about the rich mineral potential of our extraterrestrial environment. For example, the Apollo missions established that the average lunar soil contains more than 90 percent of the materials needed to construct a complicated space industrial facility. (See table 1.) The soil in the lunar highlands is rich in anorthosite, a mineral suitable for the extraction of aluminum, silicon and oxygen. Other lunar soils have been found to contain ore-bearing granules of ferrous metals like iron, nickel, titanium and chromium. Iron can be concentrated from the lunar soil

Table 1   Materials Available on the Surface of the Moon

Elements	Percentage by Weight		
	Mare[a]	Highlands[b]	Basin Ejecta[c]
Oxygen	39.7–42.3	44.6	42.2–43.8
Silicon	18.6–21.6	21.0	21.1–22.5
Aluminum	5.5–8.2	12.2–14.4	9.2–10.9
Iron	12.0–15.4	4.0–5.7	6.7–10.4
Calcium	7.0–8.7	10.1–11.3	6.3–9.2
Magnesium	5.0–6.8	3.5–5.6	5.7–6.3
Titanium	1.3–5.7	0.3	0.8–1.0
Chromium	0.2–0.4	0.1	0.2
Sodium	0.2–0.4	0.3–0.4	0.3–0.5
Manganese	0.2	0.1	0.1
Potassium	0.06–0.22	0.07–0.09	0.13–0.46
Hydrogen, carbon, nitrogen, fluorine, zirconium, nickel		100 ppm[d]	
Zinc, lead, chlorine, sulfur, other volatiles		5–100 ppm	

[a]Mare = relatively smooth, dark areas of lunar surface.
[b]Highlands = densely cratered, rugged uplands.
[c]Basin Ejecta = materials ejected out of impact crater basins.
[d]ppm = parts per million.

(called regolith) before the raw material is even refined by simply sweeping magnets over the regolith to gather the iron granules scattered within.

Some scientists have even suggested that water ice and other frozen gases (or volatiles) may be trapped on the lunar surface in perpetually shaded polar regions. If future exploration of the Moon indicates that this is true, then "ice mines" on the Moon could provide both oxygen and hydrogen—vital resources for our extraterrestrial settlements and space industrial facilities. The Moon would be able both to export chemical propellants for propulsion systems and to resupply materials for life-support systems.

Its vast mineral-resource potential, frozen volatile reservoirs and strategic location will make Mars a critical "extraterrestrial supply depot" for human expansion into the mineral-rich asteroid belt and to the giant outer planets and their fascinating collection of resource-laden moons. Smart explorer robots will assist the first human settlers on the Red Planet, enabling them to quickly and efficiently assess the full resource potential of their new world. As these early settlements mature, they will become economically self-sufficient by exporting propellants, life-support-system consumables, food, raw materials and manufactured products to feed the next wave of human expansion to the outer regions of the Solar System. Trading vessels will also travel between cislunar space and Mars, carrying specialty items to eager consumer markets in both civilizations.

The asteroids, especially Earth-crossing asteroids, represent another interesting source of extraterrestrial materials. Current Earth-based spectroscopic evidence and analysis of meteorites (which scientists believe originate from broken-up asteroids) indicate that carbonaceous (C-class) asteroids may contain up to 10 percent water, 6 percent carbon, significant amounts of sulfur and useful amounts of nitrogen. S-class asteroids, which are common near the inner edge of the main asteroid belt and among the Earth-crossing asteroids, may contain up to 30 percent free metals (alloys of iron, nickel and cobalt, along with high concentrations of precious metals). E-class asteroids may be rich sources of titanium, magnesium, manganese and other metals. Finally, chondritic asteroids, which are found among the Earth-crossing population, are believed to contain accessible amounts of nickel, perhaps more concentrated than the richest deposits found on Earth.

Using smart machines, possibly including self-replicating systems, space settlers in the next century will be able to manipulate large quantities of extraterrestrial matter and move it about to wherever it is needed in the Solar System. Some of these materials might even be refined en route, with the waste slag being used as a reaction mass in some advanced propulsion system. Many of these extraterrestrial resources will be used as the feedstock for space industries that will form the basis of interplanetary trade and commerce. For example, atmospheric ("aerostat") mining stations could be set up around Jupiter and Saturn, extracting such materials as hydrogen and helium—especially helium-3, an isotope of great value in nuclear-fusion research. Similarly, Venus could be mined for carbon dioxide in its atmosphere; Europa for water; and Titan for hydrocarbons. Large fleets of robot spacecraft could mine the Saturnian ring system for water ice while a sister fleet of robot vehicles extracts metals from the main asteroid belt. Even comets might be intercepted and mined for their frozen volatiles.

So the next time you gaze up at the night sky, don't think of space as a desolate, empty place; think of it instead as a cosmic frontier, rich in energy and material resources—

ready to be harvested through human ingenuity and advances in space technology.

See also: **asteroid; comet; Jupiter; Mars; Moon; Saturn; space industrialization; Venus**

extravehicular activity (EVA) A unique role that people play in the U.S. space program began on June 3, 1965, when Astronaut Edward H. White II left the protective environment of his *Gemini IV* spacecraft cabin and ventured into deep space. His mission—to perform a special set of procedures in a new and hostile environment—marked the start of the unique form of space technology called "extravehicular activity" or "EVA" for short. EVA may be defined as the activities conducted by an astronaut or cosmonaut outside the protective environment of his or her space capsule, aerospace vehicle or space station. With respect to the Space Transportation System or Space Shuttle, EVA is identified as the activities performed by the Shuttle astronauts outside the pressure hull or within the Orbiter payload bay when the payload bay doors are open. EVA is an optional, payload-related STS service.

The *Gemini IV* mission proved that EVA was a viable technique for performing orbital mission operations outside the spacecraft crew compartment. Then, as Gemini evolved into the Apollo Program, and Apollo into Skylab, EVA mission objectives pushed the science and art of extravehicular activity to their limit. New, more sophisticated concepts and methods were perfected, extending the capability to obtain scientific, technical and economic return from the space environment. Skylab also demonstrated the application of EVA techniques to unscheduled maintenance and repair operations—salvaging the program and inspiring its participants to new heights of aerospace accomplishment. Because of this success and usefulness, EVA capability has been incorporated into the Space Shuttle program and will also be included as an integral part of future space stations and space construction activities.

The term EVA, as applied to the Space Shuttle, includes all activities for which crewmembers don their space suits and life support systems, and then exit the Orbiter cabin into the vacuum of space to perform operations internal or external to the payload bay volume.

Shuttle-generic EVA can be divided into three basic categories: (1) Planned—that is, the EVA was planned prior to launch in order to complete a mission objective; (2) Unscheduled—that is, an EVA was not planned, but is required to achieve successful payload operation or to support overall mission accomplishments; and (3) Contingency—that is, EVA is required to effect the safe return of all crewmembers.

The following typical EVA tasks demonstrate the range of EVA opportunities that are available to space technology planners and payload designers in the Space Shuttle/Space Station era—which is also the dawn of humanity's extraterrestrial civilization.

(1) Inspection, photography and possible manual override

of vehicle and payload systems, mechanisms and components

(2) Installation, removal or transfer of film cassettes, material samples, protective covers, instrumentation and launch or entry tie-downs

(3) Operation of equipment, including tools, cameras and cleaning devices

(4) Cleaning optical surfaces

(5) Connection, disconnection and storage of fluid and electrical umbilicals

(6) Repair, replacement, calibration and inspection of modular equipment and instrumentation on the spacecraft or payloads.

(7) Deployment, retraction and repositioning of antennas, booms and solar panels

(8) Attachment and release of crew and equipment restraints

(9) Performance of experiments

(10) Cargo transfer

These EVA applications can demechanize an operational task and thereby reduce design complexity (automation), simplify testing and quality assurance programs, lower manufacturing costs and improve the probability of task success.

See also: **space construction; space station; space suit; Space Transportation System**

"faster-than-light" travel The ability to travel faster than the known physical laws of the Universe will permit. In accordance with Einstein's theory of relativity, the speed of light is the ultimate speed that can be reached in the space–time continuum. The speed of light in free space is 299,793 kilometers per second.

Concepts like "hyperspace" have been introduced in science fiction to sneak around this "speed-of-light barrier." Unfortunately, despite popular science-fiction stories to the contrary, most scientists today feel that the speed-of-light limit is a real physical law that isn't likely to change.

See also: **hyperspace; interstellar travel; relativity; tachyon**

Fermi paradox—"Where are they?" The dictionary defines the word *paradox* as an apparently contradictory statement that may nevertheless be true. According to the lore of physics, the famous Fermi paradox arose one evening in 1943 during a social gathering at Los Alamos, New Mexico, when the brilliant Italian-American physicist Enrico Fermi asked the penetrating question: "Where are they?" "Where are who?" his startled companions replied. "Why, the extraterrestrials," responded the Nobel prize-winning physicist, who was at the time one of the lead scientists on the top-secret Manhattan Project.

Fermi's line of reasoning that led to this famous inquiry has helped form the basis of much modern thinking and strategy concerning the search for extraterrestrial intelligence (SETI). It can perhaps best be summarized as follows. Our Galaxy is some 10 to 15 billion years old and contains perhaps 100 billion stars. If just *one* advanced civilization had arisen in this period of time and attained the technology necessary to travel between the stars, within 50 million to 100 million years, that advanced civilization could have diffused through or swept across the entire Galaxy—leaping from star to star, starting up other civilizations and spreading intelligent life everywhere. But as we look around, we don't see a Galaxy teeming with intelligent life, nor do we have any technically credible evidence of visitations or contact with alien civilizations, so we must perhaps conclude that no such civilization has ever arisen in the 15-billion-year history of the Galaxy. Therefore, the paradox: While we might expect to see signs of a Universe filled with intelligent life (on the basis of statistics and the number of possible "life sites," given the existence of 100 billion stars in just this Galaxy alone), we have seen no evidence of such. Are we, then, really alone? If we're not alone—where are they?

Many attempts have been made to respond to Fermi's very profound question. The "pessimists" reply that the reason we haven't seen any signs of intelligent extraterrestrial civilizations is that we really are alone. Maybe we are the first technically intelligent beings to rise to the level of space travel. Perhaps it is our cosmic destiny to be the first species to sweep through the Galaxy spreading intelligent life!

The "optimists," on the other hand, hypothesize that intelligent life exists out there somewhere and offer a variety of possible reasons why we haven't "seen" these civilizations yet. We'll discuss just a few of these proposed reasons here. First, perhaps intelligent alien civilizations really do not want anything to do with us. As a planet we may be too belligerent, too intellectually backward or simply below their communications horizon. Other optimists suggest that not every intelligent civilization has the desire to travel between the stars, or maybe they do not even desire to communicate by means of electromagnetic signals. Yet another response to the intriguing Fermi paradox is that *we* are actually *they*—the descendants of ancient astronauts who visited the Earth millions of years ago when a wave of galactic expansion passed through this part of the Galaxy.

Still another group responds to Fermi's question by declaring that intelligent aliens are out there right now but that they are keeping a safe distance, watching us either mature as a planetary civilization or destroy ourselves. A subset of this response is the extraterrestrial zoo hypothesis, which speculates that we are being kept as a "zoo" or wildlife preserve by advanced alien zookeepers who have elected to monitor our activities but not be detected themselves.

Finally, other people respond to the Fermi paradox by saying that the wave of cosmic expansion has not yet reached our section of the Galaxy—so we should keep looking! Within this response group are those who declare that the alien visitors are just now arriving among us!

If you were asked "Where are they?" just how would you respond?

See also: **ancient astronauts; Drake equation; extraterrestrial civilizations; search for extraterrestrial intelligence; unidentified flying object; Zoo hypothesis**

fission (nuclear) In nuclear fission, the nucleus of a heavy element, such as uranium or plutonium, is bombarded by a neutron, which it absorbs. The resulting compound nucleus is unstable and soon breaks apart, or fissions, forming two lighter nuclei (called fission products) and releasing additional neutrons. In a properly designed nuclear reactor, these fission neutrons are used to sustain the fission process in a controlled chain reaction. The nuclear-fission process is accompanied by the release of a large amount of energy, typically 200 million electron volts per reaction. Much of this energy appears as the kinetic (or motion) energy of the fission-product nuclei, which is then converted to thermal energy (or heat) as the fission products slow down in the reactor fuel material. This thermal energy is removed from the reactor core and used to generate electricity or as process heat.

Energy is released during the nuclear-fission process because the total mass of the fission products and neutrons after the reaction is less than the total mass of the original neutron and the heavy nucleus that absorbed it. From Einstein's famous mass-energy equivalence relationship, $E = mc^2$, the energy released is equal to the tiny amount of mass that has disappeared multiplied by the square of the speed of light.

Nuclear fission can occur spontaneously in heavy elements but is usually caused when these nuclei absorb neutrons. In some circumstances, nuclear fission may also be induced by very energetic gamma rays (in a process called photofission) and by extremely energetic (GeV-class—that is, billion-electron-volt-class)—charged particles.

The most important fissionable (or fissile) materials are uranium-235, uranium-233 and plutonium-239.

See also: **space nuclear power; space nuclear propulsion**

French Space Agency (CNES) The *Centre National d'Études Spatiales* (CNES), or French Space Agency, is the organization responsible for implementation of French space policy. It is a government scientific and technical organization operated on commercial and industrial lines. CNES began operating on March 1, 1962, and performs the following tasks: (1) prepares and implements French

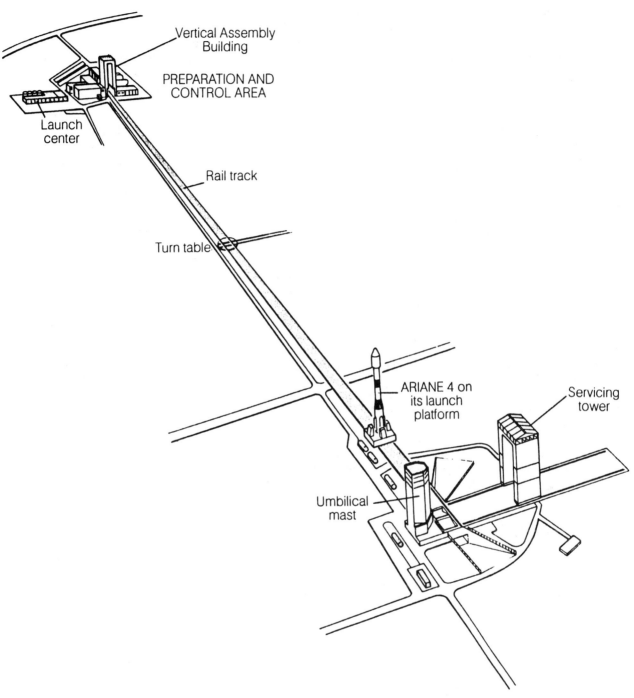

Fig. 1 The Ariane launch complex (ELA 2) at the Guiana Space Center, Kourou, French Guiana. The Ariane launch vehicle is integrated vertically in the assembly area and then moved on its launch platform to the launch site for mating of the encapsulated payload in the servicing tower. Some six hours prior to launch, the servicing tower is moved back. (Drawing courtesy of CNES and Arianespace.)

space policy, (2) stimulates and guides French industry in space matters, (3) interfaces with French administrations that are users of space services, (4) manages and supervises major operational space programs, (5) runs operational space systems, and (6) supports science and basic research that depends on space technology.

As France's national space agency, CNES manages the major national space programs carried out by industry and acts as prime contractor for the design and development phase. CNES is actively involved in a great variety of space programs through its participation in bilateral agreements and within the framework of the European Space Agency (ESA).

CNES facilities are located in Paris, Toulouse and French Guiana (South America). CNES headquarters is located in Paris along with the EVRY establishment (Evry is in the southern outskirts of Paris), which houses the Launch Vehicle Directorate responsible for the development of the Ariane family of launch vehicles on behalf of ESA.

The Toulouse Space Center (CST) serves as the main CNES engineering and technological facility, while the Guiana Space Center (CSG) serves as the Ariane vehicle launch complex. (See fig. 1.)

CNES commercial and marketing activities are handled by a number of *Groupements d'Intérêt Économique* (GIEs), or subsidiaries, such as Arianespace (the CNES subsidiary that promotes use of Ariane launch vehicles) and SPOT Image (the CNES subsidiary that promotes commercialization of SPOT remote sensing imagery). SPOT is a French Earth-observation satellite.

See also: **European Space Agency**

fusion In nuclear fusion, lighter atomic nuclei are joined together, or fused, to form a heavier nucleus. For example, the fusion of deuterium with tritium results in the formation of a helium nucleus and a neutron. Because the total mass of the fusion products is less than the total mass of the reactants (that is, the original deuterium and tritium nuclei), a tiny amount of mass has disappeared, and the equivalent amount of energy is released in accordance with Einstein's mass-energy equivalence formula:

$$E = m\,c^2$$

This fusion energy then appears as the kinetic (motion) energy of the reaction products. When isotopes of elements lighter than iron fuse together, some energy is liberated. Energy must be added to any fusion reaction involving elements heavier than iron.

The Sun is our oldest source of energy, the very mainstay of all terrestrial life. The energy of the Sun and other stars comes from thermonuclear-fusion reactions. Fusion reactions brought about by means of very high temperatures are called thermonuclear reactions. The actual temperature required to join, or fuse, two atomic nuclei depends on the nuclei and the particular fusion reaction involved. (Remember, the two nuclei being joined must have enough energy to overcome Coulombic, or "like-electric-charge," repulsion.) In stellar interiors, fusion occurs at temperatures of tens of millions of degrees Kelvin. When we try to develop useful controlled thermonuclear reactions (CTR) here on Earth, reaction temperatures of 50 million to 100 million degrees Kelvin are considered necessary.

Table 1 describes the major single-step thermonuclear reactions that are potentially useful in controlled fusion reactions for power and propulsion applications. Large space settlements and human-made "miniplanets" could eventually be powered by such CTR processes, while robot interstellar probes and giant space arks would use fusion for both power and propulsion. Helium-3 is a rare isotope of helium. Some space visionaries have already proposed mining the Jovian atmosphere or the surface of the Moon for helium-3 to fuel our first interstellar probes.

At present, there are immense technical difficulties preventing our effective use of controlled fusion as a terrestrial

Table 1   Single-Step Fusion Reactions Useful in Power and Propulsion Systems

Nomenclature	Thermonuclear Reaction	Energy Released Per Reaction (MeV) [Q Value]	Threshold Plasma Temperature (keV)[a]
(D-T)	$^2_1D + ^3_1T \rightarrow ^4_2He + ^1_0n$	17.6	10
(D-D)	$^2_1D + ^2_1D \rightarrow ^3_2He + ^1_0n$	3.2	50
	$^2_1D + ^2_1D \rightarrow ^3_1T + ^1_1p$	4.0	50
(D-³He)	$^2_1D + ^3_2He \rightarrow ^4_2He + ^1_1p$	18.3	100
(¹¹B-p)	$^{11}_5B + ^1_1p \rightarrow 3(^4_2He)$	8.7	300

where
    D is deuterium
    T is tritium
    He is helium
    n is neutron
    p is proton
    B is boron

[a]10 keV = 100 million degrees Kelvin.

or space energy source. The key problem is that the fusion gas mixture must be heated to tens of millions of degrees Kelvin and held together for a long enough period of time for the fusion reaction to occur. For example, a deuterium–tritium gas mixture must be heated to at least 50 million degrees Kelvin—and this is considered the easiest controlled fusion reaction to achieve! At 50 million degrees, any physical material used to confine these fusion gases would disintegrate, and the vaporized wall materials would then "cool" the fusion, gas mixture, quenching the reaction.

There are three general approaches to confining these hot fusion gases, or plasmas: gravitational confinement, magnetic confinement and inertial confinement.

Because of their large masses, the Sun and other stars are able to hold the reacting fusion gases together by gravitational confinement. Interior temperatures in stars reach tens of millions of degrees Kelvin and use complete thermonuclear-fusion cycles to generate their vast quantities of energy. For main-sequence stars like or cooler than our Sun (about 10 million degrees Kelvin), the proton–proton cycle, shown in table 2, is believed to be the principal energy-liberating mechanism. The overall effect of the proton–proton stellar fusion cycle is the conversion of hydrogen into helium. Stars hotter than our Sun (those with interior temperatures of 10 million degrees Kelvin and higher) release energy through the carbon cycle, shown in table 3. The overall effect of this cycle is again the conversion of hydrogen into helium, but this time with carbon (carbon-12 isotope) serving as a catalyst.

Terrestrial scientists attempt to achieve controlled fusion through two techniques: magnetic-confinement fusion (MCF) and inertial-confinement fusion (ICF). In magnetic confinement strong magnetic fields are employed to "bottle up," or hold, the intensely hot plasmas needed to make the various single-step fusion reactions occur (again, review table 1). In the inertial-confinement approach, pulses of laser light, energetic electrons or heavy ions are used to

Table 2   Main Thermonuclear Reactions in the Proton-Proton Cycle

$$^{1}_{1}H + {}^{1}_{1}H \rightarrow {}^{2}_{1}D + e^{+} + \nu$$

$$^{2}_{1}D + {}^{1}_{1}H \rightarrow {}^{3}_{2}He + \gamma$$

$$^{3}_{2}He + {}^{3}_{2}He \rightarrow {}^{4}_{2}He + {}^{1}_{1}H + {}^{1}_{1}H$$

where

$^{1}_{1}H$ is a hydrogen nucleus (that is, a proton, $^{1}_{1}p$)

$^{2}_{1}D$ is deuterium (an isotope of hydrogen)

$^{3}_{2}He$ is helium −3 (a rare isotope of helium)

$^{4}_{2}He$ is the main (stable) isotope of helium

ν is a neutrino

e^{+} is a positron

γ is a gamma ray

Table 3   Major Thermonuclear Reactions in the Carbon Cycle

$$^{12}_{6}C + {}^{1}_{1}H \rightarrow {}^{13}_{7}N + \gamma$$

$$^{13}_{7}N \rightarrow {}^{13}_{6}C + e^{+} + \nu \qquad \text{(radioactive decay)}$$

$$^{1}_{1}H + {}^{13}_{6}C \rightarrow {}^{14}_{7}N + \gamma$$

$$^{1}_{1}H + {}^{14}_{7}N \rightarrow {}^{15}_{8}O + \gamma$$

$$^{15}_{8}O \rightarrow {}^{15}_{7}N + e^{+} + \nu \qquad \text{(radioactive decay)}$$

$$^{1}_{1}H + {}^{15}_{7}N \rightarrow {}^{12}_{6}C + {}^{4}_{2}He$$

where

γ is a gamma ray

ν is a neutrino

e^{+} is a positron

very rapidly compress and heat small spherical targets of fusion material. This rapid compression and heating of an ICF target allows the conditions supporting fusion to be reached in the interior of the pellet—before it blows itself apart.

Although there are still many difficult technical issues to be resolved before we can achieve controlled fusion, it promises to provide a limitless terrestrial energy supply. Of course, fusion also represents the energy key to the full use of the resources of our Solar System and, possibly, to travel across the interstellar void.

See also: **interstellar travel; stars; Sun**

G

g   The symbol used for the acceleration due to gravity. For example, at the Earth's surface, g equals 9.8 meters per second squared.

Gaia hypothesis   The hypothesis proposed by James Lovelock (with the assistance of Lynn Margulis) that the Earth's biosphere has an important modulating effect on the terrestrial atmosphere. Because of the chemical complexity observed in the lower atmosphere, Lovelock has suggested that life-forms within the terrestrial biosphere actually help control the chemical composition of the Earth's atmosphere—thereby ensuring the continuation of conditions suitable for life. Gas-exchanging microorganisms, for example, are thought to play a key role in this continuous process of environmental regulation. Without these "co-operative" interactions, in which some organisms generate certain gases and carbon compounds that are subsequently removed and used by other organisms, the planet Earth might also possess an excessively hot or cold planetary surface, devoid of liquid water and surrounded by an inanimate, carbon dioxide-rich atmosphere.

Gaia was the goddess of Earth in ancient Greek mythology. Lovelock used her name to represent the terrestrial biosphere—namely, the system of life on Earth, including living organisms and their required liquids, gases and solids. Thus, the Gaia hypothesis simply states that "Gaia" (the Earth's biosphere) will struggle to maintain the atmospheric conditions suitable for the survival of terrestrial life.

If we use the Gaia hypothesis in our search for extraterrestrial life, we should look for alien worlds that exhibit variability in atmospheric composition. Extending this hypothesis beyond the terrestrial biosphere, a planet will either be living or else it will not! The absence of chemical interactions in the lower atmosphere of an alien world could be taken as an indication of the absence of living organisms.

While this interesting hypothesis is currently more speculation than hard, scientifically verifiable fact, it is still quite useful in developing a sense of appreciation for the complex chemical interactions that have helped to sustain life in the Earth's biosphere. These interactions among microorganisms, higher-level animals and their mutually shared atmosphere might also have to be carefully considered in the successful development of effective closed life-support systems for use on permanent space stations, lunar bases and planetary settlements.

See also: **global change**

galactic   Of or pertaining to a galaxy, such as the Milky Way Galaxy.

galactic cluster   Diffuse collection of from ten to perhaps several hundred stars, loosely held together by gravitational forces. Also called "open cluster."

See also: **globular cluster**

galaxy   A very large accumulation of from 10^6 to 10^{12} stars. By convention, when the word is capitalized (Galaxy), it refers to a particular collection of stars, such as the Milky Way Galaxy. The existence of galaxies beyond our own Milky Way Galaxy was not firmly established by astronomers until 1924.

Galaxies—or "island universes," as they are sometimes called—come in a variety of shapes and sizes. They range from dwarf galaxies, like the Magellanic Clouds, to giant spiral galaxies, like the Andromeda Galaxy. Astronomers usually classify galaxies as either elliptical, spiral (or barred spiral) or irregular.

When we talk about galaxies, the scale of distances is truly immense. Galaxies themselves are tens to hundreds of thousands of light-years across; the distance between galaxies is generally a few million light-years! For example, the beautiful Andromeda Galaxy is approximately 130,000

light-years in diameter and about 2.2 million light-years away.

See also: **Andromeda Galaxy; Magellanic Clouds; Milky Way Galaxy; stars**

gamma ray astronomy   With the arrival of the Space Age and our ability to place observation platforms above the Earth's atmosphere, scientists could collect and study gamma ray emissions from a variety of interesting cosmic sources, giving rise to the field of gamma ray astronomy. Gamma ray astronomy reveals the explosive, high-energy processes associated with such celestial phenomena as supernovae, exploding galaxies, quasars, pulsars and black hole candidates. Some of the processes associated with gamma ray emissions of interest to astrophysicists include (1) the decay of radioactive nuclei, (2) cosmic ray interactions, (3) curvature radiation in extremely strong magnetic fields, and (4) matter–antimatter annihilation.

Gamma ray astronomy is especially significant because the gamma rays being observed by spacecraft orbiting the Earth might have travelled across our entire Galaxy, or perhaps even across most of the Universe, without suffering appreciable line-of-sight alteration or loss of energy. Consequently, these energetic gamma rays reach our Solar System with the same characteristics, including directional and temporal features, as they started with at their sources. Gamma ray astronomy can provide important information on extraterrestrial phenomena not observable at any other wavelength in the electromagnetic spectrum and on spectacularly energetic events that may have occurred far back in the evolutionary history of the Universe.

See also: **astrophysics; gamma rays; Gamma Ray Observatory; stars**

Gamma Ray Observatory (GRO)   One of the major NASA space observatories to be flown in the 1990s. This spacecraft carries a variety of sensitive instruments designed to detect gamma rays over an extensive range of energies. The GRO is an extremely powerful tool for investigating some of the most puzzling mysteries in the Universe, including gamma ray bursts (short duration bursts of high-energy photons), pulsars, quasars and active galaxies.

Space scientists and astrophysicists are using GRO data to achieve the following objectives: (1) to study dynamic evolutionary forces in compact objects such as neutron stars and black holes; (2) to search for evidence of nucleosynthesis (the process of creating heavy elements in violent, supernova explosions); (3) to study gamma-ray-emitting objects whose nature is not now understood; (4) to conduct a gamma ray emission survey of our own Galaxy; (5) to study the nature of other galaxies in the energetic realm of gamma rays; and (6) to conduct a search for possible primordial black-hole emissions.

The Gamma Ray Observatory has four major instruments: the Burst and Transient Source Experiment (BATSE), Imaging Compton Telescope (COMPTEL), Oriented Scintillation Spectrometer Experiment (OSSE), and Energetic Gamma Ray Experiment Telescope (EGRET). To help achieve its overall scientific objectives, the GRO will conduct a survey of the "gamma-ray sky" over energy ranges extending from the upper end of existing X-ray observations to the highest practical observation energies.

The GRO was deployed by the Space Shuttle *Atlantis* on April 7, 1991 during Shuttle Mission STS-37 in low Earth orbit and then boosted to a higher circular orbit. This important space observatory will operate for about two years.

See also: **astrophysics; gamma ray astronomy**

gamma rays [symbol: γ] High-energy, very-short-wavelength electromagnetic radiation. Gamma-ray photons are similar to X rays, except that they are usually more energetic and originate from processes and transitions within the atomic nucleus. The processes associated with gamma-ray emissions in astrophysical phenomena include (1) the decay of radioactive nuclei, (2) cosmic-ray interactions, (3) curvature radiation in extremely strong magnetic fields and (4) matter–antimatter annihilation. Gamma rays are very penetrating and are best stopped or shielded against by dense materials, such as lead or tungsten.

See also: **astrophysics; gamma ray astronomy; Gamma Ray Observatory**

giant planets In our Solar System the large, gaseous outer planets: Jupiter, Saturn, Uranus and Neptune.

global change If we carefully explore our planet's geological record, we will discover that the Earth's environment has been subject to great change over eons. Many of these changes have occurred quite slowly, requiring numerous millennia to achieve their full impact and effect. However, other global changes have occurred relatively rapidly over time periods as short as a few decades or less. These global changes appear in response to such phenomena as the migration of continents, the building and erosion of mountains, changes in the Sun's energy output or variations in the Earth's orbital parameters, the reorganization of oceans and even the catastrophic impact of a large asteroid or comet. Such natural phenomena lead to planetary changes on local, regional and global scales, including a succession of warm and cool climate epochs, new distributions of tropical forests and rich grasslands, the appearance and disappearance of large deserts and marshlands, the advances and retreats of great ice sheets, the rise and fall of ocean and lake levels and even the extinction of vast numbers of species. The last great mass extinction (on a global basis) appears to have occurred some 65 million years ago, possibly due to the impact of a large asteroid. The peak of the most recent period of glaciation is generally considered to have occurred about 18,000 years ago when average global temperatures were about 5 degrees Celsius (9 degrees Fahrenheit) cooler than today.

Although the global changes just discussed are the in-evitable results of major natural forces currently beyond human control, it is also apparent to scientists that humans have now become a powerful agent for environmental change. For example, the chemistry of the Earth's atmosphere has been altered significantly by both the agricultural and industrial revolutions. The erosion of continents and sedimentation of rivers and shorelines have been influenced dramatically by agricultural and construction practices. The production and release of toxic chemicals have affected the health and natural distributions of biotic populations. The ever-expanding human need for water resources has affected the patterns of natural water exchange that take place in the hydrological cycle (the oceans, surface and ground water, clouds and so forth). One example is the enhanced evaporation rate from large manmade reservoirs compared to the evaporation rate from wild, unregulated rivers. As the world population grows and human civilization undergoes further technological development in the 21st century, the role of our planet's most influential animal species as an agent of environmental change will undoubtedly expand.

Over the last two decades, scientists have accumulated technical evidence that indicates that ongoing environmental changes are the result of complex interactions among a number of natural and human-related systems. For example, changes in the Earth's climate are now considered to involve not only wind patterns and atmospheric cloud populations, but also the interactive effects of the biosphere ocean currents, human influences on atmospheric chemistry, the Earth's orbital parameters, the reflective properties of our planetary system (the Earth's albedo) and the distribution of water among the atmosphere, hydrosphere and cryosphere (polar ice). The aggregate of these interactive linkages among our planet's major natural and manmade systems that appear to affect the environment has become known as *global change.*

The governments of many nations, including the United States, have now begun to address the issues associated with global change. Over the last decade, preliminary results from global observation programs (many involving space-based systems) have stimulated a new set of concerns that the dramatic rise of industrial and agricultural activities during the 19th and 20th centuries may be adversely affecting the overall Earth system. Today, the enlightened use of the Earth and its resources has become an important contemporary political and scientific issue.

The global changes that may affect both human well-being and the quality of life on this planet include global climate warming, sea-level change, ozone depletion, deforestation, desertification, drought and a reduction in biodiversity. Although, complex and dramatic phenomena in themselves, these individual global change concerns cannot be fully understood and addressed unless they are studied collectively in an integrated, multidisciplinary fashion. An effective and well-coordinated national and international research program will be required to significantly improve our knowledge of these complex Earth system processes.

This type of program will provide the technical basis through which scientists can discriminate between natural and human-influenced changes, and eventually will be able to accurately predict global change phenomena.

The overall U.S. strategy to address global change issues involves three fundamental areas: (1) research to understand the Earth's environment; (2) research and development of new technologies to adapt to, or to mitigate, environmental changes; and (3) the formulation of national and international policy responses as needed for a changing planetary environment. The overall goal of the current U.S. Global Change Research program (being developed by a committee composed of members of various federal agencies) is to provide the scientific basis for informed decision-making. More formally, the overarching goal is "to gain a predictive understanding of the interactive physical, geological, chemical, biological, and social processes that regulate the total Earth system and, hence, establish the scientific basis for national and international policy formulation and decisions relating to natural and human-induced changes in the global environment and their regional impacts."

The U.S. Global Change Research program has three parallel scientific objectives: the monitoring, understanding and predicting of global change. The Program's seven scientific elements reflect the integrated and interdisciplinary nature of this complex research effort. These elements are:

(1) *Climate and hydrologic systems*—the study of the physical processes that govern climate and the hydrological cycle, including interactions between the atmosphere, hydrosphere, cryosphere, land surface and biosphere.

(2) *Biogeochemical dynamics*—the study of the sources, sinks, fluxes, trends and interactions involving the biogeochemical constituents within the Earth system, including human activities with a focus on carbon, nitrogen, sulfur, oxygen, phosphorus and the halogens.

(3) *Ecological systems and dynamics*—the study of the responses of ecological systems, both marine and terrestrial, to changes in global and regional environmental conditions; and the study of the influence of biological communities on atmospheric, terrestrial, oceanic and climatic systems.

(4) *Earth system history*—the study and interpretation of the natural records of past environment changes that are contained in terrestrial and marine sediments, soils, glaciers and permafrost, tree rings, rocks, geomorphic features and other direct or proxy documentation of past global conditions.

(5) *Human interactions*—the study of the social factors that influence the global environment, including population growth, industrialization, agricultural practices and other land usages; and the study of human activities that are affected by the regional aspects of global change.

(6) *Solid Earth processes*—the study of geological processes (such as volcanic eruptions and erosion) that affect the global environment, especially those processes that take place at the interfaces between the Earth's surface and the atmosphere, hydrosphere, cryosphere and biosphere.

(7) *Solar influences*—the study of how changes in the near-Earth space environment and in the upper atmosphere that are induced by variability in solar output influence the Earth's environment.

In the coming decades, global change may well represent the most significant environmental, economic and societal challenges facing our planetary civilization. As we use advances in space technology to successfully build biospheres on other worlds in our Solar System in the next century, perhaps the greatest service of space technology to mankind will be a much better understanding and enlightened stewardship of our home planet.

See also: **Earth; Earth Observing System; Mission to Planet Earth**

globular cluster   Compact cluster of up to one million, generally older, stars.
See also: **galactic cluster**

gravity anomaly   A region on a celestial body where the local force of gravity is lower or higher than expected. If the celestial object is assumed to have a uniform density throughout, then we would expect the gravity on its surface to have the same value everywhere.
See also: **mascon**

Great Observatories   The four large, unmanned Earth-orbiting observatories in NASA's current astrophysics program plan that will collectively (by the late 1990s) provide an unprecedented high-resolution and high-sensitivity view of the Universe across the key portions of the electromagnetic spectrum. Included in the Great Observatories group are the Hubble Space Telescope (HST), the Gamma Ray Observatory (GRO), the Advanced X-Ray Astrophysics Facility (AXAF) and the Space Infrared Telescope Facility (SIRTF).
See also: **Advanced X-Ray Astrophysics Facility; astrophysics; electromagnetic spectrum; Gamma Ray Observatory; Hubble Space Telescope; Space Infrared Telescope Facility**

greenhouse effect   The general warming of the lower layers of a planet's atmosphere caused by the presence of "greenhouse gases," mainly carbon dioxide (CO_2) and water vapor (H_2O). As happens on Earth, the greenhouse effect occurs because our atmosphere is relatively transparent to visible light from the Sun (typically 0.3 to 0.7 micrometer wavelength), but is essentially opaque to the longer-wavelength (typically 10.6 micrometer) thermal infrared radiation emitted by the planet's surface. Because of the presence of greenhouse gases in our atmosphere—such as carbon dioxide (CO_2), water (H_2O), methane (CH_4), nitrous oxide (NO_2) and manmade chlorofluorocarbons (CFCs)—this outgoing thermal radiation from the Earth's surface is blocked from escaping to space, and the absorbed thermal energy causes a rise in the temperature of the lower atmosphere. Therefore, as the presence of greenhouse gases increases

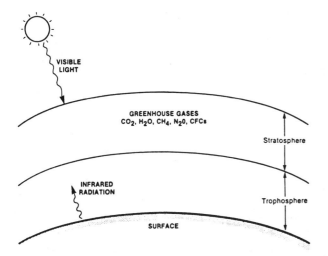

Fig. 1 The greenhouse effect in the Earth's atmosphere. (Drawing courtesy of the U.S. Environmental Protection Agency.)

in the Earth's atmosphere, more outgoing thermal radiation is trapped, and a global warming trend occurs (see fig. 1).

Scientists around the world are now concerned that human activities, such as the increased burning of vast amounts of fossil fuels, is increasing the presence of greenhouse gases in our atmosphere and upsetting the overall planetary energy balance. These scientists are further concerned that we may be creating the conditions for a *runaway greenhouse*, as appears to have occurred in the past on the planet Venus. Such an effect is a planetary climatic extreme in which all of the surface water has evaporated from the surface of a life-bearing or potentially life-bearing planet. Planetary scientists believe that the current Venusian atmosphere allows sunlight to reach the planet's surface, but its thick clouds and rich carbon dioxide content prevent surface heat from being radiated back to space. This condition has led to the evaporation of all surface water and has produced the present inferno-like surface temperatures of approximately 485 degrees Celsius (900 degrees Fahrenheit)—a temperature hot enough to melt lead.

See also: **global change**

"greening of the Galaxy" A term describing the spreading of human life, technology and culture through interstellar space and eventually across the entire Galaxy.
See also: **extraterrestrial civilizations**

H

hazards to space workers Current experience with human performance in space is mostly for individuals op-

erating in low Earth orbit (LEO). The construction of large space settlements, lunar bases, Mars surface settlements, orbiting factories and Satellite Power Systems would require human activities throughout cislunar space (the area between the Earth and the Moon) and beyond. The maximum continuous time spent by humans in space is now just several hundred days (with Soviet cosmonauts holding the duration record), and people who have experienced spaceflight are generally a small number of highly trained and highly motivated individuals.

Medical and occupational experiments performed in space and operational life-support and monitoring systems used in manned spaceflight have been extensively evaluated in preparation for the construction and operation of the U.S. Space Station *Freedom*. These evaluations and analyses have been augmented by data obtained from experiments performed under simulated space conditions on Earth.

The currently available technical data base, although limited to essentially low-Earth-orbit spaceflight, suggests that with suitable protection, people can live and work in space safely and enjoy good health after returning to Earth. Data from the 84-day *Skylab 4* mission and several long-duration Soviet Salyut and Mir space station missions are especially pertinent to the question of the ability of relatively large numbers of people to live and work in space for months at a time on permanent space stations and for perhaps years at a time at lunar bases and on human expeditions to Mars.

Some of the major cause-effect factors related to space-worker health and safety are shown in figure 1. Many of these factors require "scaling up" from current medical, safety and occupational analyses to achieve the space technologies necessary to accommodate large groups of space travelers and permanent habitats. Some of these health and safety issues include (1) preventing launch-abort, spaceflight and space-construction accidents; (2) preventing failures of life-support systems; (3) protecting space vehicles and habitats from collisions with space debris and meteoroids; and (4) providing habitats and good-quality living conditions that minimize psychological stress.

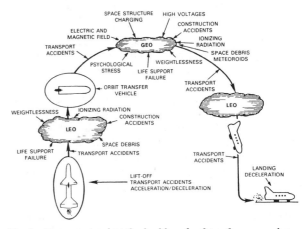

Fig. 1 Factors related to the health and safety of space workers. (Drawing courtesy of NASA and the U.S. Department of Energy.)

The biomedical effects of substantial acceleration and deceleration forces when leaving and returning to Earth, living and working in a weightless environment and the exposure to space radiation are the three main factors that must be dealt with if people are to live in cislunar space and eventually populate heliocentric (Sun-centered) space.

Astronauts and cosmonauts have adapted to weightlessness for extended periods of time in space and have experienced maximum acceleration forces equivalent to six times the Earth's gravity (6 g). No acute operational problems, significant physiological deficits or adverse health effects on the cardiovascular or musculoskeletal systems have yet been observed from these experiences.

The U.S. Space Transportation System, or Space Shuttle, can be regarded as a forerunner of more advanced "space-settler" launch vehicles. It has been designed to limit acceleration/deceleration loads to a maximum of 3 g, thereby opening space travel to a larger number of individuals.

Some physiological deviations have been observed in American astronauts and Soviet cosmonauts during and following extended space missions. Most of these observed effects appear to be related to adaption to microgravity conditions, with the affected physiological parameters returning to normal ranges either during the missions or shortly thereafter. No apparent persistent adverse consequences have been observed or reported to date. Nevertheless, some of these deviations could become chronic and might have important health consequences if they were experienced during long durations in space or in repeated long-term tours on a space station or at an orbiting construction facility.

The physiological deviations due to zero gravity have, as noted, usually returned to normal within a few days or weeks after return to Earth. However, bone calcium loss appears to require an extended period of recovery after a long-duration space mission.

Strategies are now being developed to overcome these physiological effects of weightlessness. An exercise regimen can be applied, and body fluid shifts can be limited by applying lower-body negative pressure. Antimotion medication is also useful for preventing temporary motion sickness or space sickness. Proper nutrition, with mineral supplements, and regular exercise, appear to limit other observed effects. One way around this problem in the long term, of course, is to provide acceptable levels of "artificial" gravity in larger space bases and space settlements. In fact, very large space settlements will most likely offer the inhabitants a wide variety of gravity levels, ranging from microgravity up to normal terrestrial gravity levels. This multiple-gravity-level option will not only make space-settlement life-styles more diverse than on Earth but will also prepare planetary settlers for life on their new worlds or help other space settlers adjust to the "gravitational rigors" of returning to Earth.

The ionizing-radiation environment encountered by workers in space is characterized primarily by fluxes of electrons, protons and energetic atomic nuclei. (See table

Table 1   Components of the Natural Space Radiation Environment

Galactic Cosmic Rays

Typically 85% protons, 13% alpha particles, 2% heavier nuclei
Integrated yearly fluence
1×10^8 protons/cm^2 (approximately)
Integrated yearly radiation dose:
4 to 10 rads (approximately)

Geomagnetically Trapped Radiation

Primarily electrons and protons
Radiation dose depends on orbital altitude
Manned flights below 300 km altitude avoid Van Allen belts

Solar-Particle Events

Occur sporadically; not predictable
Energetic protons and alpha particles
Solar-flare events may last for hours to days
Dose very dependent on orbital altitude and amount of shielding

1.) In low Earth orbit (LEO), electrons and protons are trapped by the Earth's magnetic fields (forming the Van Allen belts). The amount of ionizing radiation in LEO varies with solar activity. The trapped radiation belts are of concern when space-worker crews transfer from low Earth orbit to geosynchronous orbit (GEO) or to lunar surface bases. In GEO locations solar-particle events (SPEs) represent a major radiation threat to space workers. Throughout cislunar space and interplanetary space (beyond the protection of the Earth's magnetosphere), space workers are also bombarded by galactic cosmic rays. These are very energetic atomic particles, consisting of protons, helium nuclei and heavy nuclei with an atomic number (Z) greater than two (HZE particles). Shielding, solar-flare warning systems and excellent radiation dosimetry equipment should help prevent any space worker from experiencing ionizing radiation doses in excess of occupational standards established for various extraterrestrial careers.

Space workers and settlers might also experience a variety of psychological disorders, including the solipsism syndrome and the shimanagashi syndrome. The solipsism syndrome is a state of mind in which a person feels that everything is a dream and is not real. It might easily be caused in an environment (such as a small space base) where everything is "artificial," or human-made. The shimanagashi syndrome is a feeling of isolation in which individuals begin to feel left out, even though life may be physically comfortable. Careful design of living quarters and good communication with Earth should relieve or prevent such psychological disorders.

Living and working in space in the next century will present some interesting challenges and possibly even some dangers and hazards. However, the rewards of an extraterrestrial life-style, for certain pioneering individuals, will more than outweigh any such personal risks.

See also: **Earth's trapped radiation belts; people in space; "shimanagashi" syndrome; solipsism syndrome; space life sciences; space settlement; space station**

heliocentric Relative to the Sun as a center, as in "Heliocentric orbit" or "heliocentric space."

heliostat A mirrorlike device arranged to follow the Sun as it moves through the sky and to reflect the Sun's rays on a stationary collector.

Hellas (plural: Hellades) A quantity of information first proposed by the physicist Dr. Philip Morrison. It corresponds to 10^{10} bits of information—more or less the amount of information we know about ancient Greece. In considering interstellar communication with other intelligent civilizations, we would hope to send and receive something on the order of 100 Hellades of information or more at each contact.
See also: **interstellar communication**

hertz (Hz) The unit of frequency in the International or SI System of Units. One hertz is equal to one cycle per second.
See also: **electromagnetic spectrum**

HI region A diffuse region of neutral, predominantly atomic, hydrogen in interstellar space. Neutral hydrogen emits radio radiation at 1,420.4 megahertz, corresponding to a wavelength of approximately 21 centimeters. However, the temperature of the region (approximately 100 degrees Kelvin) is too low for optical emission.
See also: **HII region; nebula**

HII region A region in interstellar space consisting mainly of ionized hydrogen and existing mostly in discrete clouds. The ionized hydrogen of HII regions emits radio waves by thermal emissions and recombination-line emission, in comparison to the 21-centimeter radio-wave emission of neutral hydrogen in HI regions.
See also: **HI region; nebula**

Hubble's Law The hypothesis that the redshifts of distant galaxies and very remote extragalactic objects (such as quasars) are directly proportional to their distances (D) from us. The American astronomer Edwin Hubble (1889–1953) first proposed this relationship in 1929. Mathematically, Hubble's law can be expressed as:

$$V = H_0 D$$

where H_0 is the "Hubble constant"—the constant of proportionality in the relationship between the relative recessional velocity (V) of a distant galaxy or a very remote extragalactic object and its distance (D) from us. At present, an exact value for the Hubble Constant (H_0) is under much debate. An often encountered value is 75 kilometers per second per megaparsec [(km/sec)/Mpc], although values of from 50 to 100 (km/sec)/Mpc can be found in the technical literature.

The inverse of H_0, namely $1/H_0$, has the unit of time and is sometimes called the "Hubble time." It is a measure of the age of the Universe. It should also be noted that in an evolving Universe, the value of the Hubble "constant" will actually change with time.
See also: **galaxy; open universe; quasars**

Hubble Space Telescope (HST) The visible-light element of NASA's Great Observatories program. The Hubble Space Telescope is named for the American astronomer, Edwin P. Hubble, who revolutionized our knowledge of the size, structure and makeup of the universe through his pioneering observations in the first half of the 20th century. A fully functional HST will be used by astronomers and space scientists to observe the visible Universe to distances never before obtained and to investigate a wide variety of interesting extraterrestrial phenomena, including those associated with extragalactic astronomy and observational cosmology.

The HST is 13.1 meters (43.5 feet) long and has a diameter of 4.27 meters (14 feet). (See fig. 1.) It is designed to provide detailed observational coverage in the visible and ultraviolet portions of the electromagnetic (EM) spectrum. The HST power supply system consists of two large solar panels (unfurled on orbit), batteries and power-conditioning equipment. This power system has been designed to supply a minimum of 2,400 watts to the orbiting observatory two years after the beginning of the mission.

The 11,000 kilogram (25,500 pounds-mass) free-flying astronomical observatory was successfully placed into a 600-kilometer (380-mi) low-Earth-orbit (LEO) during the STS-31 Space Shuttle mission (April 1990). Unfortunately, previously undetected defects in the 2.4-meter diameter (94.5-in) primary mirror are now causing focusing problems that are greatly reducing mission performance, especially involving use of the wide field/planetary camera and the faint object camera. Although the HST cannot have its flawed mirror replaced on orbit and potential contamination

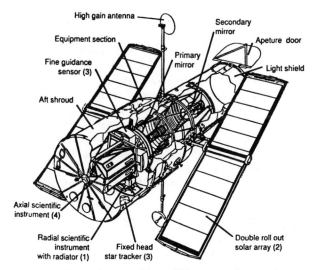

Fig. 1 A cutaway drawing of the Hubble Space Telescope. (Drawing courtesy of NASA.)

problems prevent a return to Earth of the entire HST observatory for mirror repair, the scientific instruments, which are located just behind the primary mirror at the focal plane, can be replaced on orbit by Shuttle astronauts. NASA is now planning to use a "second-generation" design of these instruments to compensate (to as great extent as possible) for the flawed mirror.

The five major HST scientific instruments are the wide field/planetary camera, faint object spectrometer, high-resolution spectograph; high-speed photometer, and faint object camera. These instruments are housed in modular units to accommodate on orbit replacement by astronauts. The wide field/planetary camera can operate in two modes. It has a wide-field capability that permits examination of large areas of space. It also can take high-resolution images of objects in our Solar System (except Mercury, which is too close to the bright light of the Sun). The faint object spectrograph gathers the spectra of extremely faint astronomical objects in the ultraviolet and visible portions of the electromagnetic spectrum. The study of such spectra tells astronomers about the nature of the celestial object being viewed, including whether it is hot or cold, dense or rarified, and even something about its chemical composition. The high-resolution spectrograph uses the full resolving capability of the telescope to view much dimmer objects than have ever been viewed by previous orbiting astronomical instruments. It observes only the ultraviolet portion of the electromagnetic spectrum, a region that cannot be seen by ground-based observatories due to the filter effects of the Earth's atmosphere. The high-speed photometer is designed to provide accurate observations of the total amount of light from a celestial object, record any fluctuations in brightness on a time scale down to microseconds and detail any fine structure related to the light source. Finally, the faint object camera (developed by the European Space Agency) takes advantage of the spatial resolution capability of the HST to capture images of faint objects in the Universe.

The difference between a fully functional Hubble Space Telescope and contemporary groundbased optical telescopes has been compared to the difference between Galileo Galilei's first telescope and its predecessor, the human eye. In time, this important orbiting observatory should allow scientists to detect objects 100 times fainter than those visible from Earth-based telescopes. It should also extend our observational reach out into the visible Universe from a present limit of about 2 billion light-years to approximately 15 billion light-years, enabling us to look back in time almost to the beginning of the Universe itself.
See also: **astrophysics, Great Observatories**

humanoid   Literally, a creature that resembles a human being. As found in the science-fiction literature, *humanoid* is frequently used to describe an intelligent extraterrestrial being, while in anthropology the term refers to an early ancestor of *Homo sapiens*.

hyperspace   A concept of convenience developed in science fiction to make "faster-than-light" travel appear credible. Hyperspace is frequently described as a special dimension or property of the Universe in which physical things are much closer together than they are in the normal space–time continuum.

In a typical science-fiction story, the crew of a spaceship simply switches into "hyperspace," and distances to objects in the "normal" Universe are considerably shortened. When the spaceship emerges out of hyperspace, the crew is where they wanted to be essentially instantly. Although this concept violates the speed-of-light barrier predicted by Einstein's special relativity theory, it is nevertheless quite popular in modern science fiction.
See also: **"faster-than-light" travel; interstellar travel; relativity; science fiction**

HZE particles   The most potentially damaging cosmic rays, with high atomic number (Z) and high kinetic energy (E). Typically, HZE particles are atomic nuclei with Z greater than 6 and E greater than 100 million electron volts. When these extremely energetic particles pass through a substance, they leave a large amount of energy deposited along their tracks. This deposited energy ionizes the atoms of the material and disrupts molecular bonds.
See also: **cosmic rays; hazards to space workers**

I

ice catastrophe   A planetary climatic extreme in which all the liquid water on the surface of a life-bearing or potentially life-bearing planet has become frozen or completely glaciated.
See also: **ecosphere; global change**

illumination from space   Just how can we presently take good advantage of the radiant energy of our parent star, the Sun? Well, in the early 1970s the late Dr. Krafft A. Ehricke and other far-thinking space engineers proposed that large, very lightweight mirrors be placed in orbit around the Earth to illuminate cities, agricultural regions, ice fields that were blocking navigation and terrestrial solar power installations. Ehricke called such mirrors "lunettas" if they reflected sunlight to localities on the dark (nighttime) side of the Earth and "solettas" if they reflected substantial amounts of solar energy (typically providing one solar constant) over limited regions of the Earth both day and night. (One solar constant is 1,371 watts per square meter of radiant solar energy.)

NASA investigations of solar-sail applications have also

Fig. 1 The Sunblazer concept. (Courtesy of NASA.)

provided the technical characteristics for a large space mirror constructed out of aluminized mylar, possibly just 0.0025 millimeter thick. We might anticipate that such giant mirrors will eventually have areal densities as low as 6 grams per square meter, which corresponds to 6 metric tons per square kilometer.

Figure 1 shows the Sunblazer concept—a proposed illumination-from-space scheme in which a series of giant, mirrorlike reflector spacecraft are placed in a equatorial, geostationary orbit over North America. These large orbiting mirrors would be so positioned that they effectively extended daylight year-round to several continental regions, as well as to Hawaii. Sixteen such reflectors, for example, could add approximately two hours of illumination during peak evening and morning rush-hour periods. One or more of these colossal reflectors could also be diverted to illuminate Alaska, especially in the Fairbanks and Anchorage areas, extending their short winter days by approximately three hours.

Entire cities or other important regions could be illuminated by a single reflector spacecraft during a time of power blackout or other emergency. Because of the costs of building and deploying such giant solar reflector spacecraft, these large mirrors would probably not be dedicated to just emergency illumination operations. However, if a series of these platforms were placed in operation (as, for example, in the Sunblazer concept), then up to four of these giant reflectors could be diverted in a nighttime emergency to illuminate any location between the Virgin Islands and the Hawaiian Islands to a level equal to street-light intensity, or 15 full Moons.

There are many exciting terrestrial applications for such colossal mirrors in space. These include nighttime illumination for urban areas; nighttime illumination for agricultural and industrial operations; nighttime illumination for disaster relief and emergency operations; frost-damage protection; local climate manipulation; increased solar flux to enhance solar energy conversion processes on the ground; ocean cell warming for climate control; enhanced agriculture through the stimulation of photosynthesis processes; and controlled snowpack melting. Can you think of any others?

Of course, such giant mirrors placed in appropriate orbits around the Moon could help illuminate mining and exploration operations during the long lunar night (approximately 14 Earth days' duration) and could be used to prevent excessive facilities and equipment "cold soak" during these extended periods of lunar darkness. In orbit around Mars, such large reflector mirrors represent one of the major tools of planetary engineering. They would help bring more sunlight to the polar regions, promoting controlled melting of the Martian polar caps. Such mirrors could also provide more benign growing environments for genetically engineered plants—some of which would be used to help "terraform" the thin, carbon dioxide-rich Martian atmosphere.

See also: **planetary engineering**

images from space How can we learn more about our extraterrestrial environment? One very effective way is to send robot spacecraft with imaging systems to target celestial objects throughout our Solar System. In the last two decades, spacecraft imaging systems have made most previous visual planetary data obsolete. Taking advantage of close flybys, orbits and landings, robot spacecraft have provided scientists with exceptionally clear and close views of alien worlds as far away as Saturn, Uranus and Neptune (see fig. 1). In fact, spacecraft have visited all the planets in our Solar System except Pluto. Pictures telemetered back from space have allowed planetary scientists to discover a Moonlike surface on Mercury and circulation patterns in the thick Venusian atmosphere. Images of Mars have shown craters, giant canyons and volcanoes on the planet's surface. Interesting details about the circulation patterns in the Jovian atmosphere have been revealed; active volcanoes have been discovered on Jupiter's moon Io; and a ring has even been found circling the King of the Planets. New moons were discovered in orbit around Saturn; known moons have been viewed close up; and rings once thought to be just four in number have now been resolved into a complex configuration consisting of more than a thousand concentric ring features (see fig. 1). Finally, the distant planets, Uranus (1986) and Neptune (1989), unknown to ancient astronomers, were recently visited by the *Voyager 2* spacecraft and extensively imaged.

These new discoveries about the planets and their moons, as well as many other exciting observations about our extraterrestrial environment, were made possible by the development of a spacecraft technology for imagery collection, processing and transmission. The spectacular views we've enjoyed of the Martian surface or of Saturn's complex rings through this spacecraft imagery technology are facsimile images and not true photographs. A scanning optical system on board the spacecraft converts sunlight reflected from a planet or moon into numerical data. The numerical

Fig. 1 This computer-assembled mosaic image of Saturn's magnificent rings was taken by *Voyager 1* on November 6, 1980, at a range of 8 million kilometers. It revealed an extraordinarily complex structure of concentric rings. (Courtesy of NASA/Jet Propulsion Laboratory.)

data are then telemetered to Earth via electromagnetic (radio) waves. On Earth, giant tracking stations collect these electromagnetic signals, and processing computers assemble the received information into useful images of alien worlds.

Most planetary spacecraft produce images of the planets and their moons by using slow-scan television cameras. These cameras take a much longer time to form and transmit images than do the commercial television system cameras used here on Earth. Although they take longer to generate, the extraterrestrial images are of a much higher quality and contain more than twice the amount of information contained in the picture appearing on your home television screen.

Since the successful flyby of Mars by the *Mariner 4* spacecraft in 1964, major improvements have occurred in spacecraft imaging systems. An entire week was required for the *Mariner 4* spacecraft to transmit enough information to create just 21 images of Mars, since the early planetary-exploration craft could only transmit its data at a rate of under 10 bits per second. (A bit is a unit of information that can be represented by a 1 or a 0.) By comparison, in 1979 the same amount of data was contained in just one *Voyager* spacecraft image of Jupiter and was transmitted in only 48 seconds.

Using the *Voyager* spacecraft imaging system as our example, the process of generating finished extraterrestrial images of alien worlds can be accomplished in five general steps:

On the spacecraft
  Step one: image scanning
  Step two: data storage and transmission
On earth
  Step three: data reception
  Step four: data storage
  Step five: image reconstruction

The *Voyager* spacecraft carries a dual television camera system on a scientific instrument platform that can be tilted in any direction for precise aiming. On command, a celestial object can be viewed with either wide-angle or narrow-angle telephoto lenses. Reflected light from the extraterrestrial target enters the lenses and falls on the surface of a selenium–sulfur vidicon television tube, 11 millimeters square. Unlike most standard television cameras, a shutter controls the amount of light reaching the tube. Exposure periods can vary from 0.005 second for very bright targets to 15 seconds or more for very faint objects or when searches are being conducted for previously unknown, and very dim, satellites. The television tube temporarily retains the image until it can be scanned or measured for brightness levels. During the scanning process, the vidicon tube surface is divided into 800 lines, each line consisting in turn of 800 points. These individual points are called pixels. (The word *pixel* is a contraction of the term *picture element*.) The total number of pixels into which a *Voyager* image is divided is then 800 × 800, or 640,000. (See fig. 2).

As each pixel is scanned for brightness, it is assigned a number from 0 to 255. The measured range from black to

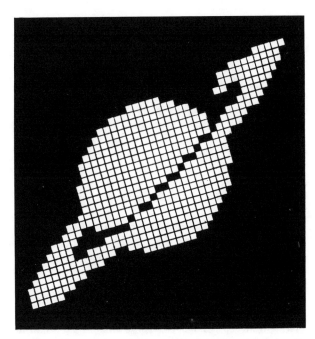

Fig. 2 A representative planetary image broken up into pixels. (Drawing courtesy of NASA /and modified by the author.)

be assigned to a pixel that had the following eight-bit binary sequence: 00101000?

If you had estimated a pixel brightness value of 20, you are correct and are well on your way to understanding how these magnificent extraterrestrial images are gathered by robot spacecraft, transmitted to Earth and computer-processed for use by scientists.

After the *Voyager* spacecraft imaging system scans the pixels and converts the light levels into binary form, the bit information is either stored on tape for later transmission or sent directly back to Earth in real time. These data are transmitted at a rate of more than 100,000 bits per second. For each *Voyager* image, 5,120,000 bits (640,000 × 8 binary bits) must be sent! Spacecraft data-storage capability is used when the vehicle passes out of sight behind a planet or moon and radio communications are temporarily eclipsed. With the *Voyager* spacecraft, data equivalent to 100 images can be stored for later transmission.

On Earth, radio signals from the spacecraft are received by one of three large radio antennas of NASA's Deep Space Network. When the spacecraft data are received on Earth, computers simultaneously store the data for future use and reassemble them into images. In generating these images, the computer converts the binary bit sequences for each pixel into small squares of light, as was shown in figure 2. The brightness of the squares is determined by the numerical value assigned to the pixel. The squares are then displayed on a television screen and assembled into a grid 800 × 800 pixels in dimension. The resulting image formed by all the lighted squares on the high-resolution television screen is a black-and-white facsimile picture of the target extraterrestrial object.

If we want color images of alien worlds, considerably more information is needed from the robot spacecraft. In the *Voyager* spacecraft, a wheel with a variety of colored filters is rotated in front of the television tube when the images are being acquired. In rapid succession, three separate images are taken of the same celestial object, through blue, green and orange filters. The television tube is scanned for each image, and the resulting binary bits are then transmitted to Earth. By the time the scanning of the third "color-filtered" image is complete, over 15 million binary

white is 256, or the number 2 to the eighth power (2^8). To express the assigned pixel brightness in terms a computer can understand, each number is converted into binary language. Eight binary digits, or bits, are needed to represent each number. Each of these eight bits, in sequence, represents a doubling of numerical value. The eight bits are arranged according to their numerical values in table 1. During information transmission, each bit is given as either a 0 or a 1. A 0 means the numerical value of that bit is zero, while a 1 means the numerical value of the bit corresponds to the bit sequence value shown in table 1. Then, to convert from binary bits back to the brightness number, we simply have to sum, or total, the values of each bit in the eight-bit sequence, or "word." Table 2 provides several examples of pixel brightness expressed in both numerical and binary form. After carefully reviewing tables 1 and 2, can you tell what brightness number would

Table 1   Binary Bit Sequence

Binary values	1	2	4	8	16	32	64	128
Binary bits	0	0	0	0	0	0	0	0

Table 2   Examples of Pixel Brightness

Binary values	1	2	4	8	16	32	64	128
Brightness no.								
0	0	0	0	0	0	0	0	0
9	1	0	0	1	0	0	0	0
58	0	1	0	1	1	1	0	0
183	1	1	1	0	1	1	0	1
255	1	1	1	1	1	1	1	1

bits $(3 \times 640{,}000 \times 8)$ are traveling toward Earth at the speed of light in the data-transmission signal.

Each color filter affects the amount of light reaching the television tube. The orange filter, for example, is transparent to orange light, but blue light appears much darker than normal. On Earth, computer processing is used to give color to the three filtered images and blend them together to form a "true" color image.

One final step may be added in the creation of these spectacular extraterrestrial images. To the unaided eye, some images might appear not to be particularly descriptive. The shading differences in planetary surfaces or cloud tops may be a little too subtle to be detected by a cursory visual examination. Scientists use a selective computer enhancement of portions of the image to bring out such subtle details. With computer processing, pixels of a particular numerical value can be assigned an unusual color to make them clearly stand out. If we have two almost identical shades of yellow, for instance, they can be colored red and blue, thereby greatly exaggerating their differences.

Images of the planets and moons of our Solar System have proved to be the most valuable source of spacecraft-acquired information available to planetary scientists. Data gathered at close range (and above the filtering effects of the terrestrial atmosphere) have produced views that are far better in quality and detail than pictures taken through telescopes on Earth. The unprecedented quality of these images from space has greatly assisted scientists in developing better theories of the nature and origin of our Solar System. Through spacecraft imaging systems in the short space of just three decades, we have "discovered" over a dozen new worlds and expanded our cosmic horizon at a rate unmatched in all previous human history!

See also: **Deep Space Network; remote sensing; Voyager**

"infective" theory of life   The belief that some primitive form of life—perhaps selected, hardy bacteria or "engineered" microorganisms—was placed on an ancient Earth by members of a technically advanced extraterrestrial civilization. This planting or "infecting" of simple microscopic life on a then-lifeless planet could have been intentional (that is, "directed panspermia") or accidental (for example, through the arrival of a "contaminated" space probe or from "space garbage" left behind by extraterrestrial visitors).

See also: **life in the Universe; panspermia**

inferior planets   Planets that have orbits that lie inside the Earth's orbit around the Sun—namely, Mercury and Venus.

infrared (IR) astronomy   The branch of astronomy dealing with infrared (IR) radiation from celestial objects. Most celestial objects emit some quantity of infrared radiation. However, when a star is not quite hot enough to shine in the visible portion of the electromagnetic spectrum, it emits the bulk of its energy in the infrared. IR astronomy, con-

sequently, involves the study of relatively cool celestial objects, such as interstellar clouds of dust and gas (typically about 100 degrees Kelvin) and stars with surface temperatures below about 6,000 degrees Kelvin.

Many interstellar dust and gas molecules emit characteristic infrared signatures that astronomers use to study chemical processes occurring in interstellar space. This same interstellar dust also prevents astronomers from viewing visible light coming from the center of our Milky Way Galaxy. However, IR radiation from the galactic nucleus is not as severely absorbed as radiation in the visible portion of the electromagnetic spectrum, and IR astronomy therefore enables scientists to study the dense core of the Milky Way.

Infrared astronomy also allows astrophysicists to observe stars (called protostars) as they are being formed in giant clouds of dust and gas (called nebula), long before their thermonuclear furnaces have completely ignited and they have "turned on" their visible light emission.

Unfortunately, water and carbon dioxide in the Earth's atmosphere absorb most of the interesting IR radiation arriving from celestial objects. There are only a few narrow IR spectral bands or windows that can be used by Earth-based astronomers in observing the Universe; and even these IR windows are distorted by "sky noise" (undesirable infrared radiation from atmospheric molecules).

However, the Space Age has provided astronomers a solution to this problem and has opened up an entirely new region of the electromagnetic spectrum to detailed observation. We can now place sophisticated IR telescopes in space, above the limiting and disturbing effects of the Earth's atmosphere.

For example, the Infrared Astronomical Satellite (IRAS), which was launched in January 1983, was the first extensive scientific effort to explore the Universe in the infrared portion of the electromagnetic spectrum. IRAS was an international effort involving the United States, the United Kingdom and the Netherlands. By the time IRAS ceased operations in November 1983, this space-based IR telescope had completed the first all-sky survey in a wide range of IR wavelengths with a sensitivity 100 to 1,000 times greater than any previous telescope. Space scientists have used IRAS data to produce a comprehensive catalog and maps of significant infrared sources in the observable Universe. These sources include stars that may possess planetary systems or at least planetary systems under formation. Survey data from IRAS have also helped scientists plan even more sophisticated IR astronomy missions.

The Infrared Space Observatory (ISO), developed by the European Space Agency (ESA) and scheduled for launch in 1993, will operate at infrared wavelengths between 3 and 200 micrometers. The ISO spacecraft will have an initial mass of 2,300 kg (5072 lbm) and will be 2.25 meters (7.38 ft) wide and 5.2 meters (17.06 ft) high. This IR-observatory will be launched by an Ariane rocket and placed into a near-equatorial, elliptical 12-hour orbit with a perigee of 1,000 km (621 mi) and an apogee of 40,000 km

(24,840 mi). This type of elliptical orbit is very suitable for an observatory satellite, because it permits continuous observations (for up to 10 hours) of selected celestial objects and also permits repeated access to all parts of the sky.

The largest part of the ISO spacecraft is the payload module, which is essentially a large thermos flask or dewar filled with liquid helium—needed to keep portions of the IR telescope at only a few degrees above absolute zero temperature. At the start of its observatory mission, the ISO will carry about 2040 liters (539 gal) of liquid helium coolant.

This liquid helium will cool some of the IR detectors to 2 degrees Kelvin and will also keep other temperature-sensitive instruments and the IR telescope mirrors at temperatures below 4 degrees Kelvin. The overall supply of liquid helium is expected to support at least 18 months of orbital observations.

Two areas of infrared astronomy that will figure extensively in ISO data collections are studies of the early phases of stellar evolution and studies of other galaxies.

Stars are formed inside dense clouds of interstellar dust and gas. As the material within these clouds condenses into protostars, the material warms up and emits IR radiation that can escape from the cloud and be detected. Some stars remain "hidden" in the visible portion of the spectrum behind veils of dust in these "stellar nurseries" even after they have reached the thermonuclear burn stage. Therefore, IR radiation is quite useful in the study of stars, because sensitive IR spectroscopic measurements can be used to probe the conditions and material abundances in regions close to evolving stars.

Many galaxies emit a large portion of their radiation—sometimes almost all of it—in the IR portion of the electromagnetic spectrum. These IR emissions can arise from dust and gas between the stars, but quasars and active galactic nuclei also emit strong IR signals.

The Space Infrared Telescope Facility (SIRTF) will be one of NASA's Great Observatory missions in support of astrophysics. Scheduled for launch in late 1998, this advanced IR observatory will collect data from distant IR-emitting galaxies, record IR images of planetary disks encircling nearby stars, examine stars in the earliest stages of their formation process, search for comets in the outer regions of the Solar System, and hunt the nearby regions of our Galaxy for telltale IR signals from brown dwarfs.

A brown dwarf is a postulated dim stellar object that possesses a mass of about ten percent or less that of our Sun. Because of this small mass, a brown dwarf star cannot sustain thermonuclear burn in its core. The only radiation emitted by such objects would be infrared, representing the residual thermal energy generated during its birthing process from a collapsing cloud of interstellar gas. With no sustained thermonuclear burn in its core regions, a brown dwarf will cool rapidly and become very dim, essentially "brown" to an observer, making it very difficult to see optically. Some astronomers now speculate that there are perhaps millions of these brown dwarfs in our Galaxy alone.

If true, the confirmation of such a brown dwarf population would help cosmologists solve the Universe's "missing mass" problem.

Maybe one of the most exciting discoveries awaiting our use of future, very sophisticated IR telescopes in space is the detection and identification of an advanced extraterrestrial civilization through its telltale astroengineering activities. In 1960 the physicist Freeman J. Dyson suggested that we could search for evidence of extraterrestrial beings by looking for artificial cosmic sources of IR radiation. He postulated that intelligent beings of an advanced civilization might eventually desire to capture all the radiant energy output of their parent star. They might subsequently construct a huge cluster of habitats and space platforms, called a Dyson sphere, around their star. This Dyson sphere, perhaps of a size comparable to the Earth's orbit around the Sun, would lie within the ecosphere of the parent star and would intercept all its radiant energy output. The intercepted starlight, after useful energy extraction, would then be reradiated to outer space at approximately ten micrometers wavelength. This infrared wavelength corresponds to a heat rejection surface temperature for the Dyson sphere of approximately 200 to 300 degrees Kelvin. Therefore, according to Dyson, if we detect an IR radiation source of about 250 degrees Kelvin that is approximately one or two astronomical units in diameter, it might just represent the astroengineering handiwork of an advanced extraterrestrial civilization!

See also: **astrophysics; black holes; cosmology; Dyson sphere; extraterrestrial civilizations; infrared radiation; nebula**

Infrared (IR) radiation That portion of the electromagnetic (EM) spectrum lying between the optical (visible) and radio wavelengths. It is frequently taken as spanning three decades of the EM spectrum, from 1 micrometer to 1,000 micrometers wavelength. The English–German astronomer Sir William Herschel (1738–1822) is credited with the discovery of infrared radiation.

See also: **astrophysics; electromagnetic spectrum**

inner planets The terrestrial planets: Mercury, Venus, Earth and Mars. These planets all have orbits around the Sun that lie inside the main asteroid belt.

intergalactic Between or among the galaxies. Although no place in the Universe is truly "empty," the space between clusters of galaxies comes very close. These intergalactic regions contain less than one atom in every 10 cubic meters. Even though the galaxies continually supply new matter to intergalactic space, the continued expansion of the Universe makes the overall effect negligible. In fact, intergalactic space is very empty and is getting more empty every moment as the Universe expands.

interplanetary Between the planets; within the Solar System.

interplanetary dust (IPD) Tiny particles of matter (typically less than 100 micrometers in diameter) that exist in space within the confines of our Solar System. By convention, the term applies to all solid bodies ranging in size from submicrometer diameter to tens of centimeters in diameter, with corresponding masses ranging from 10^{-17} gram to approximately 10 kilograms. Near the Earth the IPD flux is taken as approximately 10^{-13} to 10^{-12} gram per square meter per second [g/(m²-s)]. Space scientists have made rough estimates that the Earth collects about 10,000 metric tons of IPD per year. They also estimate that the entire IPD "cloud" in our Solar System would have a total mass of between 10^{+16} and 10^{+17} kilograms.

Recent laboratory studies of IPD materials indicate that this dust contains samples from the primitive solar nebula, preserved from destruction during the evolution of the Solar System by their residence in comets or protoplanets. Other recently examined IPD samples have yielded cores believed to consist of the interstellar dust grains that made up part of the matter from which the Solar System was originally formed. Therefore, careful laboratory examination of certain IPD samples has provided scientists with a "window" back through the entire history of our Solar System.

In another recent series of laboratory studies IPD samples attributed to stony meteorites or chondrites were compared with dust particles experienced during the comet Halley encounter by the Vega and Giotto spacecraft. These comparative studies revealed that the IPD samples examined were much richer in volatile elements (such as carbon) than any other known extraterrestrial material (e.g., meteorites) and were very similar to the dust from comet Halley. Therefore, at least some types of interplanetary dust particles appear to be of cometary origin.

One long-term goal of space scientists is the controlled orbital collection of IPD samples, permitting the determination of their velocities and trajectory information. NASA's proposed Space Station Cosmic Dust Collection Facility is one pathway being considered to achieve this goal.

See also: **comet; meteoroids; zodiacal light**

interstellar Between or among the stars.

interstellar communication One of the fundamental aspects of being human is our desire to communicate. In recent years we have begun to respond to a deep cosmic yearning to reach beyond our own Solar System to other star systems—hoping not only that someone or something is out there but that they will eventually "hear us" and perhaps even return our message.

Because of the vast distances between even nearby stars, when we say "interstellar communication" we are not talking about communication in "real time." (Communication in real time does not involve a perceptible time lag—that is, messages and responses are received immediately after they are sent.) Rather, our initial attempts at interstellar communication have actually been more like putting a message in a bottle and tossing it in the "cosmic sea," or perhaps even placing a message in a time capsule or "cosmic safety deposit box" for some future generation of human or alien beings to find and learn about life on Earth in the 20th century.

Attempts to "communicate" with alien civilizations that might exist among the stars is often called CETI, an acronym that means "communication with extraterrestrial intelligence." If, on the other hand, we quietly watch the skies for "signs" of some super-extraterrestrial civilization (for example, looking for the infrared signatures from Dyson spheres) or patiently listen for intelligent radio messages transmitted by advanced alien races (mainly in the microwave region of the spectrum), then we call the process SETI, or simply the "search for extraterrestrial intelligence."

Since 1960 there have been several serious SETI observation efforts, the vast majority involving listening to selected portions of the microwave spectrum, in hopes of detecting "radio signals" indicative of the existence of intelligent extraterrestrial civilizations among the stars. To date, none of these efforts has provided any positive evidence that such "intelligent" radio signals exist, carrying messages to other advanced or even developing galactic races. However, SETI observers have only examined a few of the billions of stars in our Galaxy and have only listened to a few rather narrow portions of the spectrum within which such intelligent signals might be transmitted. Furthermore, it is only within the last few decades that we have developed the technology, largely radio astronomy-related, to enable us to be at even a minimum "interstellar communications horizon." A century ago, for example, the Earth could have been "bombarded" with many alien signals—but no one would have had the technology to receive and interpret them.

We have also deliberately attempted to communicate with alien civilizations by sending messages out beyond the Solar System on several of our spacecraft and by sending a very powerful radio message to the stars using the world's largest radio-telescope facility, the Arecibo Observatory in Puerto Rico (see fig. 1 on page 7). Since the age of radio and television, we have also unintentionally been leaking radio-frequency signals (now about 40 light-years out) into the Galaxy. Imagine the impact some of our early television shows would have on an alien civilization capable of intercepting and reconstructing these signals!

Our three most important attempts at interstellar communication (from Earth to the stars) to date are: (1) the special message plaque placed on both the *Pioneer 10* and *11* spacecraft departing the Solar System on interstellar trajectories; (2) the "Sounds of Earth" record included on the *Voyager 1* and *2* spacecraft, which are also departing the Solar System on interstellar trajectories; and (3) the famous Arecibo Interstellar Message, transmitted on November 16, 1974, by the world's most powerful radio telescope.

See also: **Arecibo Interstellar Message; conse-**

quences of extraterrestrial contact; interstellar contact; Pioneer plaque; search for extraterrestrial intelligence; Voyager record

interstellar contact Several methods of achieving contact with intelligent extraterrestrial life-forms have been suggested. These methods include (1) interstellar travel by means of starships, leading to physical contact between different civilizations; (2) indirect contact through the use of robot interstellar probes; (3) serendipitous contact; (4) interstellar communication involving the transmission and reception of electromagnetic signals; and (5) exotic techniques involving perhaps information transfer through the modulation of gravitons, neutrinos or streams of tachyons; the use of some form of telepathy; and matter transfer through the use of hyperspace or distortions in the space–time continuum that help "beat" the speed-of-light barrier.

INTERSTELLAR TRAVEL/PHYSICAL CONTACT

The classic method of interstellar contact in science fiction is the starship. With this class of spaceship, an intelligent civilization would be capable of eventually sweeping through the Galaxy, finding and contacting other life-forms wherever they existed and planting life wherever it didn't, but could, exist. Probably nothing would be more exciting, and even a little frightening, to a technically emerging planetary society than to have its sky suddenly fill with an armada of giant starships. The inhabitants of the planet would be advanced enough to appreciate the great technology levels required to bring the alien visitors across the interstellar void. This physical contact could also prove a very humbling experience for a planetary civilization like our own, that had just struggled to achieve interplanetary spaceflight capabilities. A variety of contact scenarios can be found in science fiction. These scenarios range from a friendly welcome into a galactic community to a hostile attempt to "capture the planet." In the belligerent scenarios, those beings on the starship play the role of "invaders," while the planet's inhabitants become the "defenders." Depending on the level of technology mismatch and any literary gimmicks the S/F writers include (such as a "biological Achilles' heel" for the invading species), the battle for the planet or star system goes either way in the story.

However, even though we have successfully begun to master interplanetary flight with chemical propulsion systems and can complement these propulsion systems with more advantageous nuclear-fission- and eventually nuclear-fusion-powered propulsion systems—the energetic demands of interstellar flight simply overwhelm any propulsion technology we can extrapolate as 21st-century engineering and beyond.

One example might explain these "hard" circumstances a little better. Let's ignore *all* current engineering and materials-science limitations and construct (at least on paper) the *very best propulsion system "physics can buy"*; that is, we're going to build the most advanced propulsion

system our current understanding of physics will allow—despite the fact that the actual engineering technology to accomplish this construction task may be centuries away, if ever! We would construct a *photon rocket*, whose propellant is a mixture of equal parts of matter and antimatter. This photon rocket uses the annihilation reaction that occurs when we blend matter and antimatter. This extremely energetic reaction turns every kilogram of propellant into pure energy, mainly gamma radiation. These gamma rays would then be directed out of the rocket's special "thrust chamber" in a perfectly collimated radiation stream (that is, a radiation stream in which the rays are parallel) that provides a reaction (retrodirected) thrust to the starship and its payload. The complete conversion of just one gram of matter–antimatter in an annihilation reaction would release some 9×10^{13} joules of energy! (In comparison, a one-kiloton [(kT)] nuclear-explosion yield amounts to a release of some 4×10^{12} joules.) For the moment, we have neglected all the engineering problems of obtaining and containing antimatter and of preventing nuclear-radiation leakage into the crew compartment.

Let's now use this "best-we-can-possibly-build" photon-powered starship on a 10-year round-trip journey to a nearby star system (for example, Alpha Centauri, which is 4.23 light-years away). To further optimize this exercise, let us also assume that the entire starship, except for the matter–antimatter propellant, has a mass of only 1,000 tons and that the starship can achieve a cruising speed of 99 percent of the speed of light (0.99 c) after a reasonably short period of acceleration (about one year at a constant acceleration rate of one g). According to one set of calculations for this hypothetical round-trip interstellar mission, we would need 33,000 tons of matter–antimatter propellant (16,500 tons of each type) to annihilate en route. The total energy release associated with 33,000 tons of mass converted into pure energy is approximately 3×10^{24} joules. As a point of reference, our Sun's energy output is approximately 4×10^{27} joules per second.

A few other "engineering" details are also worth mentioning here. During the initial acceleration period, our matter–antimatter-powered starship must achieve power levels of about 10^{18} watts. If only one part in a million of this energy release leaks into the ship, it would experience a one-million-megawatt (10^{12}-watt) heat flux. A very elaborate and heavy cooling and heat-rejection system would be needed to keep the starfarers and their equipment from melting. The same, if not worse, constraints apply to radiation leakage into the crew compartment. Extensive shielding will be needed to protect the crew and their equipment both from "engine-room leakage" and also from the radiation spall (erosion of solid surfaces) that will occur when a starship moving at near-light speed hits interstellar dust and molecules.

The sobering conclusion of this paper exercise, although contrary to the bulk of popular science fiction, is that based on our current understanding of the physical laws of the Universe, interstellar starships carrying human crews on round-trip journeys within a crew's life span appear out of the question, not only for the present but for an indefinitely long time into the future. Interstellar travel is *not* a physical impossibility, but for today and, perhaps, for many tomorrows, it appears technically out of our reach. Many breakthroughs, most unimaginable at present, would have to occur before human beings can realistically think about boarding a starship and making a round-trip (or perhaps one-way) journey to another star system.

On the basis of this conclusion, we cannot initiate interstellar contact using the physical travel of human beings across the interstellar void. If contact is made by starship in the next few centuries, it will most likely be "them" visiting "us." Perhaps, some technically powerful alien society has had better luck unraveling nature's secrets and has discovered forms of energy and physical laws that are currently far beyond our intellectual horizon. If this is the case, and if these alien beings also have a societal commitment to interstellar exploration, and if they decide that our rather common G–spectral-class star is worth visiting— then perhaps . . . just perhaps, physical contact by starship will be made. But we cannot seriously include this possibility in our own attempts at interstellar contact because we have no control over the circumstances. This "starfaring" civilization either exists, or it doesn't; it is committed to interstellar exploration, or it isn't; and it either decides to explore our particular Solar System, or it bypasses us in favor of a more interesting galactic region.

In the absence of credible evidence that alien starships exist or are en route to visit us, we must be content within the next few decades to attempt interstellar contact through one of the alternative techniques suggested here.

INDIRECT CONTACT BY MEANS OF INTERSTELLAR ROBOTIC PROBES

Instead of sponsoring round-trip interstellar missions with "manned" starships, an advanced civilization might elect to send one-way robotic probes to explore neighboring star systems. The use of these interstellar probes has several potential advantages over a fully outfitted starship. First, the interstellar probe can be much smaller, since it does not require elaborate life-support equipment, crew quarters or a propellant supply for the journey back. This reduction in size greatly reduces the overall expense of the mission and eases the demands placed upon the propulsion system. Second, the probes can take a much longer time to reach their destination. A 50- or 100-year flight time for a robotic spaceship does not involve the same design complications that a mission of similar length would have on a "manned" starship. Again, this eases the technology demands on the propulsion system. We might now consider a propulsion system that reaches only one-tenth the speed of light (0.1 c) as a maximum cruising speed. While this is still a very challenging technology development, it is several orders of magnitude less of a challenge than designing a propulsion system to drive a starship at 99 percent of the speed of light or better.

A robotic interstellar probe would have to possess an advanced form of machine intelligence to execute repair functions en route, scan ahead for possible dangers in interstellar space and then execute a meaningful exploration program in a totally alien star system.

In reviewing the technical literature, three general types of interstellar-probe missions appear: (1) the flyby, (2) the sentinel and (3) the self-replicating machine (SRM). (Each of these major classes of robot-probe missions also has several possible variations, which will not be discussed here.)

The flyby interstellar robotic probe represents a one-way, one-shot attempt at interstellar contact and exploration. The probe and its complement of instruments would be launched toward the target star system and then be accelerated by the propulsion system up to a minimum velocity of about 10 percent of the speed of light (0.1 c). Because the probe does not need to decelerate when it gets to the new star system, only propellant for the initial acceleration is required. This is then the simplest, and perhaps "easiest," interstellar mission to develop. The probe would take several decades to cross the interstellar void. As it neared the target star system (perhaps at a distance of one light-year away), it would initiate an extensive long-range scanning operation to discover whether the star system contained planets that demanded special examination. If planets were detected, smaller robot scout ships would then be sent ahead. They would be launched from the mothership and powered by advanced nuclear-fission or nuclear-fusion engines. Traveling at perhaps 10 percent of the speed of light, the larger interstellar-probe mothership would only briefly encounter the new star system. It would gather as much data as possible with its on-board sensor arrays and would also collect the data transmitted from any scout ships that were sent ahead. These smaller ships might have placed themselves in orbit around the new star or landed on a planet of interest in search of life. All these data would then be transmitted back to the home civilization. The probe, its mission complete, would disappear into the interstellar void.

If any of the scout ships had detected intelligent life on a planet in the new star system, the mothership might transmit a message or even deploy a special scout ship that contained appropriate greetings and detailed information about the sponsoring civilization.

This type of one-way robot-probe mission was studied in Project Daedalus, the first detailed engineering study involving the feasibility of interstellar travel. A pulsed nuclear-fusion system, using deuterium/helium-3 as the thermonuclear fuel, was proposed as the Daedalus propulsion unit.

In a sentinel interstellar-probe mission, the propulsion system must be capable of accelerating the probe to at least 10 percent of the speed of light (0.1 c). It must also be capable of decelerating the probe spacecraft from "light speeds" to speeds that permit gravitational capture by the new star. This sentinel probe, now orbiting the target star, might spend years, decades or perhaps centuries searching the planets for signs of life. If life is detected, this probe may monitor its development, sending back information to the home civilization at selected time increments.

If the robot probe discovered an emerging technical civilization on one of the planets, it might be designed to execute a special contact protocol (procedure). For example, it could announce its presence when it was "triggered" by the detection of a certain level of technology. This robot sentinel might silently watch the planet's civilization emerge from an agrarian society to an industrial one. The development of radio-wave communication, atomic energy or spaceflight might be technology levels in the emerging planetary civilization that would trigger the sentinel to action. Suddenly, the somewhat startled younger civilization would receive a "message from the stars"! Unlike the physical arrival of "little green men" in their starship, this form of interstellar contact would be indirect, with the smart robot probe serving as a surrogate alien visitor. Some individuals have already suggested that such sentinel or monitoring probes (somewhere out there in the Solar System) are monitoring our development right now! However, our explorations of the Moon, Mars, Venus, Mercury, Jupiter, Saturn, Uranus and Neptune, have not yet discovered their suggested presence.

The third general class of robot interstellar-probe mission involves the use of a self-replicating machine (SRM). In this case, the propulsion requirements start approaching the demands of a full starship. The SRM probe must accelerate to some fraction of the speed of light, decelerate when it arrives at the target star system and then accelerate again to light-speed as it searches for another star system to explore. In the general scenario for the use of an SRM probe, the robot spaceship encounters a star system and begins the search for "suitable planets." In this case, a suitable planet is one that has the resources the probe needs to build an exact replica of itself—including computer systems, propulsion system, sensors for exploration and so on. This class of probe must also be capable of detecting and identifying life-forms, especially intelligent life-forms. If intelligent life is detected, the probe may execute a special contact protocol. This could involve the presentation of simple messages or the construction from native materials of replicas of special devices and objects from the advanced civilization. If the inhabitants of the planet have matured to the level of interplanetary spaceflight, the SRM probe might even be programmed to replicate a copy of itself to leave behind as a "technological gift." The initial self-replicating machine probe would refurbish itself with planetary resources and then make an appropriate number of robot-probe replicas. All of these SRM probes would then depart to explore other star systems. This would create an exponentially growing population of smart machine explorers passing like a wave through the Galaxy. Imagine the surprised response of the inhabitants of an emerging civilization as one of these robot probes entered their system, provided messages and replicated token objects of

"friendship," and then devoured a few choice asteroids and small moons to replicate itself several times.

To get into the interstellar-probe business, an advanced civilization would have to make a serious financial and technical commitment to interstellar exploration. For example, just to visit and possibly "monitor" all the star systems within 1,000 light-years would require about 1 million flyby or sentinel probes—or one very sophisticated and reliable self-replicating machine probe. In the case of flyby or sentinel probes, if the advanced civilization then launched one of these probes a day, just the launching process alone would take three millennia!

SERENDIPITOUS CONTACT

We cannot rule out the possibility that we might suddenly stumble on some evidence of extraterrestrial intelligence. Perhaps an archaeologist exploring ancient Mayan temples in the Yucatan will discover the wreckage of a small alien scout ship; or maybe an astronomer pondering over Hubble Space Telescope data will come across an image of an alien interstellar probe or the unmistakable signature of some great feat of astroengineering, like the construction of a Dyson sphere; or possibly, sometime in the next century, an asteroid prospector, chasing down an interesting one-kilometer-size object on her radar screen, will come face-to-face with a derelict alien starship or, even more exciting perhaps, a robot sentinel probe—which might then begin to "speak."

As the word *serendipity* implies, these would be totally unexpected but nonetheless very exciting forms of interstellar contact. However, the chances of such accidental discoveries appear astronomically small. We cannot ignore this possibility in considering interstellar contact; but we do not exercise any real control over the situation either.

TRANSMISSION AND RECEPTION OF
ELECTROMAGNETIC SIGNALS

An advanced civilization might like to minimize the expense of searching for other intelligent civilizations. One very dominant factor is the amount of energy an intelligent alien society must expend in announcing its existence to the Galaxy. As we discussed previously, sending a starship or even a robot probe across interstellar distances is a very energy-intensive activity. Today, many scientists involved in the search for extraterrestrial intelligence (SETI) believe that we cannot expect to find other intelligent life by "tossing tons of metal across the interstellar void." Even if we could build a starship or a sophisticated interstellar robot probe, the undertaking would still be extremely costly in terms of time, energy and financial resources. This is because we might have to search many star systems, perhaps out hundreds of light-years from our Sun, before we achieved contact with the nearest intelligent civilization. These SETI scientists often recommend an alternative to matter transfer: They suggest that we send some form of "radiation" instead.

Regardless of what form of radiation a civilization decides

to use in trying to achieve interstellar contact by signaling, the signal itself should have the following desirable properties: (1) the energy expended to deliver each bit or piece of information should be minimized; (2) the velocity of the signal should be as high as possible; (3) the particles or waves making up the signal should be easy to generate, transmit and receive; (4) the particles or waves should not be appreciably absorbed or deflected by the interstellar medium or planetary atmospheres; (5) the number of particles or waves transmitted and received should be much greater than the natural background; (6) the signal should exhibit some property not found in naturally occurring radiations; and (7) the radiation of such signals should be indicative of the activities of a technically advanced civilization.

Carefully reviewing these suggested requirements, SETI scientists eliminated charged particles, neutrinos, gravitons and so on in favor of spatially and temporally coherent electromagnetic waves. For example, the kinetic energy of an electron traveling at 50 percent of the speed of light (0.5 c) is about 10^8 times the total energy of a 150-gigahertz (microwave) photon. All other factors taken as equal, an interstellar communications system using electrons would need 100 million times as much power as one using microwaves (photons). More exotic particles, such as neutrinos, are not easily generated, modulated (to put a message on them) or collected.

SETI scientists have searched over the entire electromagnetic spectrum for suitable regions in which to conduct an interstellar conversation. Their generally unanimous conclusion is that the microwave region appears to be the most obvious choice for advanced civilizations to communicate with each other and even to attempt to communicate with emerging planetary civilizations. These SETI observers have also identified a special band within the microwave region called the "water hole." This narrow region lies between the spectral lines of hydrogen (1,420 megahertz [MHz]) and the hydroxyl radical (1,662 megahertz). At present, the water hole is one of the highly preferred bands used by terrestrial scientists in their search for radio signals generated by intelligent alien civilizations.

EXOTIC SIGNALING TECHNIQUES

While often suggested in science fiction as ways of rapidly transferring matter or information throughout the Universe, telepathy, the manipulation of gravitons or neutrinos, and the use of wormholes or hyperspace are "techniques" well beyond our own communications horizon. If such exotic alternatives exist and are now being used to signal the Earth, we would have no real way of knowing about it. We simply couldn't receive and interpret such messages. Remember, the terrestrial atmosphere is saturated with human-made radio and television signals—but without an antenna and the proper receiver, a person would be totally unaware of their presence. Our understanding of the Universe and "how things work in nature" is not yet

sophisticated enough for us to think about using such exotic techniques in attempting such modes of interstellar contact in the next few decades or even centuries.

See also: **ancient astronauts; consequences of extraterrestrial contact; Dyson sphere; extraterrestrial civilizations; interstellar communication; Project Daedalus; robotics in space; search for extraterrestrial intelligence**

interstellar gas and dust   Material found in space between the stars. Low-density hydrogen and other gases have been detected from their absorption and emission of specific wavelengths of light and radio waves. Fine dust particles in interstellar space can scatter starlight in much the same way as smog scatters light here on Earth.

See also: **interstellar medium**

interstellar medium (ISM)   The gas and dust particles that are found between the stars in our Galaxy. Up until about two decades ago, the interstellar medium was considered an uninteresting void. Today, through advances in radio astronomy (especially in the millimeter wave region of the electromagnetic spectrum), we now know the interstellar medium contains a rich and interesting variety of atoms and molecules as well as a population of fine-grained dust particles. There are over 100 interstellar molecules that have been discovered to date, including many organic molecules considered essential in the development of life.

These interstellar molecules are not uniformly distributed throughout interstellar space. Instead, there are essentially two basic types of interstellar clouds. The first type are very diffuse clouds, which appear to contain very little interstellar dust and in which the concentrations of gas molecules (primarily atomic hydrogen) are very low. The second type consists of dark, dense molecular clouds, often referred to as *giant molecular clouds* (GMCs) because they have total masses of perhaps millions of solar masses and dimensions that span 60 to 260 light-years. In fact, the largest of the GMCs are the most massive molecular objects yet observed in our Universe. Interstellar dust appears abundant in these molecular clouds, and molecular hydrogen (H_2) represents the dominant gas species, although there is also an interesting variety of other interstellar molecules, including carbon monoxide (CO). Gas concentrations in these molecular clouds can range from 10^3 to over 10^6 molecules per cubic centimeter. Astronomers now consider these molecular clouds as the "birthing grounds" for new stars.

Interstellar dust (which "reddens" the visible light from the stars behind it because of its preferential scattering of shorter-wavelength photons) is considered to consist of very fine silicate particles, typically 0.1 micrometers in diameter. These interstellar "sands" may sometimes have an irregularly shaped coating of water ice, ammonia ice or (solidified) carbon dioxide.

Today, the interstellar medium is considered an interesting area for astronomers, astrophysicists and exobiologists. Many of the identified interstellar molecules provide clues to the processes involved in the evolution of stellar and galactic systems and strongly suggest the essentially universal presence of the chemical building blocks of carbon-based life as we know it.

See also: **life in the Universe; radio astronomy; stars**

interstellar probes Automated interstellar spacecraft launched by advanced civilizations to explore other star systems. Such probes would most likely make use of very smart machine systems capable of operating independently for decades or centuries. Once the robot probe arrives at a new star system, it begins an exploration procedure. The target star system is scanned for possible life-bearing planets, and if any are detected, they become the object of more intense investigation. Data collected by the "mother" interstellar probe and any miniprobes (deployed to explore individual objects of interest within the new star system) are transmitted back to the home star system. There, after light-years of travel, the signals are intercepted by scientists, and interesting discoveries and information are used to enrich the civilization's understanding of the Universe.

Robot interstellar probes might also be designed to protectively carry specially engineered microorganisms, spores and bacteria. If a probe encounters ecologically suitable planets on which life has not yet evolved, then it could "seed" such barren but potentially fertile worlds with primitive life-forms or at least life precursors. In that way, the sponsoring civilization would not only be exploring neighboring star systems; it would also be spreading life itself through the Galaxy.

See also: **extraterrestrial civilizations; interstellar travel; life in the Universe; panspermia; Project Daedalus; robotics in space** (extraterrestrial impact of self-replicating systems)

interstellar travel Matter transport between star systems in a galaxy. The "matter" transported may be (1) a robot interstellar probe on an exploration or "prelifeseeding" mission, (2) an automated spacecraft that carries a summary of the cultural and technical heritage of an alien civilization, (3) a starship "manned" by intelligent extraterrestrial creatures who are on a voyage of exploration and "contact" with other life-forms or perhaps even (4) a giant interstellar ark that is carrying an entire alien civilization away from its dying star system in search of suitable planets around other stars.

See also: **interstellar contact** (interstellar travel/physical contact); **Project Daedalus; starship**

ionizing radiation Radiation capable of producing ions by adding electrons to, or removing electrons from, an electrically neutral atom, group of atoms or molecule.

J

jansky [symbol: Jy] A unit used to describe the strength of an incoming electromagnetic wave signal. The jansky is frequently used in radio and infrared astronomy. It is named after the American radio engineer, Karl G. Jansky (1905–1950), who discovered extraterrestrial radio-wave sources in the 1930s—a discovery generally regarded as the birth of radio astronomy.

$$1 \text{ jansky (Jy)} = 10^{-26} \text{ watts per meter squared per hertz}$$
$$1 \text{ Jy} = 10^{-26} \text{ W/m}^2\text{-Hz}$$

See also: **radio astronomy**

Jovian Of or relating to the planet Jupiter; (in science fiction) a native of the planet Jupiter.

Jovian planets The giant, gaseous outer planets of our Solar System. These include: Jupiter, Saturn, Uranus and Neptune.

Jupiter In Roman mythology Jupiter (called Zeus by the ancient Greeks) was lord of the heavens, the mightiest of all the gods. So, too, Jupiter is first among the planets in our Solar System. Jupiter has even been called a "near-star." If it had been only about 100 times larger, nuclear burning could have started in its core, and Jupiter would have become a star and a rival to the Sun itself. Not only is Jupiter the largest of the planets, with 318 times the mass of the Earth, but its interesting complement of 16 known satellites resemble a miniature solar system.

The largest of these moons are called the Galilean satellites. They are: Io, Europa, Ganymede and Callisto. These moons were discovered in 1610 by Galileo—an event that helped spark the birth of modern observational astronomy and the overall use of the scientific method. The discovery of these four moons provided strong support for the then-revolutionary Copernican theory (that the Sun, *not* the Earth, is at the center of our Solar System). Galileo came under bitter personal attack by ecclesiastical authorities for his "earthshaking" discoveries. Through Galileo's pioneering efforts, centered in part on his early observations of Jupiter and its four major moons, the scientific method rapidly became understood and accepted. This, in turn, resulted in the exponential growth of scientific knowledge and technology—the foundation of our modern world and the stepping-stone to our extraterrestrial civilization.

Interest in Jupiter and its moons has been greatly stimulated in recent years by the Jovian encounters of four American spacecraft: *Pioneer 10* (December 1973), *Pioneer 11* (December 1974), *Voyager 1* (March 1979) and *Voyager 2* (July 1979). These spacecraft provided a wealth of new information and spectacular photographs of Jupiter and its complement of natural satellites.

In March 1972 NASA sent the first of these four historic space probes to survey the outer planets. For each spacecraft, Jupiter was the first port of call.

Pioneer 10, which was launched from Cape Canaveral on March 2, 1972, was the first spacecraft to penetrate the main asteroid belt and travel to the outer regions of our Solar System. In December 1973 it returned the first close-up pictures of Jupiter as it flew within 130,000 kilometers of the giant planet's banded cloud tops. *Pioneer 11* followed a year later. *Voyagers 1* and *2* were launched in the summer of 1977 and returned spectacular photographs of Jupiter and its 16 satellites during flybys in 1979.

During their visits, these robot explorers found Jupiter to be a whirling ball of liquid hydrogen, topped with a uniquely colorful atmosphere, which is mostly hydrogen and helium (see fig. 1). The Jovian atmosphere also contains small amounts of methane, ammonia, ethane, acetylene, water vapor and possibly hydrogen cyanide. Exobiologists think it is possible that between the planet's frigid cloud tops and the warmer hydrogen ocean that lies below, there might be regions in the Jovian atmosphere where methane, ammonia, water and other gases could react to form organic molecules. Because of Jupiter's atmospheric dynamics, however, these organic compounds—if they exist—are probably quite short-lived. The *Galileo* probe into the Jovian atmosphere (1995) should provide some interesting data concerning these speculations.

The Great Red Spot of Jupiter has been observed for centuries by terrestrial astronomers. It is now believed to be a tremendous atmospheric storm, similar to a hurricane on Earth, which rotates counterclockwise.

The *Pioneer* and *Voyager* spacecraft detected lightning

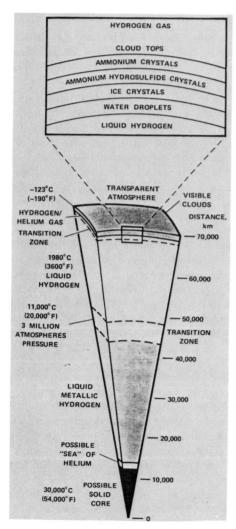

Fig. 1   A current model of Jupiter based on *Pioneer* and *Voyager* data. (Drawing courtesy of NASA.)

Table 1   Physical and Dynamic Properties of Jupiter

Diameter (equatorial)	142,982 km
Mass	1.9×10^{27} kg
Density (mean)	1.32 g/cm^3
Surface gravity	24.9 m/s^2
Escape velocity	61.0 km/s
Albedo	0.5–0.7
Atmosphere	Hydrogen (80–82%), helium (17–19%), also ammonia, methane and water vapor
Natural satellites	16
Rings	Yes (1 main ring)
Period of rotation (a Jovian day)	0.413 day
Average distance from Sun	7.78×10^8 km (5.20 AU) [43.25 light-min]
Eccentricity	0.048
Period of revolution around Sun (a Jovian year)	11.86 years
Mean orbital velocity	13.1 km/s
Magnetosphere	Yes (intense)
Radiation belts	Yes (intense)
Mean surface temperature (at cloud tops)	$-123°$C ($-190°$F)
Solar flux at planet (at top of atmosphere)	50.6 W/m^2 (at 5.2 AU)
Earth-to-Jupiter distances	
Minimum (Jupiter in opposition to Sun)	6.0×10^8 km (4.0 AU) [33.3 light-min]
Maximum (Jupiter at conjunction with Sun)	9.6×10^8 km (6.4 AU) [53.4 light-min]

SOURCE: NASA.

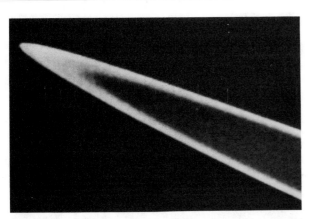

Fig. 2 An exciting view of Jupiter's main ring structure as captured by *Voyager 2*, July 1979. (Courtesy of NASA/Jet Propulsion Laboratory.)

in Jupiter's upper atmosphere and observed auroral emissions in the Jovian polar regions that appeared similar to the northern and southern lights occurring here on Earth.

Voyager 1 produced the first clear evidence of a ring encircling Jupiter. Photographs returned by this spacecraft and its sistership (*Voyager 2*) showed a narrow ring too

faint to be observed with Earth-based telescopes (see fig. 2).

Jupiter is the fifth planet from the Sun and is separated from the terrestrial planets by the main asteroid belt. The giant planet rotates at a dizzying pace—once every 9 hours, 55 minutes and 30 seconds. It takes Jupiter almost 12 Earth-years to complete a journey around the Sun. Its mean distance from the Sun is 5.2 astronomical units (AU), or 7.78×10^8 kilometers. When in opposition to the Sun, Jupiter is about 6.0×10^8 kilometers from Earth, while at conjunction this distance is about 9.6×10^8 kilometers. Table 1 provides a summary of the physical and dynamic characteristics of this giant planet.

A new NASA mission to Jupiter, called Project *Galileo*, was successfully launched on October 18, 1989 and will arrive at Jupiter in 1995. It consists of an atmospheric probe and an orbiting spacecraft to help unlock even more of the secrets of this interesting planet and its many moons.

THE MOONS OF JUPITER

Jupiter has 16 known moons, four of which (Io, Europa, Ganymede and Callisto) are called the Galilean satellites. Very little was actually known about the Jovian moons until the *Pioneer 10* and *11* and *Voyager 1* and *2* spacecraft

Table 2 Properties of the Moons of Jupiter

Moon	Diameter (km)	Semimajor Axis of Orbit (km)	Period of Rotation (days)
Sinope	30 (approx)	23,700,000	758 (retrograde)
Pasiphae	40 (approx)	23,300,000	735 (retrograde)
Carme	30 (approx)	22,350,000	692 (retrograde)
Ananke	20 (approx)	20,700,000	617 (retrograde)
Elara	80 (approx)	11,740,000	260.1
Lysithea	20 (approx)	11,710,000	260
Himalia	180 (approx)	11,470,000	251
Leda	10 (approx)	11,110,000	240
Callisto[a]	4,820	1,880,000	16.70
Ganymede[a]	5,270	1,070,000	7.16
Europa[a]	3,130	670,900	3.55
Io[a]	3,640	422,000	1.77
Thebe (1979J2)	80 (approx)	222,000	0.675
Amalthea	270 (approx)	181,300	0.489
Adrastea (1979J1)	40 (approx)	129,000	0.298
Metis (1979J3)	40 (approx)	127,900	0.295

[a] Galilean satellite.

SOURCE: NASA.

Table 3 Physical Data for the Galilean Moons

Galilean Moon	Diameter (km)	Mass (kg)	Density (g/cm³)	Albedo
Callisto	4,820	1.07×10^{23}	1.8	0.17
Ganymede	5,270	1.49×10^{23}	1.9	0.43
Europa	3,130	0.487×10^{23}	3.0	0.64
Io	3,640	0.891×10^{23}	3.5	0.63

SOURCE: NASA.

encountered Jupiter between 1973 and 1979. These flybys provided a great deal of valuable information and imagery, including the discovery of three new moons (Metis, Adrastea and Thebe), active volcanism on Io, and a possible liquid water ocean underneath the icy surface of Europa. Tables 2 and 3 provide physical and dynamic data for these Jovian moons.

The outermost eight Jovian satellites are quite tiny, rocky objects, ranging in diameter from about 10 to 180 kilometers (6.2 to 112 miles). Four of these satellites (Leda, Himalia, Lysithea and Elara) orbit Jupiter in the same direction as the Galilean satellites, but inclined to the giant planet's equator by between 25 and 29 degrees. The outermost four moons (Ananke, Carme, Pasiphae and Sinope) are in a retrograde orbit around the planet (i.e., they orbit in a direction opposite to Jupiter's direction of rotation). Scientists currently think that all eight of these outer moons are fragments of larger celestial bodies that were captured (by Jupiter's gravitational field) from the asteroid belt.

Callisto is the outermost and least reflective of the Galilean moons. Its very heavily cratered surface indicates that this moon has probably undergone very little change since it was formed (see fig. 3). Its most prominent surface feature is a huge impact basin, named Valhalla. Valhalla has a bright central zone some 600 kilometers (372 mi) across, which is surrounded by numerous concentric rings extending outward for nearly 2,000 kilometers (1,242 mi) from the center.

Ganymede is not only Jupiter's largest moon, but the

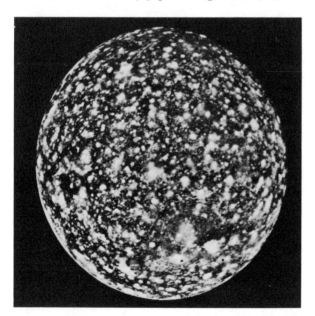

Fig. 3 An image of Callisto taken by *Voyager 2* at a range of 1.1 million kilometers. Scientists think that Callisto's surface is a mixture of ice and rock that dates back to the final stages of planetary formation (over 4 billion years ago)—when its surface was bombarded by a torrent of meteorites. Younger craters show up as bright spots, probably because they expose fresh ice. (Courtesy of NASA.)

Fig. 4 Europa, smallest of the Galilean moons, as imaged by *Voyager 2* at a distance of 241,000 kilometers. (Courtesy of NASA.)

largest moon in our Solar System. The Saturnian moon, Titan, is slightly smaller with a diameter of 5,150 kilometers (3,200 mi). Ganymede's estimated density of 1.9 grams per centimeter-cubed (g/cm^3) suggests that it, like Callisto, has a bulk composition of about 50 percent water by weight. *Voyager* images indicate that approximately half of Ganymede's surface is exposed (water) ice and half is darker rock.

Europa is the smallest and most reflective of the Galilean satellites. This Jovian moon has a very smooth, uncratered surface composed primarily of water ice (see fig. 4). In fact, the elevation difference between the lowest and the highest points on this unusual world is estimated to be less than 100 meters (330 ft).

One of the most exciting possibilities, strongly hinted at by the *Voyager* data, is that Europa may have an ocean of liquid water beneath its surface layer of ice and frost. Figure 5 shows a possible structural model for Europa, including this liquid water ocean. The *Voyager* imagery depicted Europa as a world covered with cracked ice sheets. This moon is composed of rock and ice, with the denser silicate rock thought to form a core. Remnant internal heat from radioactive decay or tidal pumping (or both) may allow this water ocean to exist as a liquid at some depth beneath the protective ice surface. As shown in figure 5, when a crack in the ice occurs, the liquid water will flash-boil into the vacuum of space, and then it will fall back on Europa as frost.

Some exobiologists now speculate that if the Europan ocean exists, it may also contain extraterrestrial life-forms! They assume that Europa, like its parent planet Jupiter, had methane, ammonia and water all present as primordial

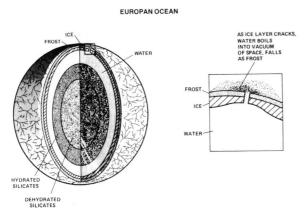

EUROPAN OCEAN

Fig. 5 A model of Europa—including a possible ocean of liquid water beneath its outer layer of ice. (Courtesy of NASA.)

volatiles. An earlier, much warmer Jupiter may have stimulated the chemical evolution of life in Europa's ancient ocean. Then, as Jupiter cooled and the environmental conditions on Europa changed, these alien life-forms, once started, may have tenaciously evolved into hardier creatures that could now be lurking in the depths of this extraterrestrial ocean. These Europan life-forms—should they exist—might cluster around the cracks or thin spots in the surface ice, desperately trying to gather the feeble but life-supporting rays of the Sun and even "planetshine" from nearby Jupiter. (At Jupiter's distance from the Sun, the solar flux is only about 1/27th what it is above the Earth's atmosphere.)

Perhaps submarine volcanic activity on Europa provides the energy necessary to heat sulfur compounds that are then used by extraterrestrial microorganisms in a process called chemosynthesis. (Chemosynthesis is a chemical parallel to photosynthesis.) Exobiologists are quick to point out that life, once started on Earth, has now spread over the planet to even the most hostile locations, such as under the polar ice pack and in the darkest depths of the oceans. They therefore reason that it is possible that alien life, once initiated billions of years ago in a more benign ancient Europan ocean, may still cling to existence in the current dark, watery world beneath the moon's protective layer of surface ice. Of course, only additional exploration will resolve this most intriguing line of speculation.

The innermost of the Galilean satellites, Io, is the most colorful and active of the Jovian moons. Like the Earth and Venus, Io has active volcanoes, which send sulfur-laden plumes as high as 300 kilometers (186 mi) above its surface. Scientists estimate that upwards of 10 billion tons of material erupt from Io's interior each year, enough to coat Io's entire surface annually with a fresh layer of sulfur-rich materials—creating its bright red, orange, yellow and black ("pizza-colored") appearance. This volcanism on Io is driven by the interior heating caused by gravitationally induced tidal stresses within its crusts. These stresses result from Io's close elliptical orbit around Jupiter and, as some sci-

entists now speculate, may eventually lead to Io's complete melting "from the inside out."

Finally, three of the four innermost Jovian moons (Metis, Adrastea and Thebe) were discovered as a result of the *Voyager 1* spacecraft encounter in 1979. The other tiny inner moon, Amalthea, was discovered in 1892 by Earth-based telescopic observation. Metis and Adrastea orbit just outside Jupiter's recently discovered "main ring" (see fig. 2), while Thebe orbits Jupiter outside of Amalthea. All of these inner satellites are small, nonspherical rocky objects. Amalthea is the largest, with irregular dimensions of 270 by 165 by 150 kilometers (168 by 103 by 93 mi).

See also: **Pioneer 10, 11; Project Galileo; Voyager**

K

Kardashev civilizations The Soviet astronomer, N. Kardashev, in describing the possible technology levels of various alien civilizations around distant stars, distinguished three types or levels of extraterrestrial civilizations. He used the civilization's overall ability to manipulate and harness energy resources as the prime comparative figure of merit. A Kardashev Type I civilization would be capable of harnessing the total energy capacity of its home planet; a Type II civilization, the energy output of its parent star; while a Type III civilization would be capable of using and manipulating the energy output of their entire galaxy.

See also: **extraterrestrial civilizations**

Kirkwood gaps Gaps or "holes" in the main asteroid belt between Mars and Jupiter where essentially no asteroids are located. These gaps, initially explained in 1857 by Daniel Kirkwood, an American astronomer (1814–95), are caused by the gravitational attraction of Jupiter. They occur at distances from the Sun that correspond to orbital periods that are harmonics (that is, $\frac{1}{2}$, $\frac{2}{5}$, $\frac{1}{3}$, $\frac{1}{4}$, and so on) of Jupiter's orbital period.

See also: **asteroid**

L

Laboratory hypothesis A variation of the Zoo hypothesis response to the Fermi paradox. This particular hypothesis postulates that the reason we cannot detect or interact with technically advanced extraterrestrial civilizations in the Galaxy is because they have set the Solar System up as a "perfect" laboratory. These hypothesized extraterrestrial experimenters want to observe and study us but do not want us to be aware of or influenced by their observations.

See also: **Fermi paradox; Zoo hypothesis**

Lagrangian libration points   The five points in outer space where a small object can have a stable orbit in spite of the gravitational attractions exerted by two much more massive celestial objects when they orbit about a common center of mass. The existence of such points was first postulated by the French mathematician Joseph Louis Lagrange (1736–1818). The Trojan group of asteroids, which occupy such Lagrangian points 60 degrees ahead of and 60 degrees behind the planet Jupiter in its orbit around the Sun, are one example.

In cislunar space—that is, the region of space associated with the Earth–Moon system—the five Lagrangian points arise from a balancing of the gravitational attractions of the Earth and Moon with the centrifugal force that an observer would feel in the rotating Earth–Moon coordinate system. The main feature of these points or regions in cislunar space (see fig. 1) is that an object placed there will keep a fixed relation with respect to the Earth and Moon as the entire system revolves around the Sun.

The Lagrangian points called L_1, L_2 and L_3 in figure 1, are saddle-shaped "gravity valleys," with the interesting property that if you move an object at right angles (that is, perpendicularly) to a line connecting the Earth and the Moon (called the Earth–Moon axis), this object will slide back toward the axis; however, if you displace the object along this axis, it will move away from the Lagrangian point indefinitely. Because of this, these three Lagrangian points are called points of unstable equilibrium.

In contrast, the Lagrangian points L_4 and L_5 present bowl-shaped "gravity valleys." If an object at L_4 or L_5 is moved slightly in any direction, it returns to the Lagrangian point. These two points are therefore called points of stable

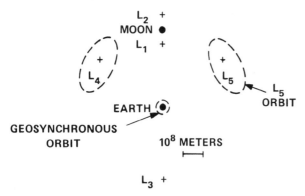

Fig. 1   Five Lagrangian libration points in the Earth–Moon system. (Drawn by the author.)

equilibrium. Lagrangian points L_4 and L_5 are located on the Moon's orbit about the Earth, at equal distance from both the Moon and the Earth. They have been proposed as the sites for large human settlements in cislunar space.

large space structures   The building of very large and complicated structures in space will be one of the hallmarks of our extraterrestrial civilization. Although enormous, these structures will actually help to shrink the total cost of using space, and their construction and operation will support the "humanization of space." Space transportation costs are assessed by both the mass of the payload and its volume. So it is very wise to design space hardware that is both light and modest in dimensions. Structures that would have been too fragile to stand up under their own weight on Earth can now be compactly stored in a compact payload container and then safely deployed in their final, extensive configuration in the microgravity environment of space.

The new capability to supervise deployment and construction operations in orbit is a crucial factor in the effective use of space. The Space Shuttle can carry a work force of up to seven astronauts per flight and will remain close at hand while the early construction jobs are performed.

All of this creates exciting new possibilities for the engineering of space hardware and poses an entirely new set off technical challenges to space technologists. What are the strongest, lightest and most stable materials to use in space construction? How do you load a launch vehicle so as to build these colossal objects with the fewest trips into space? What are the best ways to assemble these structures once the materials are delivered to the orbiting "construction sites"? And perhaps the most obvious question: What kinds of structures will we want to build?

Television viewers in the remote regions of Alaska, the Rocky Mountains and Appalachia rapidly joined the world of modern space-based electronic communications and entertainment in 1974. The reason was simple—ATS-6 (standing for Advanced Technology Satellite-6), a nine-meter-wide dish antenna that relayed TV signals down to small receivers in previously isolated areas of the United States, was successfully launched. ATS-6 had the largest civilian communications antenna launched to date.

Some of the new superantennas now being considered will be 10 times that size, or bigger than a football field! This tremendous size means a boost in transmitting power as well as increased sensitivity to weak signals from the ground. As a result, instead of needing massive antenna dishes on the Earth straining to hear weak messages from space as we do today, the roles will be reversed. (This reversal of roles is also called a technology inversion.) A few giant antennas placed in geostationary orbits covering the globe will replace numerous smaller communications satellites now in space. And millions of inexpensive satellite-receiving dishes on the rooftops of homes and businesses will receive signals now picked up by very large and powerful ground stations. The true "electronic cottage" will have been born.

The implications of this coming boom in antenna performance for change in our daily lives are profound. Many exciting ideas and technologies consistent with this "large-antenna revolution" already exist. These include working models of personal "Dick Tracy" wrist radios, designs for electronic mail systems with direct home delivery via satellite and 300-channel television sets tuned to stations all over the globe.

And there will be other exciting and vital space-based information services economically emerging as new space industries as a result of large antenna farms and platforms in orbit. For example, figure 1 shows the large antenna system of one proposed design for an Earth Observation Spacecraft (EOS) being assembled in orbit. When folded, this system would be only 4.1 meters wide by 17.8 meters long and could easily be carried in the Space Shuttle payload bay. However, when erected in orbit, it is actually some 120 meters by 60 meters and has a 116-meter-long mast. This advanced antenna system is designed to observe Earth resources, including measuring soil moisture, sea-surface conditions, and ice and its boundaries.

The first large space antennas will most likely be deployables. They will fold into compact containers on Earth, go up whole in one launch vehicle flight, then deploy automatically in space in a single operation. The key, obviously, is to have the largest possible antenna dish unfolding from the smallest and lightest possible package.

One type of deployable—the hoop-column or "may-pole," antenna—would open up once in orbit much like an umbrella does. A cylinder no bigger than a school bus can be transformed within an hour into a huge antenna dish 100 meters across. Depending on the length of the various strings that stretch the fabric taut inside its stiff outer hoop, this type of antenna can be designed into many shapes—that is, the bowl of the dish could be made flat, more hollowed out or even made of four different surfaces, each

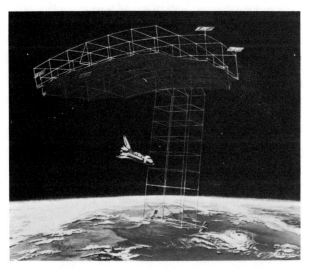

Fig. 1   In-orbit assembly of an intricate antenna system for one proposed Earth Observation Spacecraft (EOS). (Courtesy of NASA.)

focusing a beam in its own different direction. Multibeam feeds could also allow one antenna to do the work of several by pointing signals toward different areas of the Earth's surface.

In another type of deployable antenna, called the offset wrap-rib type, the dish fabric is attached t flexible ribs that wrap around a central hub. The whole package is quite compact initially, but once in space, another marvelous transformation in size takes place. A long (some 150 meters for a 100-meter-diameter antenna dish) mast telescopes out from the core and turns a corner so that the dish is offset and not blocked by the mast. This is an advantage in sensitive radar and radiometry missions. Then, like a pinwheel coming to life, the ribs unfurl and straighten until they fully extend to stretch and support a round dish.

Whatever their ultimate shape, these large space structures will place great demands on the materials from which they are made. Even though they will be free from the weight stresses imposed by gravity on the surface of the Earth, there will be other strains from their tight packaging and from the space environment itself. Space technologists and engineers will need to build these structures with new materials—materials that are at the same time light, very strong, thermally stable and either flexible or rigid (depending on their application). Telescoping masts must be light, yet remain very stiff. Antenna ribs, on the other hand, need to be strong but should be flexible enough to wrap around their hub. Furthermore, the configuration needs to remain fixed in position equally well in the hot Sun as in the frigid shadows encountered in orbit, because if a structure were to expand with heat or shrink with cold, it would upset the extremely precise shape of an antenna (some of which can be out of tolerance no more than a few millimeters in a total diameter of 100 meters).

One substance that appears to meet these rigorous demands quite well is the graphite–epoxy composite now used in lightweight tennis rackets, golf clubs, airplane parts and the Space Shuttle. A three-meter-long hollow tube of this material can be lifted with one finger yet, in its particular applications, is ten times stronger than steel.

Other materials are tailored for specific jobs. The hundreds of threads that pull and stretch a hoop-column antenna into shape might be made of quartz filament, because quartz is very stable. The antenna dishes themselves can be made of fabrics that fold like cloth before they are deployed. These would be metal meshes woven like nylon stockings or soft patio screening and coated with a very thin layer of gold for reflectivity. A finer mesh will be used for dishes that handle smaller wavelengths. For very small wavelengths there are ultrathin membranes made of transparent films coated with metals that look and feel like sheets of Christmas tinsel.

Let us now imagine that six different groups want to fly different remote sensing instruments in Earth orbit, all at about the same altitude and inclination. Instead of crowding and cluttering low Earth orbit with six different spacecraft, why not just build one large platform to which all six instruments are attached? They could then share the cost of the power and communications systems, stability control and cooling devices. Shuttle astronauts or space station workers would need to visit just one place in space (instead of six) to repair and maintain the systems. This space platform appears to be a technically sound and economically good idea.

Some of these large space platforms, especially those dedicated to communications and information services, will need to hover in geostationary Earth orbit (GEO) about 35,900 kilometers high in order to look down on large sections of the globe or to stay fixed in one spot (as seen by an observer on Earth). Since the Space Shuttle Orbiter ascends no higher than a few hundred kilometers above the surface of the Earth, orbital transfer vehicles must be attached either to an undeployed package (as taken right out of the Orbiter's cargo bay) or to an already assembled structure to boost it to a higher orbit.

Eventually, no matter how cleverly the platforms and antennas are packed, they will be too large to unfold in a single deployable unit. At that point, we will have to send up "erectable" space structures in pieces. These pieces can be loaded in the Shuttle's cargo bay on Earth, lifted into low Earth orbit, unfolded and finally assembled by space workers into a single, giant structure.

What kinds of building blocks will we use on these floating construction devices? Ideally, they should be basic, simple and adaptable to many different types of structure. These "erectables" have their roots in common household objects—in collapsible cardboard boxes, folding chairs, telescoping car radio antennas, accordion baby gates—anything we have tried to make smaller and more portable. Masts for dish antennas will telescope into their full lengths from small cylinders. Latticed trusses will store as flat packages, unfold first into diamond shapes and then finally into tetrahedrons.

But in each case, no matter how flexible their hinges when stored, the modules must hold stiff when deployed in space, as would the hexagonal pieces for large antennas. Looking a bit like minitrampolines when unfolded, these hexagons will be attached precisely and rigidly to form great reflecting surfaces many city blocks in area.

Not all of these building blocks will need to unfold. Some of them will store quite easily just as they are, such as the light graphite-epoxy tubes that will stack inside one another like ice cream cones and sit on racks like arrows in a quiver. These tubes would then be attached to form struts that can themselves be joined to build larger beams or trusses. Or they might be used to form a thin hoop for a space antenna.

Unfolding with the push of a button, deployable antennas will, in a sense, build themselves. "Erectables," on the other hand, will not. Space workers or very smart automated machines controlled by astronauts will have to snap the separate pieces together. Ongoing assembly projects will therefore mean having the first construction sites in space, giving rise to a new type of work for human beings.

Many important factors must be taken into account in planning and conducting a space construction project. These include safety and fatigue of the space workers; speed in moving from one space location to another; the requirement for simple tools and the need to restrain them so they don't float away; and how much time the Space Shuttle or manned orbital transfer vehicle would lose lingering at the construction site, if the site is distant from a space station or space base.

In one method of assembly, space workers tethered to the Space Shuttle or space station would simply move from beam to column to module, snapping, locking or latching everything together. Their travel time could be shortened by wearing the jet-packs called manned maneuvering units (MMUs). It is not yet, however, entirely certain how we will combine space worker and sophisticated machine in space deployment and assembly operations. For some projects it might be more efficient to move the astronauts around on a scaffold in some type of mobile work station instead of having them free-flying all over the construction site. The scaffold would rest on a frame in the Shuttle Orbiter's cargo bay or as a part of the space station and move either up-down or right-left. As sections of the structure were completed, they would be moved away from the work station so that the part to be built is always in reach. Astronauts could also stand in open cherry pickers attached to the Shuttle's 15-meter remote manipulator arm or a similar manipulator system on the space station and be moved from beam joint to beam joint like telephone line workers working on high wires. Even more sophisticated operations would involve the use of closed cherry pickers, where space workers inside a comfortable chamber would operate sophisticated manipulator arms in a highly mobile space system.

Or, for repetitive and perhaps very dangerous construction tasks, unmanned free-flying teleoperators—essentially smart robots—could do the work with their own dexterous manipulator arms. There could also be assembler devices to form three-dimensional structures from struts by following simple, repeatable steps and maneuvering television units that would transmit pictures to technicians in the Shuttle or space-station control room so that they could direct the assembly work by remote control. These advanced construction devices would most likely be used early in the next century, perhaps in conjunction with large, permanently inhabited space platforms. In the meantime, astronauts will have to learn to erect structures the size of large stadiums in the unusual and challenging world of microgravity. Seemingly easy tasks will become complicated. For example, workers trying to turn ordinary bolts will be as likely to turn themselves as the bolts, thanks to the lack of leverage that accompanies the free-fall condition of orbiting objects.

After deployable and erectable systems, the next logical step will be to build large structures completely from scratch by fabricating the construction elements in space. A machine for that very purpose has already been considered. Called the automated beam builder, this device would heat, shape and weld the material into meter-wide triangular beams that might be cut to any length, then latched together to build large structures. With the beam builder, the dreams for humanity's extraterrestrial civilization as found in the science-fiction literature would be converted to practical blueprints for colossal structures.

As such platforms and structures grow in size, they will become even more complicated. The need to control and maintain a perfectly fixed attitude is crucial to antennas and remote sensing instruments, which would be useless unless pointed precisely. This means that these mammoth structures will not be able to wobble or bend out of alignment. Several things will combine to distort their orbital positions, because as large as these structures will be, they will also be relatively light and delicate. For example, a 50-meter-diameter space antenna would have a mass of approximately 4.5 metric tons. An entire large space structure could easily be pushed out of alignment by the steady, streaming pressure of the solar wind; in addition, every time a spacecraft docked or made physical contact with one of these large platforms, the delicate balance of forces controlling an orbiting object would again be upset. Some future structures will be so very large and extensive that they will even experience "tidal effects," as if they were minimoons with gravity (Earth's in particular) tugging harder on one edge than on the other.

Obviously, to successfully build and use such large structures in space, we will need precise and sophisticated controls for stability, starting with sensors to indicate just when the structure is moving out of alignment. On-board computers could then determine how to compensate; and finally, small gas jets located around the structure would fire to make the necessary corrections. All of this would be a constant, self-regulating process in an age when mammoth space structures served humanity throughout cislunar space.

See also: **space construction; space industrialization; space station**

life in the Universe   The history of life in the Universe can be explored in the context of a grand, synthesizing scenario called *cosmic evolution* (see fig. 1). This sweeping scenario links the development of galaxies, stars, planets, life, intelligence and technology and then speculates on where the ever-increasing complexity of matter is leading.

And where *does* all this lead to in the cosmic-evolution scenario? Well, we should first recognize that all living things are extremely interesting pieces of matter. Lifeforms that have achieved intelligence and have developed technology are especially interesting and valuable in the cosmic evolution of the Universe! Intelligent creatures with technology, including human beings here on Earth, can exercise conscious control over matter in progressively more effective ways as the level of their technology grows. Ancient cave dwellers used fire to provide light and water. Modern humans harness solar energy, control falling water and split atomic nuclei to provide energy for light, warmth, industry and entertainment. People in the next century

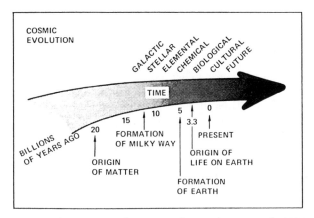

Fig. 1   The scenario of cosmic evolution. (Courtesy of NASA [from the work of Eric J. Chaisson].)

will "join atomic nuclei" (controlled fusion) to provide light, warmth, industry, entertainment and interplanetary communications for their settlements on the Moon and Mars! The trend should be obvious. Some scientists even speculate that if technologically advanced civilizations throughout the Galaxy can learn to live with the awesome powers unleashed in such advanced technologies, then it may be the overall destiny of advanced intelligent life-forms (including perhaps humans) to ultimately exercise control over all the matter in the Universe!

According to modern scientific theory, living organisms arose naturally on the primitive Earth through a lengthy process of *chemical evolution* of organic matter. This process began with the synthesis of simple organic compounds from inorganic precursors in the atmosphere; continued in the oceans, where these compounds were transformed into increasingly more complex organic substances; and then culminated with the emergence of organic microstructures that had the capability of rudimentary self-replication and other biochemical functions.

Human interest in the origins of life extends back deep into antiquity. Throughout history, each society's "creation myths" seem to reflect that particular people's view of the extent of the Universe and their place within it. Today, in the Space Age, the scope of those early perceptions has expanded well beyond the reaches of our Solar System to the stars and interstellar clouds that populate the seemingly limitless expanse of outer space. Just as the concept of *biological evolution* implies that all living organisms have arisen by divergence from a common ancestry, so too the concept of *cosmic evolution* implies that all matter in our Solar System has a common origin. Following this line of reasoning, scientists now postulate that life may be viewed as the product of countless changes in the form of primordial stellar matter—changes brought about by the interactive processes of astrophysical, cosmochemical, geological and biological evaluation.

If we use the even larger context of cosmic evolution, we can further conclude that the chain of events that led to the origins of life here on Earth extends well beyond planetary history: to the origin of the Solar System itself, to processes occurring in ancient interstellar clouds that spawned stars like our Sun, and ultimately to the very birth within these stars (through nucleosynthesis) of the elements that make up living organisms—the *biogenic elements*. The biogenic elements are those generally judged to be essential for all living systems. Scientists currently place primary emphasis on the elements hydrogen (H), carbon (C), nitrogen (N), oxygen (O), sulfur (S) and phosphorous (P). The compounds of major interest are those normally associated with water and with organic chemistry, in which carbon is bonded to itself or to other biogenic elements. The essentially "universal" presence of these compounds throughout interstellar space gives exobiologists the scientific basis for forming the important contemporary hypothesis that *the origin of life is inevitable throughout the cosmos wherever these compounds occur and suitable planetary conditions exist.* Present-day understanding of life on Earth leads modern scientists to the conclusion that life originates on planets and that the overall process of biological evolution is subject to the often chaotic processes associated with planetary and solar system evolution (e.g., the random impact of a comet on a planetary body or the unpredictable breakup of a small moon).

Scientists now use four major epochs in describing the evolution of living systems and their chemical precursors. These are:

(1) The cosmic evolution of biogenic compounds—an extended period corresponding to the growth in complexity of the biogenic elements from nucleosynthesis in stars, to interstellar molecules, to organic compounds in comets and asteroids.

(2) Prebiotic evolution—a period corresponding to the development (in planetary environments) of the chemistry of life from simple components of atmospheres, oceans and crustal rocks, to complex chemical precursors, to initial cellular life forms.

(3) The early evolution of life—a period of biological evolution from the first living organisms to the development of multicellular species.

(4) The evolution of advanced life—a period characterized by the emergence of progressively more advanced life forms, climaxing perhaps with the development of intelligent beings capable of communicating, using technology and exploring and understanding the Universe within which they live.

As scientists now unravel the details of this process for the chemical evolution of terrestrial life, we must also ask ourselves another very intriguing question: If it happened here, did or can it happen elsewhere? In other words, what are the prospects for finding extraterrestrial life—in this Solar System or perhaps on Earthlike planets around distant stars?

According to the principle of mediocrity (frequently used by exobiologists), there is nothing "special" about the Solar System or the planet Earth. Therefore, as these scientists speculate, if similar conditions have existed or are now present on "suitable" worlds around alien suns, the chemical evolution of life will also occur.

Perhaps a more demanding question is: Does alien life develop to a level of intelligence? And even more interesting: Do intelligent alien life-forms acquire advanced technologies and learn to live with these vast powers over nature? It is just possible that we here on Earth are the only life-forms anywhere in the Galaxy to acquire intelligence and high technology! Just think, for a moment, about the powerful implications of this simple conjecture. Are we the best the Universe has been able to produce in over 15 billion years of cosmic evolution? If so, a human being—any human being—is something very special.

Our preliminary search on the Moon and on Mars for extraterrestrial life in the Solar System has to date been unsuccessful. However, the giant, gaseous outer planets and their constellations of intriguing moons present some tantalizing possibilities. Who cannot get excited about a possible ocean of liquid water beneath the Jovian moon Europa and the (remote) chance that this extraterrestrial ocean might contain alien life-forms! Of course, even the final verdict concerning life on Mars (past or present) will not be properly resolved until more detailed investigations of the Red Planet have occurred. Perhaps a terrestrial explorer will stumble upon a remote exobiological niche in some deep Martian canyon; or possibly a team of miners, searching for certain ores on Mars, will uncover the fossilized remains of an ancient creature that roamed the surface of the Red Planet in more hospitable environmental eras! Speculation, yes—but not without reason.

All we can say now with any degree of certainty is that our overall understanding of the cosmic prevalence of life will be significantly influenced by the exobiological discoveries (pro and con) that will occur in the next few decades. In addition to looking for extraterrestrial life on other worlds in our Solar System, exobiologists can also search for life-related molecules in space in order to determine the cosmic nature of prebiotic chemical synthesis.

Recent discoveries, for example, show that comets appear to represent a unique repository of information about chemical evolution and organic synthesis at the very outset of the Solar System. (For example, after reviewing Comet Halley encounter data, space scientists think that comets have remained unchanged since the formation of the Solar System.) Exobiologists now have evidence that the organic molecules considered to be the molecular precursors to those essential for life are prevalent in comets. These discoveries have provided further support for the hypothesis that the chemical evolution of life has occurred and is now occurring widely throughout the Galaxy. Some scientists even suggest that comets have played a significant role in the chemical evolution of life on Earth. They hypothesize that significant quantities of important life-precursor molecules could have been deposited in an ancient terrestrial atmosphere by cometary collisions.

Meteoroids are solid chunks of extraterrestrial matter. As such, they represent another source of interesting information about the occurrence of prebiotic chemistry beyond the Earth. In 1969, for example, meteorite analysis provided the first convincing proof of the existence of extraterrestrial amino acids. (Amino acids are a group of molecules necessary for life.) Since that time, a large amount of information has been gathered showing that many more of the molecules considered necessary for life are also present in meteorites. As a result of this line of investigation, it now seems clear to exobiologists that the chemistry of life is not unique to the Earth. Future work in this area should greatly help our understanding of the conditions and processes that existed during the formation of the Solar System. These studies should also provide clues concerning the relations between the origin of the Solar System and the origin of life.

The basic question, Is life—especially intelligent life—unique to the Earth? lies at the very core of our concept of self and where we fit in the cosmic scheme of things. If life is precious and rare, then we have a truly serious obligation to the entire (as yet "unborn") Universe to carefully preserve the organic heritage that has taken over 4 billion years to evolve on this tiny planet. If, on the other hand, life (including intelligent life) is abundant throughout the Galaxy, then we should eagerly seek to learn of its existence. Fermi's famous question, "Where are they?" then takes on even more significance in the Age of Space.

See also: **amino acid; Drake equation; exobiology; extraterrestrial civilizations; extraterrestrial contamination; Fermi paradox; Viking Project**

light flash A momentary flash of light seen by astronauts in space, even with their eyes closed. Scientists believe that there are probably at least three causes for these light flashes. First, energetic cosmic rays passing through the eye's "detector" (the retina) ionize a few atoms or molecules, resulting in a signal in the optic nerve. Second, HZE particles can produce Cerenkov radiation in the eyeball. (Cerenkov radiation is the bluish light emitted by a particle traveling very near the speed of light when it enters a medium in which the velocity of light is less than the particle's speed.) Finally, alpha particles from nuclear collisions caused by very energetic Van Allen belt protons can produce ionization in the retina, again triggering a signal in the optic nerve. Astronauts have reported seeing these light flashes in a variety of sizes and shapes.

See also: **cosmic rays; Earth's trapped radiation belts; hazards to space workers; HZE particles**

light-minute (lm) A unit of length equal to the distance traveled by a beam of light (or any electromagnetic wave) in the vacuum of outer space in one minute. Since the speed of light (c) is 299,792.5 kilometers per second (km/sec) in free space, a light-minute corresponds to a distance of approximately 18 million kilometers.

light-second (ls) A unit of length equal to the distance traveled by a beam of light (or any electromagnetic wave) in the vacuum of outer space in one second. Since the speed of light (c) in free space is 299,792.5 kilometers per second (km/sec), a light-second corresponds to a distance of approximately 300,000 kilometers.

light-year (ly) The distance that light travels at 3×10^8 meters per second in one year (3.15×10^7 seconds). One light-year is equal to a distance of 9.46×10^{12} kilometers, or about 63,000 times the distance from the Earth to the Sun.

Local Group A small grouping of about 20 galaxies, of which the Milky Way (our Galaxy) and the Andromeda Galaxy are dominant members.
See also: **cluster of galaxies; galaxy**

lunar Of or pertaining to the Moon.

lunar bases and settlements When human beings return to the Moon, it will not be for a brief moment of scientific inquiry as occurred in the Apollo program, but rather as permanent inhabitants of a new world. They will build bases from which to completely explore the lunar surface, establish science and technology laboratories that take advantage of the special properties of the lunar environment and exploit the Moon's mineral resources in support of humanity's extraterrestrial expansion.

In the first stage of one possible lunar-development scenario, men and women, along with their smart machines, would return to the Moon to conduct more extensive site explorations and resources evaluations. These efforts would pave the way for the first permanent lunar base.

The next critical stage in humanity's use of the Moon will be the establishment of the first permanent lunar base camp. In this base a team of from 10 up to perhaps 100 "permanent" lunar workers will set about the task of fully investigating the Moon. They will take particular advantage of the Moon as a "science in space platform" and perform the fundamental scientific and engineering studies needed to confirm the specific roles that the Moon will play in the full development of cislunar space. For example, the discovery of frozen volatiles (including water) in the perpetually frozen recesses of the Moon's polar regions could change strategies for orbital-transfer-vehicle and space-station resupply as well as open up new, more rapid pathways for the development of a full-scale lunar civilization.

Many lunar base and settlement applications, both scientific and industrial, have been proposed since the Apollo program. Some of these concepts include (1) a lunar scientific laboratory complex; (2) a lunar industrial complex to support space-based manufacturing; (3) an astrophysical observatory for Solar-System and deep-space surveillance; (4) a "fueling" station for orbital transfer vehicles that travel throughout cislunar space; (5) a training site and assembly point for manned Mars expeditions; (6) a nuclear waste repository for the very long-lived radioisotopes (such as the transuranic nuclides) originating in terrestrial nuclear fuel cycles and for spent space nuclear power plants used throughout cislunar space; and (7) the site of innovative political, social and cultural developments—essentially rejuvenating humanity's concept of itself and its ability to change its destiny.

All these lunar base and settlement applications are very exciting and definitely deserve expanded study—especially in the context of a permanent space station in low Earth orbit. The question space planners are already beginning to ask is: Where do we go from a space station in low Earth orbit? The two most popular responses are the Moon or Mars. A human expedition to Mars can well be served by the capabilities and technologies associated with a flourishing lunar base complex.

As lunar activities mature, the initial lunar base will grow into an early settlement of about 1,000 more or less "permanent" residents. Then, as the lunar industrial complex expands and lunar raw materials, food and manufactured products start to support space industrialization throughout cislunar space, the lunar settlement itself will expand to a population of around 10,000, with electric energy demands on the order of several megawatts.

Mature settlements on the Moon will continue to grow, reaching a population of about 500,000 and attaining a social and economic "critical mass" that supports true self-sufficiency from Earth. This moment of self-sufficiency for the lunar civilization will also be a very historic moment in human history. For from that time on, the human race will exist in two distinct and separate "biological niches"—we will be terran and nonterran, or extraterrestrial.

With the rise of a self-sufficient, autonomous lunar civilization, future generations will have a "choice of worlds" on which to live and prosper. Of course, such a major social development will also produce its share of "cultural backlash." Citizens of the 21st century may start seeing personal ground vehicles with such bumper-sticker slogans as: "This is my world—love it or leave it!"; "Terran go home"; or perhaps "Protect terrestrial jobs—ban lunar imports."

All major lunar-development strategies include the use of the Moon as a platform from which to conduct "science in space." Scientific facilities on the Moon will take advantage of its unique environment to support platforms for astronomical, solar and space science (plasma) observations. The unique environmental characteristics of the lunar surface include low gravity (one-sixth that of the Earth), high vacuum (about 10^{-12} torr [a torr is a unit of pressure equal to 1/760 of an atmosphere]), seismic stability, low temperatures (especially at the poles) and a low radio noise environment on the far side. More advanced astronomical observations will be made from space in the future, mainly to escape from the distortional effects of the Earth's atmosphere and ionosphere.

Astronomy from the lunar surface offers the distinct advantages of a low radio noise environment and a stable platform in a low-gravity environment. The far side of the Moon is permanently shielded from direct terrestrial radio emissions. As future radio-telescope designs approach their ultimate (theoretical) performance limits, this uniquely quiet lunar environment may be the only location in all cislunar space where sensitive radio-wave detection instruments can be used to full advantage, both in radio astronomy and in our search for extraterrestrial intelligence (SETI). In fact, radio astronomy, including extensive SETI efforts, may represent one of the main "lunar industries" of the next century. In a figurative sense, SETI performed by lunar-based scientists will be "extraterrestrials" searching for other extraterrestrials.

The Moon also provides a solid, seismically stable, low-gravity, high-vacuum platform for conducting precise interferometric and astrometric observations. For example, the availability of ultrahigh-resolution (micro-arc-second) optical, infrared and radio observatories will allow us to search for extrasolar planets encircling nearby stars.

A lunar scientific base also provides life scientists with a unique opportunity to extensively study biological processes in reduced gravity (1/6 g) and in low magnetic fields. Genetic engineers can conduct their experiments in comfortable facilities that are nevertheless physically isolated from the Earth's biosphere. Exobiologists can experiment with new types of plants and microorganisms under a variety of simulated alien-world conditions. Genetically engineered "lunar plants," grown in special facilities, will become a major food source and also supplement the regeneration of the atmosphere in the lunar habitats.

The true impetus for large, permanent lunar settlements will most likely arise from the desire for economic gain—a time-honored stimulus that has driven much technical, social and economic development on Earth. The ability to provide useful products from native lunar materials will have a controlling influence on the overall rate of growth of the lunar civilization. Some "early lunar products" can now easily be identified. These products would support overall space-industrialization efforts. They include (1) the production of oxygen for use as a propellant for orbital transfer vehicles used throughout cislunar space (for example, low Earth orbit to lunar orbit or lunar orbit to geostationary Earth orbit); (2) the use of "raw" lunar soil and rock materials for radiation shielding—a critical, mass-intensive component of space stations, space settlements and personnel transport vehicles; and (3) the production of ceramic and metal products to support the construction of large structures and habitats in space.

The initial lunar bases can be used to demonstrate industrial applications of native Moon resources and to operate small pilot factories that provide selected raw and finished products for use both on the Moon and in Earth orbit.

The Moon has large supplies of silicon, iron, aluminum, calcium, magnesium, titanium and oxygen. Lunar soil and rock can be melted to make glass—in the form of fibers, slabs, tubes and rods. Sintering (a process whereby a substance is formed into a coherent mass by heating [but without melting]) can produce lunar bricks and ceramic products. Iron metal can be melted and cast or converted to specially shaped forms using powder metallurgy. These lunar products would find a ready "market" as shielding materials, in habitat construction, in the development of large space facilities and in electric power generation and transmission systems.

Lunar mining operations and factories can then be expanded to meet growing demands for lunar products throughout cislunar space. With the rise of lunar agriculture, the Moon may even become our "extraterrestrial breadbasket"—providing the majority of all food products consumed by humanity's extraterrestrial citizens.

One interesting space-industrialization scenario involves an extensive lunar surface mining operation that provides the required quantities of materials in a preprocessed condition to a giant space manufacturing complex located at Lagrangian libration point 4 or 5 (L_4 or L_5). These exported lunar materials would consist primarily of oxygen, silicon, aluminum, iron, magnesium and calcium locked into a great variety of complex chemical compounds. It is anticipated by space-technology visionaries that the Moon will become the chief source of materials for space-based industries in the latter part of the 21st century.

Numerous other tangible and intangible advantages of lunar settlements will accrue as a natural part of their creation and evolutionary development. For example, the high-technology discoveries originating in a complex of unique lunar laboratories could be channeled directly into appropriate economic and technical sectors on Earth, as new ideas, techniques, products and so on. The existence of "another human world" will create a permanent open-world philosophy for all human civilization. Application of space technology, especially lunar-base-generated technology, will trigger a terrestrial renaissance, leading to an overall increase in the creation of wealth, the search for knowledge and even the creation of beauty by all humanity.

Our present civilization—as the first to venture into cislunar space and to create permanent lunar settlements—will long be admired, not only for its great technical and intellectual achievements but also for its innovative cultural accomplishments. Finally, it is not too remote to speculate that the descendants of the first lunar settlers will become first the interplanetary, then the interstellar, portion of the human race! The Moon and its emerging civilization will become our gateway to the Universe.

See also: **Lagrangian libration points; Moon; Satellite Power System; space industrialization; space settlement**

lunar crater A depression, usually circular, on the surface of the Moon. It frequently occurs with a raised rim called a ringwall. Craters range in size up to 250 kilometers in diameter. The largest lunar craters are sometimes called "walled plains." The smaller craters—say, 15 to 30 kilometers across—are often called "craterlets"; while the very smallest, just a few hundred meters across, are called "beads." Many lunar craters have been named after famous people, usually astronomers.

See also: **Moon**

lunar day The period of time associated with one complete orbit of the Moon about the Earth. It is equal to 27.322 "Earth"-days. The lunar day is also equal in length to the sidereal month.

Lunar Observer Are there volatile materials such as water ice trapped in permanently darkened, frigid polar regions of the Moon? The exploitation of the lunar resource base and the development of our extraterrestrial civilization could be greatly enhanced by such a marvelous discovery! NASA's Lunar Observer will help scientists assess the Moon's resources, including a search for frozen volatiles at the poles. It would also measure the Moon's elemental and mineralogical surface composition, surface topography and gravity field on a global basis. This mission can be launched at almost any time from Earth and inserted into an initial elliptical orbit around the Moon for certain very close remote sensing measurements, such as gamma-ray spectroscopy. This would be followed by a one-year operating period in which the spacecraft would travel in a 50- to 100-kilometer circular polar orbit, allowing all the regions of the Moon to be observed and assessed. A 1996 launch date is currently envisioned for this mission.

See also: **lunar bases and settlements.**

lunar rovers Manned and automated (robot) vehicles used to help explore the Moon's surface. The Lunar Rover Vehicle (LRV), shown in figure 1, was also called a space buggy and the "Moon car." It was used by American astronauts during the *Apollo 15, 16* and *17* expeditions to the Moon. This vehicle was designed to climb over steep slopes, go over rocks and move easily over sandlike lunar surfaces. It was able to carry more than twice its own mass (about 210 kilograms) in passengers, scientific instruments and lunar material samples. This electric-powered vehicle could travel about 16 kilometers per hour (10 miles per hour) on level ground. The vehicle's power came from two 36-volt silver-zinc batteries that drove independent ¼-horsepower electric motors in each wheel. Apollo astronauts used their space buggies to explore well beyond their initial lunar-landing sites. With these vehicles, they were able to gather Moon rocks and travel much farther and quicker across the lunar surface than if they had had to explore on foot. For example, during the *Apollo 17* expedition, the rover traveled 19 kilometers (12 mi) on just one of its three excursions. The informal four-wheeled-vehicle lunar speed record is now approximately 17.6 kilometers per hour (11 miles per hour) and was set by the *Apollo 17* astronauts.

Figure 2 shows a design concept for an automated, or robot, lunar rover vehicle. Automated surface rovers such as these would gather soil and rock samples from remote areas of the Moon—for example, the polar regions—and return these materials for analysis at an automated lunar

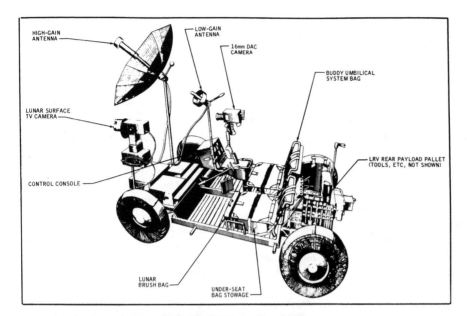

Fig. 1   The Lunar Rover Vehicle, or "Moon car," used by the Apollo astronauts. (Courtesy of NASA.)

Fig. 2   Artist's concept of an automated lunar rover exploring the remote regions of the Moon's south pole. Soil and rock samples gathered by the robot vehicle would be returned to an automated lunar base *(seen in the background)* for analysis. (Courtesy of NASA.)

base (shown in the background of fig. 2) or at a manned settlement. These surface rovers would help identify the location of valuable mineral deposits and pave the way for lunar mining operations.

See also: **lunar bases and settlements**

Magellanic Clouds   The two dwarf galaxies that are closest to our Milky Way Galaxy. The Large Magellanic Cloud

(LMC) is about 150,000 light-years away and the Small Magellanic Cloud (SMC) approximately 170,000 light-years distant. Both are visible to observers in the Southern Hemisphere and resemble luminous clouds several times the size of the Full Moon. Their presence was first recorded in 1519 by the Portuguese explorer Ferdinand Magellan, after whom they are named.

Figure 1 is a photograph of the large Magellanic Cloud taken by the United Kingdom's Schmidt Telescope in Australia. The white rectangle superimposed on this picture identifies the region also scanned by the Infrared Astronomical Satellite (IRAS) in 1983.

See also: **dwarf galaxy; Local Group; Milky Way Galaxy**

Magellan mission   A NASA Solar System exploration mission to the planet Venus. On May 4, 1989, the 3,550 kilogram (7,825 pound-mass) *Magellan* spacecraft was delivered to Earth orbit by the Space Shuttle *Atlantis* during the STS 30 mission, and then sent on an interplanetary trajectory to the cloud-shrouded planet by a solid-fueled inertial upper stage (IUS) rocket system. On August 10, 1990, the *Magellan* spacecraft (named for the famous 16th century explorer Ferdinand Magellan) was inserted into orbit around Venus and began initial operations for its 243-day-duration radar mapping mission (see fig. 1).

The prime scientific goal of the Magellan mission is to improve our overall knowledge of the geologic history of Venus. The *Magellan* spacecraft has acquired high-resolution radar image maps of over 80 percent of the cloud-shrouded surface of Venus. The *Magellan's* synthetic aperture radar (SAR) system can penetrate the planet's opaque atmosphere and provide photographic quality surface images of sufficient resolution to identify surface features as small as 100 meters (300 feet). The spacecraft has also gathered altimetry and gravity data to help planetary sci-

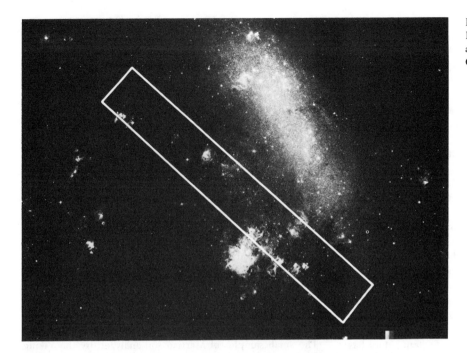

Fig. 1 A photograph of the Large Magellanic Cloud. (Courtesy of NASA and the United Kingdom's Schmidt Observatory in Australia.)

entists accurately map the planet's gravity field and to detect internal stresses and density variations. By providing planetary scientists data on the morphology of Venus and by identifying its small-scale geologic characteristics, the highly successful Magellan mission is now allowing the evolutionary history of Venus to be compared with that of Earth.

See also: **Venus**

magnetosphere The region around a planet in which charged atomic particles are influenced by the planet's own magnetic field rather than by the magnetic field of the

Fig. 1 This mosaic image of the Venusian surface consists of adjacent pieces of two Magellan radar image strips obtained on August 16, 1990. With a resolution of 120 meters (400 ft), this high-resolution radar image reveals many new geologic features about a region near the east flank of a major volcanic upland called Beta Regio. For example, the bright line across the center of the image is a fracture or fault zone. (Courtesy of NASA/JPL.)

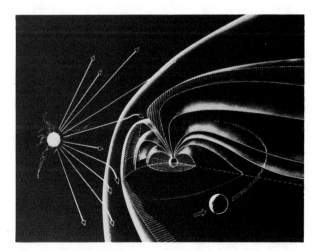

Fig. 1 Spacecraft and space probes have helped us map much of the Earth's magnetosphere—the region of magnetic field structures and streams of trapped radiation belts around our planet. (Courtesy of NASA.)

parent star, as projected by the stellar (solar) wind. The magnetosphere generally includes any trapped radiation belts that might encircle the planet. As shown in figure 1, studies by spacecraft and space probes have now mapped much of the region of magnetic field structures and streams of trapped particles around the Earth. The solar wind, streaming out from the Sun, shapes the Earth's magnetosphere into a teardrop, with a long magnetotail stretching out opposite the Sun.

See also: **Earth's trapped radiation belts**

Mars Throughout human history Mars, the Red Planet, has been at the center of astronomical thought. The ancient Babylonians, for example, followed the motions of this wandering red light across the night sky and named it after Nergal, their god of war. In time, the Romans, also honoring their own god of war, gave the planet its present name. The presence of an atmosphere, polar caps and changing patterns of light and dark on the surface caused many pre-Space-Age astronomers and scientists to consider Mars an "Earthlike planet"—the possible abode of extraterrestrial life.

Of all the other planets in the Solar System, Mars has been the leading candidate for possessing extraterrestrial life-forms. Pre-Space-Age scientists and astronomers observed the Red Planet through Earth-based telescopes and saw what appeared to be straight lines crisscrossing its surface. These observations (later determined to be optical illusions) led to the very popular notion that an intelligent race of Martians had constructed a large system of irrigation canals to support life on a "dying planet." In fact, when Orson Welles broadcast a radio drama in 1938 based on H. G. Wells' science-fiction classic *War of the Worlds*, enough people believed the report of invading Martians to create a near-panic in some areas.

Another reason for exobiologists to anticipate the existence of life on Mars was the apparent seasonal color changes of the planet's surface. These observed changes led to speculation that environmental conditions on Mars might cause certain Martian vegetation to bloom in the warmer months and become dormant during the colder periods.

Within the last three decades, however, sophisticated robot spacecraft—flybys, orbiters and landers—have shattered these romantic myths of a race of ancient Martians struggling to bring water to the more productive regions of a dying world. Spacecraft-derived data have shown instead that the Red Planet is actually a "halfway" world. (See table 1.) Part of the Martian surface is ancient, like the surfaces of the Moon and Mercury, while part is more evolved and Earthlike. Contemporary information about Mars is presented in tables 2 and 3.

In August and September 1975, two Viking spacecraft were launched on a mission to help answer the question, Is there life on Mars? Each Viking spacecraft consisted of an orbiter and a lander. While scientists did not expect these spacecraft to discover Martian cities bustling with intelligent life, the exobiology experiments on the lander were designed to find evidence of primitive life-forms, past or present. Unfortunately, the results sent back by the two robot landers were teasingly inconclusive.

Scientists still don't know whether there is life on Mars. The small samples of Martian soil analyzed by the Viking Landers were specially treated in three different experimental protocols (procedures) designed to detect biological processes. While some of the tests indicated that biological activities were occurring, the same responses could be explained by Martian soil chemistry. There was also a very notable absence of evidence that organic molecules (life precursors) exist on Mars.

Despite the inconclusive results of the Viking project exobiology experiments, we now know more about Mars than any other planet (except Earth) in our Solar System. A large number of missions to Mars have been conducted by the United States and the Soviet Union (review table 1). Before the Viking project missions, for example, four Mariner spacecraft missions were sent to Mars by NASA. Three of these Mariner missions were flybys, while one involved an orbiter that conducted an extensive initial survey of the planet.

Table 1   Missions to Mars (1962–1990)

Mission (Payload)	Country of Origin	Launch Date	Remarks
Mars 1	USSR	Nov. 1, 1962	Passed Mars at 193,000 km on June 19, 1963, but communications failed on March 21, 1963.
Mariner 3	USA	Nov. 5, 1964	Mission failed (shroud malfunction).
Mariner 4	USA	Nov. 28, 1964	Successful planetary and interplanetary exploration; encounter (flyby) occurred on July 14, 1965; 22 images taken of Mars.
Zond 2	USSR	Nov. 30, 1964	Passed Mars at 1,500 km on Aug. 6, 1965; however, communications/batteries failed in May 1965.
Mariner 6	USA	Feb. 25, 1969	Successful planetary exploration; Mars flyby within 3,300 km on July 31, 1969; studied atmosphere of Mars; returned TV images of surface.

Mission (Payload)	Country of Origin	Launch Date	Remarks
Mariner 7	USA	March 27, 1969	Successful planetary exploration; Mars flyby at 3,518 km on Aug. 5, 1969; spacecraft identical to *Mariner 6*.
Mars 2 (Orbiter and Lander)	USSR	March 28, 1971	Orbiter circled Mars on Nov. 27, 1971; capsule (lander) ejected at 47°E and landed.
Mars 3 (Orbiter and Lander)	USSR	May 28, 1971	Orbited Mars on Dec. 2, 1971; capsule ejected and soft-landed at 45°S 158°W.
Mariner 9	USA	May 30, 1971	Orbited Mars on Nov. 13, 1971; transmitted over 6,900 images of Martian surface; all scientific instruments operated succesfully; mission terminated on Oct. 27, 1972.
Mars 4	USSR	July 21, 1973	Passed Mars at 2,200 km on Feb. 10, 1974, but failed to enter Mars orbit as planned.
Mars 5	USSR	July 25, 1973	Orbited Mars on Feb. 2, 1974; gathered Martian data and served as relay station.
Mars 6 (Orbiter and Lander)	USSR	Aug. 5, 1973	Lander soft-landed on Mars at 24°S 25°W on March 12, 1974; returned atmospheric data during descent.
Mars 7 (Orbiter and Lander)	USSR	Aug. 9, 1973	Missed Mars by 1,300 km (aimed at 50°S 28°W); contact lost on March 9, 1974.
Viking 1 (Orbiter and Lander)	USA	Aug. 20, 1975	Nuclear-powered lander successfully soft-landed on July 20, 1976, on the Plain of Chryse at 47.97°W 22.27°N. First on-site analysis of surface material of another planet.
Viking 2 (Orbiter and Lander)	USA	Sept. 9, 1975	Nuclear-powered lander successfully soft-landed on Sept. 3, 1976, on the Plain of Utopia at 47.67°N 225.75°W; returned scientific data.
			Viking 1 and 2 Orbiter spacecraft returned over 40,000 high-resolution images of the surface of Mars; also collected gravity-field data, monitored atmospheric water levels and thermally mapped selected sites on the surface.
Phobos 1	USSR	July 7, 1988	Lost in interplanetary space on Aug 29, 1988, due to telemetry error.
Phobos 2	USSR	July 12, 1988	Orbited Mars in late January 1989; acquired images of Martian moon Phobos in February 1989; then, lost in space in orbit around Mars on March 27, 1989 just prior to Phobos rendezvous mission; computer error in control system suspected cause of failure.

SOURCE: Author, NASA, ESA, Glaskosmos.

Mariner 4 was launched by NASA late in 1964. It flew past the Red Planet on July 14, 1965, and approached to within 9,656 kilometers of its surface. Looking at the 22 close-up photographs of Mars taken by *Mariner 4*, scientists found no evidence of Martian cities, Martian canals or even flowing water!

The *Mariner 6* and 7 missions to Mars followed during the summer of 1969. These flyby spacecraft returned approximately 200 photographs that showed a diversity of surface conditions. *Mariner 9* was launched on May 30, 1971, and arrived at Mars 5½ months after lift-off. *Mariner 9* orbited the planet, returning over 7,000 images that revealed previously unknown Martian surface features, including evidence of ancient rivers and possible seas that could have existed at one time on the Red Planet (see fig. 1.)

Table 2   Physical and Dynamic Data for Mars

Diameter (equatorial)	6,794 km
Mass	6.42×10^{23} kg
Density (mean)	3.9 g/cm^3
Surface gravity	3.73 m/sec^2
Escape velocity	5.0 km/sec
Albedo	0.15
Atmosphere (main components by volume)	
Carbon dioxide (CO_2)	95.32%
Nitrogen (N_2)	2.7%
Argon (Ar)	1.6%
Oxygen (O_2)	0.13%
Carbon monoxide (CO)	0.07%
Water vapor (H_2O)	0.03%[a]
Natural satellites	2 (Phobos and Deimos)
Period of rotation (a Martian day)	1.026 days
Average distance from Sun	2.28×10^8 km (1.523 AU)
Eccentricity	0.093
Period of revolution around Sun (a Martian year)	687 days
Mean orbital velocity	24.1 km/sec
Solar flux at planet (at top of atmosphere)	590 W/m^2 (at 1.52 AU)

[a]variable

SOURCE: Developed by author based on NASA data.

Table 3   *Viking 1* Lander Site Chemistry

Location: Chryse Planitia 47.97°W 22.27°N

Local topography: low, rolling hills, mostly covered by fine-grained debris.

Remarks: Surface samples analyzed at both *Viking 1* and *2* landing sites are similar; it appears that these are representative of fine debris everywhere on the planet.

Compound[a]	Percentage Mass
SiO_2	44.7
Al_2O_3	5.7
Fe_2O_3	18.2
MgO	8.3
CaO	5.6
K_2O	<0.3
TiO_2	0.9
SO_3	7.7
Cl	0.7

[a]Except for chlorine (Cl), the elements silicon (Si), aluminum (Al), iron (Fe), magnesium (Mg), calcium (Ca), potassium (K), titanium (Ti), and sulfur (S) are all present in the Martian soil as their common oxides, such as silicon dioxide (SiO_2).

SOURCE: NASA.

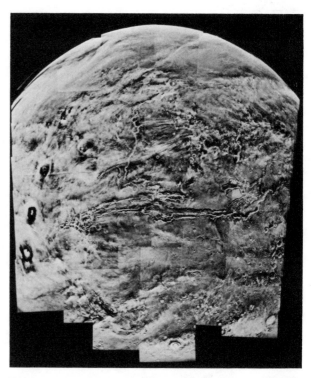

Fig. 1   A composite image of the Martian surface, including the Valles Marineris, which is as long as the North American continent from coast to coast. (Taken by *Viking Orbiter 1,* February 1980.) (Courtesy of NASA/Jet Propulsion Laboratory).

The Viking project was the first mission to successfully soft-land a robot spacecraft on another planet (excluding here, of course, the Earth's Moon). All four Viking spacecraft (two orbiters and two landers) exceeded by considerable margins their design goal lifetime of 90 days. The four spacecraft were launched in 1975 and began to operate around or on the Red Planet in 1976. When the *Viking 1* Lander touched down on the Plain of Chryse on July 20, 1976, it found a bleak landscape. Several weeks later, its twin, the *Viking 2* Lander, set down on the Plain of Utopia and discovered a more gentle, rolling landscape. One by one these robot explorers finished their highly successful visits to Mars. The *Viking Orbiter 2* spacecraft ceased operation in July 1978; the *Lander 2* fell silent in April 1980; *Orbiter 1* managed at least partial operation until August 1980; and the *Viking Lander 1* made its final transmission on November 11, 1982. NASA officially ended the Viking mission May 21, 1983.

As a result of these interplanetary missions, we now know that Martian weather changes very little. For example, the highest atmospheric temperature recorded by either Viking Lander was minus 21 degrees Celsius (midsummer at the *Viking 1* site); while the lowest recorded temperature was minus 124 degrees Celsius (at the more northerly *Viking 2* site during winter).

The atmosphere of Mars was found to be primarily carbon dioxide. Nitrogen, argon and oxygen are present in small percentages, along with trace amounts of xenon and krypton. The Martian atmosphere contains only a wisp of water (about 1/1000th as much as found in the Earth's atmosphere). But even this tiny amount of water can condense out and form clouds that ride high in the Martian atmosphere or form local patches of morning fog in valleys. There is also evidence that Mars had a much denser atmosphere in the past—one capable of permitting liquid water to flow on the planet's surface. Physical features resembling riverbeds, canyons and gorges, shorelines and even islands hint that large rivers and maybe even small seas once existed on the Red Planet.

THE SEARCH FOR EXTRATERRESTRIAL LIFE ON MARS

Each Viking Lander had cameras capable of "capturing" any large Martian life-forms that might inhabit the regions near either landing site. In thousands of images taken on the surface of Mars, no large life-forms were detected. The Viking Landers also carried three biology experiments to search for extraterrestrial life (especially at the microorganism level). These experiments were (1) the gas-exchange experiment, which was designed to detect metabolization and related gaseous-exchange products in the presence of a nutrient solution; (2) the labeled-release experiment, designed to detect the release of radioactivity from labeled carbon (using radioisotope-tracer techniques) bound in a nutrient solution; and (3) the pyrolitic-release experiment, based on the exobiological hypothesis that any Martian life-form would be able to incorporate atmospheric carbon (tagged with a radioisotope tracer) in the presence of sunlight—that is, by photosynthesis.

All three exobiology experiments exhibited signs of activity, but many scientists now feel that this activity was probably due to chemical, and not biological, reactions in the Martial soil. In addition, the Viking's gas chromatograph/mass spectrometer (GCMS) failed to detect any organic material, even at very low concentrations.

Therefore, despite intensive initial investigations, the greatest Martian mystery of all—Is there life?—still remains a mystery. More sophisticated expeditions to Mars are needed in the next few decades before exobiologists can state that life does (or perhaps did) exist on Mars—or else positively conclude that Mars is a barren, lifeless world. In either circumstance, the implications for our models depicting the cosmic prevalence of life will be profound.

THE MOONS OF MARS

Mars has two small, irregularly shaped moons called Phobos ("fear") and Deimos ("terror"). These natural satellites were discovered in 1877 by Asaph Hall. They both have ancient, cratered surfaces with some indication of regoliths to depths of possibly five meters or more. (See figs. 2 and 3.) The physical properties of these two Martian moons are presented in table 4. Infrared spectrometer data from the recent Soviet *Phobos 2* probe suggest that the Phobian rocks hold much less water in their crystalline

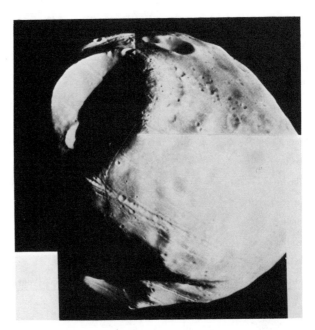

Fig. 2   The Martian satellite Phobos, photographed from a range of 612 kilometers by the *Viking Orbiter 1*. Stickney is the largest crater on Phobos; it measures about 10 kilometers across and can be seen at the left near the morning terminator (the border between the illuminated and unilluminated areas of Phobos's surface). (Courtesy of NASA/Jet Propulsion Laboratory.)

Fig. 3   The tiny Martian satellite Deimos, as imaged by the *Viking Orbiter 1*. (Courtesy of NASA.)

structure than do rocks found on the surface of Mars. This observation would be consistent with the long held hypothesis that both Phobos and Deimos are "captured" asteroids.

THE U.S. SPACE EXPLORATION INITIATIVE (SEI)

In a dramatic and important speech on July 20, 1989, (the 20th anniversary of the Apollo 11 landing on the Moon), President George Bush called for an evolutionary human expansion into the Solar System (first a permanent

Table 4   Physical and Dynamic Properties of the Martian Moons, Phobos and Deimos

Property	Phobos	Deimos
Characteristic dimensions (both are irregularly shaped)		
Longest dimension	27 km	15 km
Intermediate dimension	21 km	12 km
Shortest dimension	19 km	11 km
Mass	9.8×10^{15} kg	2.0×10^{15} kg
Density	2.0 g/cm^3	1.9 g/cm^3
Albedo	0.05	0.06
Surface gravity	1 cm/sec^2	0.5 cm/sec^2
Rotation	Synchronous	Synchronous
Semimajor axis of orbit	9,378 km	23,459 km
Eccentricity	0.018	0.0008
Sidereal period (approx.)	7 hr, 39 min	30 hr, 18 min

SOURCE: Developed by author based on NASA data.

Table 5   Scenario for the Exploration and Settlement of Mars in the 21st Century

Stage 1: Advanced exploration with sophisticated spacecraft
• Mars Observer

Stage 2: Robotic surface exploration
• Mars Network mission (penetrators)
• Mars airplane
• Mars surface rovers
• Mars Sample Return mission (MSRM)

Stage 3: Initial human expeditions
• Nuclear electric propulsion (NEP)
• Nuclear thermal rockets

Stage 4: Development of Martian resource base
• Base site preparation (automated)
• Initial surface base (10–100 persons)
• Early Martian settlements (100–1,000 persons)
  —Initiation of planetary engineering projects
• Autonomous Martian civilization (population over 100,000)
  —Full-scale planetary engineering projects
  —Permanent human presence in heliocentric space
  —Independent of Earth–Moon planetary system

SOURCE: Developed by the author based on National Space Council, NASA and aerospace industry data.

presence in low Earth orbit, then bases on the surface of the Moon and finally a human expedition to and bases on Mars). This statement is often called the President's space exploration initiative (SEI).

The planet Mars serves as a long-term goal and objective in this multidecade-long initiative. Why should we go to Mars? Here are some of the most important reasons now being identified in support of a U.S. (and quite possibly international) manned expedition to Mars in the second decade of the next century: (1) to fulfill the fundamental human imperative to explore; (2) to understand planetary evolution (e.g., why did Mars and Venus turn out so differently than Earth); (3) to find out if life once existed on Mars; and (4) to continue our Nation's journey into space.

WILL THERE BE A MARTIAN IN YOUR
FUTURE?

As we enter the next millennium, human development will be highlighted by the establishment of our extraterrestrial civilization. We will first learn how to permanently occupy near-Earth space and then expand human activities throughout cislunar space (the area between Earth and the Moon). With the rise of self-sufficient space settlements on the Moon and in Earth orbit, humankind will then expand into heliocentric (Sun-centered) space. Mars will become the focus of extensive exploration and resource-evaluation missions, using robotic spacecraft and human expeditions. Table 5 describes one possible grand scenario for the exploration and development of the Red Planet in the next century and beyond.

NASA has proposed that a Mars Observer be placed into a polar orbit around the Red Planet. From a Sun-synchronous, nearly circular 360-kilometer (224-mi) orbit, it would observe the planet for almost a full Martian year (approximately two Earth-years), making global maps of elemental and mineralogical surface compositions. It would also measure the seasonal cycles of carbon dioxide, water and dust in the Martian atmosphere and study the interactions between volatile reservoirs (such as the polar caps) and the Martian atmosphere. This spacecraft would pave the way for more extensive exploration programs, robotic and manned, that would take place in the 21st century.

Planetary scientists believe that experiments performed from a network of surface penetrators can provide the important facts needed to start understanding the evolution, geologic history and nature of a planetary body such as Mars. "Network science" can, in fact, be defined as a set of systematic measurements made over a relatively long period of time (such as one Martian year) at widely distributed locations on a planet's surface. The measurements typically would include seismic, meteorologic, heat flow, water content, geochemical and imagery—as needed to fully characterize each site. A Mars Probe Network would help establish the chemical composition of near-surface materials, evaluate general atmospheric-circulation patterns and investigate the internal structure and seismicity of the planet.

Some planetary scientists think that a ligthweight, robotic Mars airplane represents a very versatile tool for exploring the Red Planet. This type of airplane could perform high-resolution surface mapping missions, carry and distribute special sensor packages or play an important scouting role for a manned expedition or robotic Mars Sample Return mission. These hydrazine-engine-powered aircraft would be capable of flying at altitudes of 500 meters up to 15,000 meters, with respective ranges of 26 kilometers and 6,700 kilometers.

Smart surface rovers also have a very useful role to play in the overall exploration of Mars in the next century. These smart autonomous rovers can gather samples of surface materials, deploy instruments on or beneath the Martian surface and perform detailed site investigations. A rover can work independently, or perhaps be operated in teams of two or four, transmitting data back to an orbiting mothership, which in turn relays the findings back to Earth.

An automated Mars Sample Return mission would, as the name implies, return one or more samples of Martian surface material to Earth or to an Earth-orbiting quarantine facility, where extensive scientific investigations can be made by terrestrial scientists.

A human expedition to Mars early in the next century would represent our physical penetration into heliocentric space. One proposed mission scenario, using a nuclear-electric-propelled spacecraft, involves a single spacecraft with a Mars lander that would carry a crew of five on a 2.6-year duration mission to the Red Planet. The nuclear-electric propulsion (NEP) system would be powered by twin megawatt-class advanced-design space nuclear reactors. Closed air and water life-support systems and artificial gravity would sustain the crew throughout the flight through interplanetary space.

The Mars expedition vehicle could be assembled at Space Station *Freedom* in low Earth orbit. It could then be transferred to geosynchronous Earth orbit, where its crew would board it for the long, spiraling outward journey to Mars (about 510 days). It would take another 39 days to perform a capture spiral maneuver around Mars, ending up in a circular, 3,000-kilometer-altitude orbit above the Red Planet. The crew would then engage in a 100-day reconnaissance mission, including a 30-day surface exploration mission by three of the five crew members. A 23-day Mars departure spiral would start the electrically propelled vehicle back to Earth. Mars-to-Earth transfer would take about 229 days under optimum coast conditions. The vehicle would then execute a 16-day capture spiral to geosynchronous orbit around the Earth, and the crew would transfer to a special quarantine (if necessary) and debriefing facility on the space station before eventually returning to Earth. In another Mars expedition scenarios under study, a permanent lunar base would play a major role, especially in crew training and space technology demonstrations.

Exactly what happens after the first human expedition to Mars is, of course, open to wide speculation at present. People here on Earth could simply marvel at "another outstanding space-technology first" and then settle back to their "more pressing" terrestrial pursuits. (This pattern unfortunately followed the spectacular Apollo Moon landing missions of 1969–72.) On the other hand, if this first human expedition to Mars were widely recognized and accepted as the technical precursor to our permanent occupancy of heliocentric space, then Mars would truly become the central object of greatly expanded space activities (perhaps complementing the rise of a self-sufficient lunar civilization).

Very sophisticated surface rovers could be used to prepare suitable sites for the first permanent bases, each housing perhaps 10 to 100 people. These early bases on the Red Planet would focus their activities on detailed exploration and resource identification and would most likely be supported by a Mars-orbiting space station (possibly a "natural" space station using Phobos as a base location). Another important objective for the first "Martians" will be to conduct basic science and engineering projects that take advantage of the Martian environment and native Martian materials. Then, as prospects for Martian-based industries grow, these early bases will also expand, reaching the size of modest settlements, with upwards of 1,000 Martians each—all committed to discovery, adventure and profit on the Red Planet.

As the early settlements mature and are economically nourished by the planet's resource base, Martian fuels, food, water, metals and manufactured products will support the wave of human expansion to the mineral-rich asteroid belt and to the giant outer planets and their intriguing complement of moons.

During this growth and expansion process, a point will be reached when the Martian population, for all practical purposes, becomes fully self-sufficient. The very thought of a neighboring planet inhabited by intelligent beings has always been stimulating to the human race.

An autonomous Martian civilization will most likely be characterized by the initiation of planetary engineering projects. Planetary engineering involves the large-scale modification or reconfiguration of a planet to provide a more habitable ecosphere for the terrestrial settlers. On Mars the human settlers would most likely first seek to make the atmosphere more dense (and eventually even breathable) and to alter its temperature extremes to more "Earthlike" ranges. These planetary engineering efforts could include melting the polar caps and transporting large quantities of liquid water to the equatorial regions of the planet—perhaps by means of a large series of open canals. Those who speculated in the early 20th century that Mars was inhabited by intelligent creatures who constructed canals to transport water from the polar regions may only have been a century or two early in their bold assumptions!

Mars, properly explored and used, not only opens up the remainder of heliocentric space for human development; it also establishes the technological pathways for the first interstellar missions. The development of very smart machines, the ability to modify the ecosphere of a planet and the technology to control and manipulate large quantities of energy are all necessary if human explorers are ever to venture across the interstellar void in search of new worlds around distant suns. The conquest of Mars in the 21st century will provide a large portion of the technology base needed to seriously consider flights beyond our own Solar System.

See also: **exobiology; Mars Observer; Mars base; Mars penetrator; Mars Sample Return Mission; Mars surface rovers; planetary engineering; Viking Project**

Mars airplane   How would we use an unmanned airplane in the further exploration of Mars? Some scientists think that such lightweight robot vehicles represent a very versatile means of transportation around the Red Planet. They could either carry experiment packages or play an important support role for special missions like a Mars Sample Return mission (MSRM). This Mars airplane could be used, for example, to deploy a network of science stations, such as seismometers or meteorology stations, at selected Martian sites within an accuracy of a few kilometers. With a 40-kilogram payload capacity, the flying platform could perform high-resolution imagery (say, less than 0.5-meter ground-spot size) or conduct magnetic, gravity and geochemical surveys. It would be capable of flying at altitudes of 500 meters to 15 kilometers, with ranges of 25.5 to 6,700 kilometers.

The Mars airplane has many characteristics of a terrestrial competition glider. It would have a very lightweight airframe made of carbon fiber composites and weigh less than 40 kilograms. The wings, fuselage and tail sections would fold, allowing it to fit into a protective aeroshell for its initial descent into the Martian atmosphere after deployment from its "aircraft-carrier" spaceship. This Mars airplane would be powered by a 15-horsepower hydrazine airless engine that is used to drive a three-meter propeller.

This type of extraterrestrial airplane offers two basic options. First, it can be employed as a powered flyer that performs aerial surveys, atmospheric soundings and so on and then crashes when its fuel supply is exhausted. Or else it can be built as a plane equipped with a variable-thrust rocket so that it may soft-land and then take off again.

One basic mission concept would be to send several of these "aircraft-carrier" spaceships to Mars on an extensive exploration program. Each spaceship would carry up to four Mars airplanes (folded and tucked in their respective aeroshells). While the carrier spacecraft orbits Mars, individual airplanes would be deployed on command from Earth into the Martian atmosphere. Each Mars airplane would enter the atmosphere of the Red Planet in its protective aeroshell. While descending through the upper Martian atmosphere, this aeroshell would be discarded, parachutes deployed and the plane unfolded. It would continue to descend to its operational altitude and then fly off on its mission of exploration, instrument deployment or sample collection. Once its hydrazine fuel was exhausted, each Mars airplane would acquire an eternal resting place among the shifting red dunes of the planet. Some would land softly and be found nearly intact by later teams of terrestrial explorers, while others would crash and become crumpled piles of "extramartian" debris. Communications with scientists on Earth would be maintained through a network of strategically located relay or communications

satellites orbiting around Mars. This Mars airplane, or bigger versions of it, could also be used in direct support of a human expedition to Mars. The terrestrial explorers would use such robot scouts to find suitable sites for more detailed scientific investigation. In time, the thin Martian atmosphere could become host to a squadron of such flyers—each gliding across the surface of the Red Planet in response to scientific targeting instructions from a human explorer of the 21st century.

See also: **Mars; Mars base; Mars Sample Return Mission; Mars surface rovers**

Mars base   For automated Mars missions, the spacecraft and robotic surface rovers will generally be small and self-contained. For human expeditions to the surface of the Red Planet, however, two major requirements must be satisfied: life support (habitation) and surface transportation (mobility). Habitats, power supplies and life support systems will tend to be more complex in a permanent base on the Martian surface that must sustain human beings for years at a time. Surface mobility systems will also have to grow in complexity and sophistication as early Martian explorers and settlers travel several hundred kilometers or so away from their base camp. At a relatively early time in any permanent Mars surface base program, the use of Martian resources to support the base must be tested vigorously and then quickly integrated in the development of an eventually self-sustaining surface infrastructure.

Most likely, the initial Martian habitats will resemble standardized lunar base (or space station) pressurized modules and will be transported from cislunar space to Mars in a prefabricated condition by interplanetary nuclear-electric-propulsion (NEP) cargo ships. On the surface of Mars these modules will be configured and connected as needed and then covered with a meter or so thickness of Martian soil for protection against the lethal effects of solar flare radiations or continuous exposure to cosmic rays on the planet's surface. (Unlike the Earth's atmosphere, the very thin Martian atmosphere does not shield very well against ionizing radiations from the space environment).

See also: **Mars; Mars airplane; Mars surface rovers**

Mars Observer (MO)   NASA's Mars Observer will follow up on the earlier discoveries of the Mariner (4, 6, 7 and 9) and Viking (1 and 2) programs, emphasizing the geologic and climatic evolution of this interesting, complex planet. This program will provide a spacecraft in orbit around Mars from which the entire Martian surface and atmosphere will be mapped for at least one Martian year (equivalent to 687 Earth-days). The main objectives of this mission to the Red Planet are to (1) determine on a global scale the elemental and mineralogical character of the planet's surface materials; (2) define the global topography and gravitational fields; (3) establish the nature of the Martian magnetic field; (4) determine the time and space distribution, abundance, sources and sinks of volatile material and dust over a full seasonal cycle; and (5) explore the structure and circulation patterns of the Martian atmosphere.

The Mars Observer will be the first mission to use a *planetary-observer* class NASA spacecraft. This mission is scheduled for launch in September 1992 onboard an expendable Titan III commercial launch vehicle. The interplanetary trajectory to Mars from Earth orbit will be provided by a transfer orbit stage (TOS) vehicle. After an 11-month cruise through interplanetary space, the Mars Observer will arrive at Mars in August 1993. The spacecraft will then be inserted into a circular, near-polar orbit with an average altitude of about 360 kilometers (224 mi). A plane change will place this spacecraft in a 93-degree sun-synchronous mapping orbit for the remainder of the mission. Thorough mapping of the entire Martian surface will continue for a period of one Martian year. Important "seasonal" data will be collected on the climatology, surface composition, topography, gravity field and magnetic field of Mars. At the end of the mission, this spacecraft will be boosted to a higher *quarantine* orbit in order to prevent atmospheric entry before the year 2019.

The Mars Observer will carry seven major instruments; (1) a gamma ray spectrometer that measures the abundance of chemical elements (such as iron, silicon, potassium, thorium and uranium) on the surface of Mars; (2) a thermal-emission spectrometer that maps the mineral content of surface rocks, ices, frosts and the composition of clouds; (3) a line-scan camera that takes low-resolution images of Mars on a daily basis to support climate studies, and also takes medium- and high-resolution images of selected areas to support studies of surface geology and the interactions between the surface and the atmosphere; (4) a laser altimeter that determines the topographic relief of the Martian surface; (5) a pressure-modulator infrared radiometer that measures dust and condensates in the Martian atmosphere, as well as provides profiles of temperature, water vapor, and dust opacity as they change with latitude, longitude and season; (6) a radio-science investigation that measures atmospheric refractivity as it varies with altitude to determine the temperature profile of the Martian atmosphere and that uses tracking data to measure the Martian gravity field; and (7) a magnetometer and electron reflectometer that determine the nature of the Martian magnetic field and its overall interactions with the solar wind.

The Mars Observer will also participate in an ambitious international investigation of Mars through an agreement with France and the Soviet Union. This participation involves the Mars balloon relay experiment.

The Soviet Union's planned Mars '94 mission will deploy balloon-borne instrument packages in the Martian atmosphere. During their operating lifetimes, these balloon-borne packages will transmit data to both a Soviet orbiter spacecraft and the U.S. Mars Observer. Special equipment on the Mars Observer spacecraft, supplied by the French Space Agency, CNES (Centre National d'Étude Spatiales),

will receive these balloon package data for transmission back to Earth.

The Mars Observer will, therefore, study the surface, atmosphere, interior and magnetic fields of Mars for a full Martian year—providing truly global data that will greatly improve our understanding of the evolution of Mars. One subject of particular interest to planetary scientists is the role that water once played on Mars. Although there is no liquid water on the surface of Mars today, the Mariner and Viking missions found ample evidence that liquid water once flowed there long ago. Through *comparative planetology*, scientists are examining the planetary neighbors Venus, Earth and Mars, and are trying to understand why each has evolved so differently. (In comparative planetology, scientists compare the differences and similarities in the evolutionary histories of the planets in order to better understand the processes that formed and modified each of them.)

Finally, the information collected by the Mars Observer will also provide an essential data base for planning more ambitious future missions to the Red Planet in the next century, including sophisticated robot surface rovers and sample return missions, the initial human expeditions and the establishment of permanent surface bases and settlements.

See also: **Mars; Mars airplane; Mars base; Mars penetrator; Mars Sample Return Mission**

Mars penetrator   The overall goal of planetary-exploration missions is to gather data that will help in understanding the formation, evolution and present state of planetary bodies and prepare the way for human expeditions, resource exploitation and permanent habitation, if appropriate. Planetary bodies are complex and are characterized by both surface and interior irregularities. Precise measurements of key physical properties of these celestial bodies, when taken from a network of sites distributed over the planet's surface, can provide a database for constructing a composite picture of that planet. In addition, such a network provides simultaneous measurements of transient phenomena of the planet's interior, surface and atmosphere. In fact, a network of complementary stations is one very effective way of obtaining meaningful data as a function of time and location. These data will provide the information necessary to develop sound working hypotheses and models about planetary processes and eventually pave the way for more extensive exploration and possible exploitation by human explorers.

Space scientists have concluded that experiments performed from a network of penetrators can provide essential facts needed to begin understanding the evolution, history and nature of a planetary body, such as Mars. "Network science," in fact, can be defined as a set of systematic measurements over a relatively long time duration (one Martian year at a minimum) at locations distributed widely over the planet's surface. The scientific measurements should necessarily include seismic, meteorologic and, if possible, other experiments to characterize the local site—such as heat flow, water content, geochemistry and imaging.

There are several main reasons for using penetrators to deploy science experiments on Mars and other celestial bodies. First, the penetrators establish a global network of stations to measure transient phenomena over widely scattered regions on the planet's surface. Second, the penetrators actually place some of the experiments beneath the planet's surface. Finally, the penetrators provide an effective way of characterizing local sites in regions that could not be safely reached by larger, more sophisticated surface rovers. In fact, penetrators can be useful in any future Mars mission, since their deployment immediately establishes a planetary monitoring network.

When such penetrators are used in a planetary or global monitoring network, they possess inherent advantages for certain science experiments. For example, penetrators provide a much better coupling to the ground, thereby supporting a broad spectrum of seismic-event measurements free from noise introduced by meteorologic activity. The acquisition of time-varying measurements of wind direction, velocity, atmospheric pressure and humidity can be made at many more sites than would be economically possible using sophisticated surface vehicles. When such meteorological data are combined with image observations of transient events over widely separated regions, a more general understanding of the planet's atmospheric processes can be gained.

A typical penetrator system consists of four major subassemblies: (1) the launch tube, (2) the deployment motor, (3) the decelerator (usually a two-stage device) and (4) the penetrator itself. The launch tube attaches to the host spacecraft and houses the penetrator, deployment motor

and two-stage decelerator. The deployment motor is based on well-proven solid-rocket motor technology and provides the required derobit velocity. The two-stage decelerator includes a furlable umbrella heat shield for the first stage of hypersonic deceleration. The second stage consists of a small drogue (parachute), which ensures proper impact conditions. The penetrator itself is a steel device, shaped like a rocket, with a blunt ogive (curved) nose and conical-flared body. The afterbody of the penetrator remains at the planet's surface, with the forebody penetrating the subsurface material. Figure 1 shows the penetrator with typical network instruments—here, seismic and meteorologic sensors. Penetrator subsystems include structure, data processing and control, communications, power, thermal control and umbilical cable in addition to scientific instruments and sensors. The penetrator structure is designed to penetrate a variety of soils and rocks and to withstand the effects of the way it enters the ground (called inclination and angle of attack) at impact. The afterbody includes a deployable boom for the meteorologic instruments and an antenna. Penetrator power is provided by a nuclear energy source, called a radioisotope thermoelectric generator (or RTG, for short).

Before being attached to the "carrier spacecraft," each penetrator system will be assembled and checked out. If sterilization is necessary to support a planetary quarantine protocol (procedure), each penetrator system will be subjected to terminal sterilization procedures after assembly. It will then be attached to the carrier, or "mother spacecraft," in a biologically shielded launch tube. During the flight to Mars, housekeeping information and instrument-status data will be sent to the spacecraft through an umbilical connection and then transmitted back to Earth.

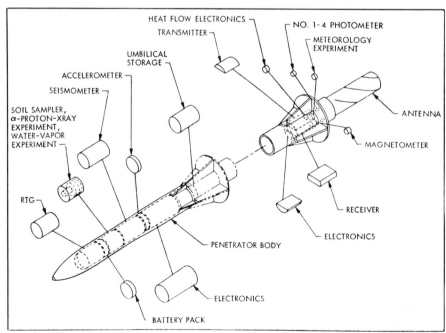

Fig. 1   Components of a typical Mars penetrator. (Courtesy of NASA.)

The penetrators will be monitored during this time by scientist on Earth, who will review the housekeeping data. When the carrier spacecraft arrives at Mars, the penetrators will be individually targeted, with the spacecraft positioned so as to properly "aim" the launch tube for propulsive separation. Separation from the mothership involves a sequence of actions that include venting pressure, opening the launch-tube covers and firing the deployment motor.

After separation from the mother spacecraft, one by one each penetrator will independently enter the Martian atmosphere behind a deployable heat shield and then float down on its parachute. Upon impact, the probe will bury itself in the Martian soil, leaving some instrumentation and an antenna at the surface. Communications with scientists on Earth from these surface/subsurface extraterrestrial monitoring sites will be accomplished by means of the orbiting mothership, which now interrogates each penetrator at least once a Martian day. A very large network of penetrators is considered necessary to obtain a general atmospheric-circulation model of Mars; but many other planetary science objectives can be satisfied, at least partially, with a minimum of three to six probe stations.

The overall scientific objection in creating a Mars Probe Network would be to establish a planetary network of seismic stations, meteorological stations, and geochemical and geophysical observation sites that can remotely and automatically operate on Mars for an extended period of time. The penetrator and its afterbody would contain a wide variety of instruments emplaced in the Martian soil and on its surface. Specific scientific objectives for this penetrator network on Mars include (1) a determination of the chemical composition of Martian near-surface materials, (2) a study of the internal structure and seismicity of Mars, (3) an evaluation of the general circulation patterns of the Martian atmosphere and (4) a characterization of local atmospheric conditions in a variety of Martian locales—many inaccessible to larger, surface rover vehicles. Individual penetrators as well as an extensive network can be used as part of any future Mars mission involving widespread exploration, robotic or human.

See also: **Mars; Mars base; Mars Observer; space nuclear power**

Mars Sample Return Mission (MSRM) The purpose of an automated Mars Sample Return mission (as the name implies) is to use a robot rover vehicle to traverse the surface of Mars collecting rock and soil samples and then return these collected specimens to Earth for analysis. In one contemporary sample-return-mission scenario, four expendable launch vehicles would be used to deliver the four major mission components to Mars: an imaging orbiter spacecraft, a communications and data relay orbiter spacecraft, a lander and an Earth-return spacecraft.

The imaging orbiter would support Martian surface operations by providing images for a landing site survey and then images of the area ahead of the robot rover vehicle to assist terrestrial scientists in planning its overall travels during exploration and sample collection.

The communications and data relay orbiter would serve as a relay link between the robot rover on the Martian surface and the Earth. However, this surface robot rover must also be equipped with stereo computer vision and a sophisticated "computer brain" that would allow it to make many immediate maneuvering and mission science decisions (e.g., this rock or that one?) by itself. This is necessary because the round-trip speed-of-light time of up to 40 minutes from Earth to Mars precludes any significant "real-time" teleoperation of rover vehicles actions by a human ground controller on Earth.

The lander vehicle would include this "smart" robot rover (at least one and possibly a pair) as well as an ascent rocket stage to take the sample canister back up into Martian orbit. As presently envisioned, this robot surface rover vehicle would have a mass of 1,500 kilograms (3,300 lb-mass) and would be powered by a 300-watt-electric radioisotope thermoelectric generator (RTG). (See fig. 1.) After about one year of exploring across the Martian surface, the robot rover would return to the lander and transfer the special canister containing the collected rock and soil specimens to the ascent vehicle, which would then take this canister into orbit around Mars. Once relieved of its sample canister, the robot rover could still continue exploring other portions of the surface of Mars.

After lift-off from the Martian surface, the ascent vehicle portion of the original lander system would then rendezvous and dock with the Mars-orbiting Earth return spacecraft. While coupled in orbit around the Red Planet, the sample canister would be automatically transferred from the ascent vehicle to the Earth return vehicle, which would then depart for Earth. After an interplanetary journey of about one year, this automated spacecraft, carrying its

Fig. 1 A smart robot surface rover collecting samples during the automated Mars Sample Return mission. (Artist rendering courtesy of NASA/JPL.)

precious cargo of Martian soil and rocks would achieve orbit around the Earth.

To avoid any potential problems of extraterrestrial contamination of the Earth's biosphere by alien microorganisms that might possibly be contained in the Martian soil or rocks, the sample canister would most likely be first quarantined and analyzed in a special Earth-orbiting facility, possibly at Space Station *Freedom*. Then, if no hazards were discovered, these Martian soil and rock samples (still encapsulated) would be allowed to enter the terrestrial biosphere, so that more extensive investigations could be performed at special laboratories on the Earth's surface.

The Mars Sample Return mission will provide a wealth of important and unique information about the Red Planet. Scientists consider this mission the next logical step beyond the Mars Observer mission and a significant and necessary step toward eventual human expeditions to Mars in the next century.

See also: **extraterrestrial contamination; Mars; Mars Observer**

Mars surface rovers   Automated rovers and mobility systems can be used to satisfy a number of exploration needs on the surface of Mars. For example, they can acquire specific samples of surface materials for automated evaluation in on-board laboratories; they can deploy instruments on or beneath the Martian surface; or they can perform extensive site investigations. The mobility of such surface rovers can also be used to extend the range of landers, Mars airplanes or a Mars Sample Return Mission. In addition, surface rovers could also serve as an independent mission, operating perhaps in teams of two or four and transmitting data back to Earth via a Mars-orbiting mothership. Two pairs of surface rovers might travel up to five kilometers each Martian day and also assist one another as needed.

Planetary surface rovers can be designed to meet a full range of mission requirements. They can vary in mass from 20 to approximately 2,000 kilograms. Large, autonomous full-capacity rovers would typically have a total mass of between 400 and 500 kilograms, including a scientific-payload capability of 80 to 100 kilograms. For operation in hostile planetary environments (including operations in darkness), these robot rovers could be powered by radioisotope thermoelectric generator (RTG) systems. They would be capable of autonomously traveling approximately 400 meters per Martian day and would have a total range of several hundred kilometers. Full-capacity, automated rovers would have a mission design lifetime of at least one Martian year.

See also: **lunar rovers; Mars airplane; Mars Base; Mars Sample Return Mission; robotics in space** (contemporary space robots); **space nuclear power**

Martian   Of or relating to the planet Mars; (currently in science fiction and once a permanent settlement is established) a native of the planet Mars.

mascon   A term meaninig "mass concentration." An area of mass concentration or high density within a celestial body, usually near the surface. In 1968, data from five U.S. lunar orbiter spacecraft indicated that regions of high density or mass concentration existed under circular maria (extensive dark areas) on the Moon. The Moon's gravitational attraction is somewhat higher over such mascons, and their presence perturbs (causes variations in) the orbits of spacecraft around the Moon.

See also: **Moon**

mass driver   An electromagnetic device that can accelerate payloads to a very high terminal velocity. Small, magnetically levitated vehicles, called "buckets," are used to carry the payloads. These "buckets" would contain superconducting coils and be accelerated by pulsed magnetic fields along a linear trace or guideway. When these buckets reach an appropriate terminal velocity, they release their payloads and are then decelerated for reuse.

materials research and processing in space (MRPS)   Materials research and processing in space, first using the Space Shuttle/Spacelab configuration and later through modules on permanent space stations and dedicated space industrial platforms, offers a great potential for significantly advancing materials science and for creating new products for use on Earth and in space.

Materials-science experiments conducted on Skylab, the Apollo–Soyuz Test Project (ASTP) and Space Shuttle missions have demonstrated that new knowledge can be gained through the effective use of the microgravity environment found in orbit. Metals, for example, can be solidified without the disturbing effects of gravity-driven convection. Crystallographers can watch crystals form in microgravity and learn why certain dopant materials (small amounts of impurities) that are needed to improve electronic characteristics distribute themselves far more uniformly in orbit then when crystals are formed on Earth. These valuable materials-science insights can be used to improve terrestrial manufacturing techniques or to establish the framework for space-based manufacturing.

For the field of materials science and technology, one of the main advantages of the Space Age is the ability to escape from terrestrial gravitational effects that adversely impact materials-processing operations on Earth. These gravitational effects include convection, sedimentation and buoyancy. In addition, the microgravity environment of space provides the opportunity to conduct "containerless processing." By using the long-duration microgravity environment of an orbiting space platform to overcome these adverse gravitational effects, both new materials and new manufacturing processes can be developed.

Convection is the spontaneous mixing or stirring in a liquid or gas as (fluid) currents flow between temperature gradients. Convective phenomena are unpredictable and chaotic and often lead to undesirable structural and compositional differences in a material after it has solidified. Both crystal growth and solidification processes are enhanced if convective disturbances are suppressed. A microgravity environment gives materials scientists and engineers the opportunity to reduce or completely eliminate such undesirable convective phenomena.

On Earth, gravity causes heavier components of a mixture to settle to the bottom (sedimentation) while less dense materials rise to the top (buoyancy). As a result, sedimentation or buoyancy effects complicate terrestrial manufacturing processes involving different-density alloys or composite materials. There are hundreds of potentially interesting metallic combinations that, like oil and water, just won't properly mix on Earth. As long as these metallic mixtures remain essentially separated through sedimentation or buoyancy effects, they are not particularly useful. But when combined in microgravity, the lighter-density components of such mixtures or alloys will remain suspended for indefinitely long periods of time and therefore permit the formation of essentially uniform solid composites or alloys whose constituents have large density differences. When uniformly mixed and properly solidified, many of these new composite materials take on unusual properties, such as very high strengths, excellent semiconductor behavior or perhaps outstanding performance as a superconductor.

Hydrostatic pressure places a strain on materials during solidification processes on Earth. Certain crystals are sufficiently dense and delicate that they are subject to strain under the influence of their own weight during growth. Such strain-induced deformations in crystals degrade their overall performance. In microgravity, heat-treated, melted and resolidified alloys can be developed free of such deformations.

Containerless processing in microgravity eliminates the problems of container contamination and wall effects. These are often the greatest source of impurities and imperfections when a molten material if formed here on Earth. But in space a material can be melted, manipulated and shaped, free of contact with a container or crucible, by acoustic, electromagnetic or electrostatic fields. In microgravity the surface tension of the molten material helps hold it together, while on Earth this cohesive force is overpowered by gravity.

One promising new space-based materials-manufacturing technology is the manufacture of certain high-purity pharmaceuticals. One example is the production of urokinase, lifesaving drug in blood-clot treatments. Urokinase is an enzyme produced from a specific kidney cell. However, until recently, it has been virtually impossible to isolate this kidney cell from others. Cells may be separated from one another by a process called electrophoresis. The basis on which electrophoresis works is that living cells suspended in a liquid have a small negative electric charge on their surfaces. Cells vary not only in electric charge but also in size and weight. When an electric charge is passed through the liquid, various cells start drifting at different rates because of their varied charges, sizes and shapes. Individual cells of the same type flock together in a band, or layer.

One difficulty with electrophoresis on Earth is that the electric current heats the water, causing convection currents in the solution. This prevents clear lines of demarcation between bands of nearly identical cells, such as kidney cells.

Electrophoresis experiments performed on board the Apollo spacecraft (in the ASTP) and on board the Space Shuttle have demonstrated that in microgravity, where gravity-induced convection currents do not occur, the urokinase-producing cell can be isolated from others. By improving techniques for separation of the urokinase-producing cell, manufacturers hope to extract many times as much urokinase as they could when the cell was not isolated.

Alloys are mixtures of two or more metallic or metallic and nonmetallic materials that provide a new material with useful properties of both or all constituents. For example, steel is an alloy of iron and carbon, while brass is an alloy of copper and zinc. Many alloys cannot be made on Earth or at least are extremely difficult to make on Earth, because the heavier element tends to settle too fast when the mixture is solidifying. An example of this situation is aluminum antimonide, which is an alloy of aluminum and antimony. One attribute of aluminum antimony is that it may be 30 to 50 percent more efficient than silicon when used in solar cells or computer circuit chips. Antimony is more than three times denser and tends to settle faster than aluminum when the two are mixed in a molten state. This causes a lack of uniformity in the alloy, severely limiting its usage in solar cells or computer circuit chips. In microgravity experiments, however, a largely uniform mixture has been obtained.

Crystals are solids in which most of the atoms are arranged in a regular pattern. Crystals such as silicon and germanium have extensive applications in the electronics and computer industries. These crystals are chemically grown under carefully controlled conditions. In experiments on board Skylab, the Apollo spacecraft used in the ASTP mission and the Space Shuttle/Spacelab configuration, higher-quality crystals were obtained then those made under similar conditions here on Earth.

Monodisperse (identically sized) latex spheres, whose diameters are smaller then the width of a human hair, are used in both medicine and industry to conduct research and to calibrate instruments. A variety of sizes are required to meet different needs. However, terrestrial gravity has limited the sizes that could be produced to about three micrometers. Beyond this size, gravitational influences give the particles a tendency to settle or clump. Experiments in microgravity have resulted in the development of larger latex spheres than is possible on Earth.

Materials research and processing in space offers the potential of not only stimulating space commerce but also positively influencing the quality of life for all on Earth in the next century:

See also: **microgravity; space commerce; space industrialization**

Megasphere A truly enormous, Dyson-sphere-like structure, tens of parsecs across, that captures the energy output of millions of stars clustered tightly together at the center of a galaxy. (A parsec is 3.26 light-years.) While a Dyson sphere represents the astroengineering feat of a mature Kardashev Type II civilization (solar-system level), the Megasphere represents the spectacular astroengineering accomplishment of an extremely powerful, galactic-level, Kardashev Type III extraterrestrial civilization.

See also: **Dyson sphere; extraterrestrial civilizations**

Mercurian Of or relating to the planet Mercury; (in science fiction) a native of the planet Mercury.

Mercury The innermost planet in our Solar System, orbiting the Sun at approximately 0.4 astronomical unit (AU). This planet, named for the messenger god of Roman mythology, is a scorched, primordial world that is only 40 percent larger in diameter than the Earth's Moon.

Before the Space Age, astronomers had attempted to identify its surface features, but because Mercury is so small and so close to the Sun, and is usually lost in its glare, this innermost planet remained only a featureless white blue in their telescopes. The first detailed observa-

tions were made in the late 1960s, when scientists bounced radar signals off its surface. Analysis of the returning radar signals revealed a rough surface and permitted the first accurate determination of Mercury's rotation rate. Prior to these pioneering radar observations, astronomers believed that the planet always kept the same face toward the Sun. These scattered signals indicated, however, that Mercury actually turns on its axis within a period of approximately 59 days. It takes the planet approximately 88 days to orbit around the Sun.

NASA's *Mariner 10* spacecraft provided the first close-up views of Mercury (see fig. 1). This spacecraft was launched from Cape Canaveral in November 1973. After traveling almost five months (including a flyby of Venus), this spacecraft passed within 805 kilometers (500 mi) of Mercury on March 29, 1974. *Mariner 10* then looped around the Sun and made another rendezvous with Mercury on September 21, 1974. This encounter process was repeated a third time on March 16, 1975, before the control gas used to orient the spacecraft was exhausted. This triple flyby of the planet Mercury by *Mariner 10* is sometimes referred to as Mercury I, II and III in the technical literature.

The images of Mercury transmitted back to Earth by *Mariner 10* revealed an ancient, heavily cratered world that closely resembled the Earth's Moon. Unlike the Moon, however, Mercury has huge cliffs (called lobate scarps) that crisscross the planet. These great cliffs were apparently formed when Mercury's interior cooled and shrank, com-

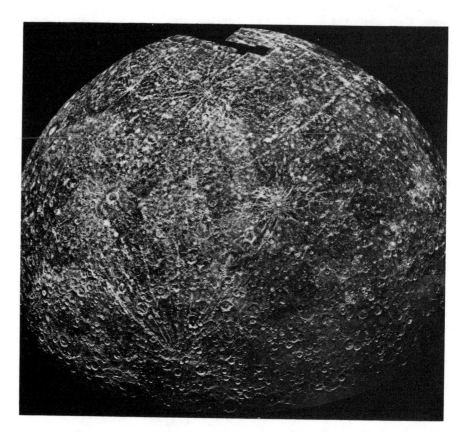

Fig. 1 This mosaic of several hundred images from *Mariner 10* provides an enhanced view of Mercury's southern hemisphere. (Courtesy of NASA/Jet Propulsion Laboratory.)

pressing the planet's crust. The cliffs are as high as 2 kilometers (1.2 mi) and as long as 1,500 kilometers (932 mi).

To the surprise of scientists, instruments on board *Mariner 10* discovered that Mercury has a weak magnetic field. It also has a wisp of an atmosphere—a trillionth of the density of the Earth's atmosphere and made up mainly of traces of helium, hydrogen, sodium and potassium.

Temperatures on the sunlit side of Mercury reach approximately 700 degrees Kelvin (427 degrees Celsius)—a temperature that exceeds the melting point of lead; while on the dark side, temperatures plunge to a frigid 100 degrees Kelvin (minus 173 degrees Celsius). Quite literally, Mercury is a world seared with intolerable heat in the daytime and frozen at night.

The "days" and "nights" on this planet are quite long by terrestrial standards, since it takes about 59 Earth-days for Mercury to make a single rotation about its axis. The planet spins at a rate of approximately 10 kilometers per hour (6

Table 1   Physical and Dynamic Properties of the Planet Mercury

Radius (mean equatorial)	2,439 km
Mass	3.30×10^{23} kg
Mean density	5.44 g/cm³
Acceleration of gravity (at the surface)	3.70 m/sec²
Escape velocity	4.25 km/sec
Normal albedo (averaged over visible spectrum)	0.125
Surface temperature extremes	$-173°$ C to $+427°$ C
Atmosphere	negligible
Number of natural satellites	none
Flux of solar radiation	
Aphelion	6,290 W/m²
Perihelion	14,490 W/m²
Semimajor axis	5.79×10^7 km (0.387 AU)
Perihelion distance	4.60×10^7 km (0.308 AU)
Aphelion distance	6.98×10^7 km (0.467 AU)
Eccentricity	0.20563
Orbital inclination	7.004 degrees
Mean orbital velocity	47.87 km/sec
Sidereal day (a Mercurean "day")	58.646 Earth-days
Sidereal year (a Mercurean "year")	87.969 Earth-days
Earth-to-Mercury distance	
Maximum	20.25×10^7 km (11.26 light-min)
Minimum	8.94×10^7 km (4.97 light-min)

SOURCE: NASA.

miles per hour), measured at its equator. For comparison, the Earth spins at about 1,600 kilometers per hour (1,000 miles per hour) at its equator.

As shown in figure 1, Mercury's surface features include large regions of gently rolling hills and numerous impact craters like those found on the Moon. A large number of these craters are surrounded by blankets of ejecta (material thrown out at the time of a meteorite impact) and secondary craters that were created when chunks of ejected material fell back down to the planet's surface. Because Mercury has a higher gravitational attraction than the Moon, these secondary craters are not spread as widely from each primary crater as occurs on the Moon. One major surface feature discovered by *Mariner 10* is a large impact basin called "Caloris," which is about 1,300 km (810 mi) in diameter. Scientist now believe that Mercury has a large, iron-rich core—the source of its weak, but detectable, magnetic field.

Table 1 presents some contemporary physical- and dynamic-property data about the Sun's closest planetary companion.

Because Mercury lies deep in the Sun's gravity field, its detailed exploration with sophisticated orbiters and landers will require the development of advanced planetary spacecraft that take advantage of intricate "gravity-assist" maneuvers involving both the Earth and Venus.

metagalaxy   The entire system of galaxies, including our Milky Way Galaxy; the entire contents of the Universe together with the region of space it occupies.

meteoroids   An all-encompassing term that refers to solid objects found in space, ranging in diameter size from micrometers to kilometers and in mass from less than 10^{-12} gram to more than 10^{+16} grams. If these pieces of extraterrestrial material are less than 1 gram, they are often called micrometeoriods.

When objects of more than approximately 10^{-6} grams reach the Earth's atmosphere, they are heated to incandescence (that is, they glow with heat) and produce the visible effect popularly called a meteor).

If some of the original meteoroid survives its glowing plunge into the Earth's atmosphere, the remaining unvaporized chunk of space matter is then called a meteorite.

Scientists currently think that meteoroids originate primarily from asteroids and comets that have perihelia (portions of their orbits nearest the Sun) near or inside the Earth's orbit. The parent celestial objects are assumed to have been broken down into a collection of smaller bodies by numerous collisions. Recently formed meteoroids tend to remain concentrated along the orbital path of their parent body. These "stream meteoroids" produce the well-known meteor showers that can be seen at certain dates from Earth.

Meteoroids are generally classified by composition as stony meteorites (chondrites), irons and stony-irons. Of the meteorites that fall on Earth, stony meteorites make up about 93%, irons about 5.5% and stony-irons about 1.5%. Astronomers use the composition of a meteoroid to make inferences about the parent celestial body. Meteoroids are attracted by the Earth's gravitational field, so that the meteoroid flux from allowed directions in near-Earth space is actually increased up to approximately 1.7 over the interplanetary meteoroid flux value. The Earth also shields certain meteoroid arrival directions. Both of these factors—the defocusing factor and the shielding factor—must be considered.

How much extraterrestrial material falls on the Earth each year? Space scientists estimate that about 10^{+7} kilograms (or 10,000 metric tons) of "cosmic rocks" now fall on our planet annually.

What is the meteoroid hazard to an astronaut or cosmonaut in Earth orbit? In June 1983, during the STS-7 Space Shuttle mission, the right-hand middle windshield (windshield no. 5) of the Orbiter *Challenger* was struck by either a micrometeoroid or a tiny piece of manmade space debris. Although the astronaut crew was not endangered by this collision, the outer windshield pane suffered a 0.5-cm (0.20-in) wide damage area (including a small impact crater 0.227 cm [0.0892 in] wide and 0.0452 cm [0.0178 in] deep) and had to be replaced. Similarly, on July 27, 1983, a micrometeoroid or small fragment of space debris struck and damaged a window on the Soviet *Salyut 7* space station. The impact caused a loud crack that was heard by both cosmonauts onboard. Soviet space officials characterized this collision as an "unpleasant surprise." However, the 0.38-cm (0.15-in) diameter crater that was formed on the *Salyut 7*'s window did not threaten the pressure seal integrity of the window. Micrometeoroids large enough to cause such damage are considered rare by space scientists. For example, table 1 presents a contemporary estimate for the time between collisions between an object the size of a Space Shuttle Orbiter in low Earth orbit and a meteoroid of mass greater than a given meteoroid mass.

On a much larger "collision scale," meteoroid impacts are now considered by planetary scientists to have played a basic role in the evolution of planetary surfaces in the early history of the Solar System. Although dramatically evident in the cratered surfaces on many planets and moons, here on Earth this stage of surface evolution has essentially been lost due to later crustal recycling and weathering processes.

Table 1 Time Between Meteoroid Collisions for a Space Shuttle Orbiter in Low Earth Orbit (300-km Altitude)

Minimum Meteoroid Mass (g)	Estimated Time Between Collisions (yr)
10	350,000
1	25,000
0.1	1,800
0.01	130

SOURCE: NASA data.

Space scientists now believe that meteorites typically spend from 10 to 500 million years exposed to the space environment. This suggests relatively recent collisions and breakups have occurred in the asteroid belt. Some meteorites are basalts, aged about 4.5 billion years, indicating that early melting (perhaps due to decay heat from primordial, short-lived radioactive elements) may have occurred on their parent asteroids. The formative ages of most meteorites (typically 4.5 to 4.6 billion years) provides us with a firm estimate for the age of our Solar System. Solar wind gases have been found trapped within some meteorites, providing space scientists with a "snapshot" of past solar activity. Finally, amino acids and other organic compounds of extraterrestrial origin have also been discovered in several carbon-rich meteorites.

LUNAR METEORITES

Although it is now generally accepted by planetary scientists that most meteorites found on Earth have come from asteroids, at least eight meteorites found in Antarctica have now been identified as rocks ejected from the lunar surface during very energetic asteroid or comet impacts. Analysis of the chemical composition of these "lunar meteorites" indicated that the impact sites were most likely quite distant from the Apollo (U.S.-manned) and Luna (Soviet-robot) landing sites. If an asteroid or comet of sufficient mass and velocity hits the surface of the Moon, a small fraction of the impacted lunar material could depart from the Moon's surface with velocities greater than its escape velocity (2.4 km/sec). Recent computer simulations of such highly energetic collisions show that a fraction of the ejected material will eventually reach the Earth's surface, with Moon-to-Earth transit times ranging from under 1 million to upwards of 100 million years.

MARTIAN METEORITES

If impact material ejected from the Moon has reached the Earth, then planetary scientists also speculate that very energetic asteroid or comet impacts on Mars (5.0 km/sec escape velocity) could be the source for other types of interesting meteorites recently found in Antarctica. Current investigation of one suspected "Martian meteorite," an 8-kilogram (17.6-lb) meteorite called EETA 79001, has revealed the presence of a small quantity of the type of carbon-bearing material considered necessary for the emergence of life. A few of the scientists examining this "piece of Mars" have even very cautiously hinted that they might be looking at some type of extraterrestrial fossil life-form. Other scientists prefer to wait and compare these data with the data provided by Mars rocks collected during future robot sample return missions before commenting further on the possibility of life (past or present) on the Red Planet.

See also: **asteroid; comet; hazards to space workers; space debris**

microgravity Because the inertial trajectory of a spacecraft (for example, the Space Shuttle) compensates for the force of the Earth's gravity, an orbiting spacecraft and all its contents approach a state of free-fall. In this state of free-fall, all objects inside the spacecraft appear "weightless."

It is important to understand how this condition of weightlessness, or the apparent lack of gravity, develops. Newton's universal law of gravitation tells us that any two objects have a gravitational attraction for each other that is proportional to their masses and inversely proportional to the square of the distance between their centers of mass. It is also interesting to recognize that a spacecraft orbiting the Earth at an altitude of 400 kilometers is only 6 percent farther from the center of the Earth than it would be if it were on the Earth's surface. Using Newton's law, we find that the gravitational attraction at this particular altitude is only 12 percent less than the attraction of gravity at the surface of the Earth. In other words, an Earth-orbiting spacecraft and all its contents are very much under the influence of the Earth's gravity! The phenomenon of weightlessness occurs because the spacecraft and it contents are in a continual state of free-fall.

Figure 1 describes the different orbital paths a falling object may take when "dropped" from a point above the Earth's sensible atmosphere. With no tangential-velocity component, an object would fall straight down (trajectory 1) in our simplified demonstration. As we give the object an increasing tangential-velocity component, it still "falls" toward the Earth under the influence of terrestrial gravitational attraction, but the velocity component now gives the object a trajectory that is a segment of an ellipse. As shown in trajectories 2 and 3 in figure 1, as we give the object a larger tangential velocity, the point where it finally hits the Earth moves farther and farther away from the release point. If we keep increasing this velocity component, the object eventually "misses the Earth" completely (trajectory 4). As the tangential velocity is further increased, the object's trajectory takes the form of a circle (trajectory

5) and then a larger ellipse, with the release point representing the point of closest approach to the Earth (or "perigee"). Finally, when the initial tangential-velocity component is about 41 percent greater than that needed to achieve a circular orbit, the object follows a parabolic, or escape, trajectory and will never return (trajectory 6).

Einstein's principle of equivalence tells us that the physical behavior inside a system in free-fall is identical to that inside a system far removed from other matter that could exert a gravitational influence. Therefore, the term *zero gravity* (also called "zero g") or *weightlessness* is frequently used to describe a free-falling system in orbit.

Sometimes people ask what is the difference between mass and weight. Why do we say, for example, "weightlessness" and not "masslessness"? *Mass* is the physical substance of an object—it has the same value everywhere. *Weight*, on the other hand, is the product of an object's mass and the local acceleration of gravity (in accordance with Newton's second law of motion, $F = ma$). For example, you would weigh about one-sixth as much on the Moon as here on Earth, but your mass remains the same in both places.

A "zero-gravity" environment is really an ideal situation that can never be totally achieved in an orbiting spacecraft. The venting of gases from the space vehicle, the minute drag exerted by a very thin terrestrial atmosphere at orbital altitudes and even crew motions create nearly imperceptible forces on people and objects alike. These tiny forces are collectively called "microgravity." In a microgravity environment, astronauts and their equipment are almost, but not entirely, weightless.

Microgravity represents an intriguing experience for space travelers. You can perform slow-motion somersaults and handsprings. You can float with ease through a space cabin (see fig. 2). You can push off one wall of a space station and drift effortlessly to the other side. You can lift or move "heavy" objects, which are essentially weightless. And if you're just a little bit clumsy, you don't need to worry about dropping things—whatever slips from your hand will simply float away.

However, life in microgravity is not necessarily easier than life on Earth. For example, the caloric (food-intake) requirements for people living in microgravity are the same as those on Earth. Living in microgravity also calls for special design technology. A beverage in an open container, for instance, will cling to the inner or outer walls and, if shaken, will leave the container as free-floating droplets or fluid globs. Such free-floating droplets are not merely an inconvenience. They can annoy crew members (no one wants to get "slimed" in orbit), and they represent a definite ha ard to equipment, especially sensitive electronic devices and computers.

Therefore, water is usually served in microgravity through a specially designed dispenser unit that can be turned on or off by squeezing and releasing a trigger. Other beverages, such as orange juice, are typically served in sealed containers through which a plastic straw can be inserted.

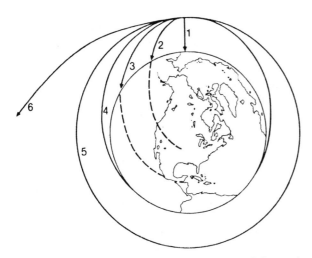

Fig. 1 Various orbital paths of a falling body around the Earth. (Drawing courtesy of NASA.)

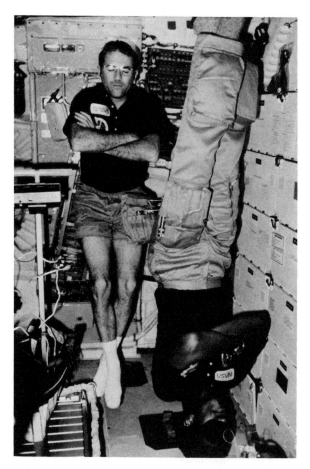

Fig. 2   STS-8 astronauts (left) Truly and (right) Bluford enjoy a "floating" rest session in the Shuttle *Challenger's* cabin. (Courtesy of NASA.)

When the beverage is not being sipped, the straw is simply clamped shut.

Microgravity living also calls for special considerations in handling solid foods. Crumbly foods are provided only in bite-sized pieces to avoid having crumbs floating around the space cabin. Gravies, sauces and dressings have a viscosity (stickiness) that prevents them from simply lifting off the food trays and floating away. Typical space food trays are equipped with magnets, clamps and double-adhesive tape to hold metal, plastic and other utensils. Astronauts are provided with forks and spoons. However, extraterrestrial diners must learn to eat without sudden starts or stops if they expect the solid food to stay on their utensils.

Personal hygiene is also a bit challenging in microgravity. Waste water in the Shuttle's galley from utensil cleanup or an astronaut's washing (sponge bath) is directed away by a flow of air (which provides a force substituting for gravity) to a drain that then leads to a sealed tank. Shuttle astronauts must take sponge baths rather than showers or regular baths. However, space-station crews will most likely have

more "Earthlike" shower facilities, perhaps something like the apparatus used on Skylab.

Because water adheres to the skin in microgravity, perspiration can also be annoying, especially during strenuous activities. In the absence of proper air circulation, perspiration can accumulate layer by layer on an astronaut's skin.

Waste elimination in microgravity represents another challenging design problem. To help Shuttle astronauts go to the bathroom, a special toilet device has been engineered to closely resemble the normal sanitary procedures performed here on Earth. The main differences are that the astronaut must use a seat belt and foot restraints to keep from drifting. The wastes themselves are flushed away by a flow of air and a mechanical device.

The Space Shuttle's waste-collection system has a set of controls that are used to configure the system for various operational modes, including urine collection only, combined urine and feces collection and emesis collection (vomit collection). The overall microgravity toilet system consists of a commode (or waste collector) to handle solid wastes and a urinal assembly to handle fluids.

The urinal is used by both male and female astronauts—with the individual either holding the urinal while standing or sitting on the commode with the urinal mounted to the waste-collection system. Since the urinal has a contoured cup with a spring assembly, it provides a good seal with the female crew member's body. During urination, a flow of air creates a pressure differential that draws the urine off into a fan separator/storage tank assembly.

The microgravity commode is used for collecting both feces and emesis. When properly functioning, it has a capacity for storing the equivalent of 210 person-days of vacuum-dried feces and toilet tissue. This device may be used up to four times per hour, and it may be used simultaneously with the urinal. To operate the waste collector during defecation, the astronaut positions himself or herself on the commode seat. Handholds, foot restraints and waist restraints help the individual maintain a good seal with the seat. The crew member uses this equipment like a normal terrestrial toilet, including tissue wipes. Used tissues are disposed of in the commode. Everything stored in the waste collector—feces, tissues and fecal and emesis bags—are then subjected to vacuum-drying in the collector.

Shaving can also cause problems in microgravity, if whiskers end up floating around the cabin. These free-floating whiskers could damage delicate equipment (especially electronic circuits and optical instruments) or else irritate the eyes and lungs of space travelers. One solution is to use a safety razor and shaving cream or gel. The whiskers will adhere to the cream until wiped off with a disposable towel. Another approach is to use an electronic razor with a built-in vacuum device that sucks away and stores the whiskers as they are cut.

For long-duration space missions, other personal hygiene tasks that might require some special procedure or device in microgravity include nail trimming and hair cutting. Special devices have also been developed for female astronauts to support personal hygiene requirements associated with the menstrual cycle.

Microgravity living is definitely different from the lifestyles permitted at the bottom of a one-g gravity well on the surface of the Earth. Furniture, for example, must be bolted in place—or else it will simply float around the cabin. Tether lines, belts, Velcro anchors and handholds enable astronauts to move around and to keep themselves and other objects in place.

Sleeping in space is also an interesting experience. Shuttle astronauts can sleep either horizontally or vertically while in orbit. Their fireproof sleeping bags attach to rigid padded boards for support. But the astronauts themselves literally sleep "floating in air."

Special tools (such as torqueless wrenches), handholds and foot restraints help an astronaut turn a nut or tighten a screw while in orbit. These devices are needed to balance or neutralize reaction forces. If these special devices were not available, a space worker might find himself or herself helplessly rotating around a "work piece" or work station.

Exposure to microgravity also causes a variety of physiological (bodily) changes. For example, space travelers appear to have "smaller eyes," because their faces have become puffy. They also get rosy cheeks and distended veins in their foreheads and necks. They may even be a little bit taller than they are on Earth, because their body masses no longer "weigh down" their spines.

Leg muscles shrink, and anthropometric (measurable postural) changes also occur. Astronauts tend to move in a slight crouch, with head and arms forward.

Some space travelers suffer from a temporary condition resembling motion sickness. This condition is called space sickness or space-adaptation syndrome. In addition, sinuses become congested, leading to a condition similar to a cold.

Many of these microgravity-induced physiological effects appear to be caused by fluid shifts from the lower to the upper portions of the body. So much fluid goes to the head that the brain may be fooled into thinking that the body has too much water. This can then result in an increased production of urine.

Extended stays in microgravity tend to shrink the heart, decrease production of red blood cells and increase production of white blood cells. A process called resorption occurs. This is the leaching of vital minerals and other chemicals (such as calcium, phosphorus, potassium and nitrogen) from the bones and muscles into the body fluids that are then expelled as urine. Such mineral and chemical losses can have adverse physiological and psychological effects. In addition, prolonged exposure to a microgravity environment might cause bone loss and a reduced rate of bone-tissue formation.

While a brief stay (say, up to 30 days) in microgravity may prove to be an exhilarating, nondetrimental experience for most space travelers, permanent space bases and large space settlements will resort to "artifical gravity" (created through the slow rotation of these habitats and facilities) to provide a more "Earthlike" home in cislunar space and to

avoid any serious health effects that might arise from permanent exposure to a microgravity environment. These space habitats might even offer the inhabitants the very exciting possibility of life in a multiple-gravity-level world, with different modules or zones simulating gravity conditions from microgravity up to normal terrestrial gravity.

Besides providing an interesting new dimension for human experience, the microgravity environment of an orbiting space system offers the ability to manufacture new and improved materials and products that simply cannot be made on Earth. Although microgravity can be simulated from here on Earth using drop towers, special airplane trajectories and sounding rocket flights, these simulations are only short-duration activities (seconds to minutes) that are frequently "contaminated" by vibrations and other undesirable effects. However, the long-term microgravity environment found in orbit provides an entirely new dimension for materials-science research, life-science research and manufacturing processes. Today, we can only partially speculate on the overall impact that access to microgravity will have on our 21st-century life-styles. Through the use of permanent space stations and platforms, we will be the first human generation that can regularly examine material behavior, physical processes, manufacturing techniques and life processes in the absence of Earth's loving but firm one-g grasp! The potential for revolutionary breakthroughs, unanticipated discoveries and unusual developments in a great number of technical areas is simply astounding!

See also: **materials research and processing in space; Newton's laws of motion; orbits of objects in space; space commerce; space industrialization; space settlement; space station; Space Transportation System**

microorganism   A very tiny plant or animal, especially a protozoan or a bacterium.
See also: **extraterrestrial contamination**

microwave   A comparatively short-wavelength electromagnetic (EM) wave in the radio-frequency portion of the EM spectrum. The term *microwave* is usually applied to those EM wavelengths that are measured in centimeters, approximately 30 centimeters to 1 millimeter (with corresponding frequencies of 1 GHz [gigahertz] to 300 GHz).
See also: **electromagnetic spectrum; Satellite Power System**

Milky Way Galaxy   Our home Galaxy. The immense band of stars stretching across the night sky represents our "inside view" of the Milky Way. Classified as a spiral galaxy, the Milky Way is characterized by the following general features: a spherical *central bulge* at the *galactic nucleus;* a thin *disk* of stars, dust and gas formed in a beautiful, extensive pattern of spiral arms; and a *halo* defined by an essentially spherical distribution of globular clusters. This disk is between 2,000 and 3,000 light-years thick and is some 100,000 light-years in diameter. It contains primarily younger, very luminous, metal-rich stars (called Population I stars), as well as gas and dust. Most of the stars found in the halo are older, metal-poor stars (called Population II stars); while the galactic nucleus appears to contain a mixed population of older and younger stars. Some astrophysicists now speculate that a massive black hole, containing millions of "devoured" solar masses, may lie at the center of many galaxies, including our own. Current estimates suggest that our Galaxy contains between 200 and 600 billion solar masses. (A solar mass is a unit used for comparing stellar masses, with one solar mass being equal to the mass of our Sun.) Our Solar System is located about 30,000 light-years from the center of the Galaxy.

See also: **black holes; galaxy; stars**

mirror matter   A popular name for antimatter, which is the "mirror image" of ordinary matter. For example, an antielectron (also called a positron) has a positive charge, while an ordinary election has a negative charge.

See also: **antimatter**

Mission to Planet Earth   A proposed NASA initiative to understand our home planet, how forces shape and affect its environment, how that environment is changing, and how those planetary changes will affect current and future generations of human beings. This initiative was first formalized in a NASA report entitled *Leadership and America's Future in Space* (1987) by Dr. Sally K. Ride (America's first woman in space). The overall goal of Mission to Planet Earth is to obtain a comprehensive scientific understanding of the entire *Earth System*, by describing how its various components function, how they currently interact and how they might evolve in the future. Comprehensive observations of the Earth from the unique vantage point of space have given scientists the capability to effectively study, for the first time in human history, the complex, highly interactive processes of the Earth as a synergistic system that is constantly changing and evolving.

Interactive physical, chemical and biological processes connect the oceans, continents, atmosphere and biosphere of the Earth in a very complex way. Oceans, ice-covered regions (called the cryosphere) and the atmosphere are closely linked and shape the Earth's climate; volcanism links the solid (inner) Earth (called the lithosphere) with the atmosphere; and biological activity contributes substantially to the cycling of chemicals important to life, such as carbon, oxygen and nitrogen. It is also quite clear as a result of observing the Earth from space that human activity is also a making a major impact on the Earth System.

In the original NASA initiative, Mission to Planet Earth would:

(1) Establish and maintain a global observational system in space that would include both experimental and operational observation platforms in low-inclination, polar and geostationary orbits. These free-flying observation platforms (built by the United States and other nations) would perform integrated, long-term measurements of our planet and its key interactive processes.

(2) Use the large volume of data from these space platforms in coordination with data obtained on Earth (ground-truth data) and reliable global modeling numerical techniques to document, understand and eventually predict the major consequences of global change.

This integrated system would measure the full complement of our planet's characteristics, including global cloud cover, vegetation cover and ice cover; global rainfall and moisture; ocean chlorophyll content and ocean topography; the motions and deformations of the Earth's tectonic plates; and atmospheric concentration of gases such as carbon dioxide (CO_2), methane (CH_4) and ozone (O_3).

Space-based observations would also be coordinated with ground-based measurements, and the data from all these observations would then be combined and integrated by state-of-the-art information management systems. This highly automated, integrated database would then support the development of diagnostic and predictive Earth System models.

See also: **Earth Observing System; global change**

moon   A natural satellite of any planet.

Moon   The Moon (the term is capitalized when used in this sense) is the Earth's only natural satellite and closest celestial neighbor. While life on Earth is made possible by the Sun, it is also regulated by the periodic motions of the Moon. For example, the months of our year are measured by the regular motions of the Moon around the Earth, and the tides rise and fall because of the gravitational tug-of-war between the Earth and the Moon. Throughout history, the Moon has had a significant influence of human culture, art and literature. Even in the Age of Space, it has proved to be a major sociotechnical stimulus. It was just far enough away to represent a real technological challenge to reach it; yet it was still close enough to allow us to be successful on the first concentrated effort. The Apollo expeditions to the Moon from 1969 to 1972 made us extraterrestrial travelers.

Scientific interest in the Moon dates back to the earliest periods of recorded history, but it wasn't until very recently, with the birth of the Space Age, that we finally developed the technical tools needed to examine our celestial companion firsthand. As evidenced by the "sensational" news clipping describing the pioneering rocket work of Dr. Robert Goddard (the father of American rocketry), the Moon was considered totally inaccessible by the vast majority of people. Today, however, it has become a "planet" to visit, to explore and eventually to inhabit.

The Moon was the first extraterrestrial object surveyed by spacecraft. Table 1 provides a detailed summary of all the major unmanned missions to the Moon, while table 2 provides information about the manned Apollo program expeditions into cislunar space and on to the lunar surface itself.

Fig. 1   The crater Tsiolkovsky as photographed from the *Apollo 15* Command Module, 1971. (Courtesy of NASA.)

From evidence gathered by the early unmanned missions (such as Ranger, Surveyor and the Lunar Orbiter spacecraft) and by the Apollo missions, lunar scientists have learned a great deal more about the Moon and have been able to construct a geologic history dating back to its infancy. Table 3 provides selected physical and dynamic properties of the Moon.

Because the Moon does not have any oceans or other free-flowing water and lacks a sensible atmosphere, appreciable erosion, or "weathering," has not occurred there. In fact, the Moon is actually a "museum world." The primitive materials that lay on its surface billions of years ago in an excellent state of preservation. (See fig. 1.) Scientists believe that the Moon was formed over four billion years ago and then differentiated quite early, perhaps only 100 million years later. Tectonic activity ceased eons ago on the Moon. The lunar crust and mantle are quite thick, extending inward to more than 800 kilometers. However, the deep interior of the Moon is still unknown. It may contain a small iron core at its center, and there is some evidence that the lunar interior may be hot and even partially molten. Moonquakes have been measured within the lith-

osphere and interior, most being the result of gravitational stresses. Chemically, the Earth and the Moon are quite similar, though compared to the Earth, the Moon is depleted in the more easily vaporized materials. The lunar surface consists of highlands composed of alumina-rich rocks that formed from a globe-encircling molten sea and maria made up of volcanic melts that surfaced about 3.5 billion years ago. However, despite all we have learned in the past three decades about our nearest celestial neighbor, lunar exploration has really only just started. Several puzzling mysteries still remain, including an explanation for the remnant magnetism measured in the rocks despite the absence of a lunar dynamo (a hot, molten core that conducts electric currents) and the origin of the Moon itself.

There are three general theories concerning the origin of the Moon that are currently popular. The least "exotic" hypothesis suggests that the Moon and the Earth accreted (accumulated) from a nebula of gas and dust that surrounded the primordial Sun in much the same relative positions that they occupy today. Another theory proposes that the Moon formed in a different part of the solar nebula but was later captured by the Earth. In this lunar-origin

Table 1   A Summary of Moon Missions

SPACECRAFT	LAUNCH DATE; NATIONALITY; WEIGHT; MISSION/RESULTS
Pioneer 0	Aug. 17, 1958; USA; 38 kg; attempt to orbit Moon/launch vehicle exploded at an altitude of 16 km.
Pioneer 1	Oct. 11, 1958; USA; 38 kg; attempt to orbit Moon/launch vehicle reached an altitude of 113,830 km, then fell back into the South Pacific.
Pioneer 2	Nov. 8, 1958; USA; 39 kg; attempt to orbit Moon/launch vehicle reached an altitude of 1,550 km, then fell back to Earth near Africa.
Pioneer 3	Dec. 6, 1958; USA; 6 kg; attempt to fly-by Moon/launch vehicle reached an altitude of 102,320 km, then fell back to Earth over Africa.

Table 1   A Summary of Moon Missions (1958–1989) (continued)

SPACECRAFT	LAUNCH DATES; NATIONALITY; WEIGHT; MISSION/RESULTS
Luna 1[a]	Jan. 2, 1959; USSR; 361 kg; attempt to impact Moon/partial success—missed Moon by some 5,000 km, then entered solar orbit.
Pioneer 4	March 3, 1959; USA; 6 kg; lunar flyby/successful—passed Moon at 60,500 km, then entered solar orbit.
Luna 2	Sept. 12, 1959; USSR; 390 kg; impact Moon/successful—first lunar impact, struck 335 km from visible center.
Pioneer P-1	Sept. 24, 1959; USA; 170 kg; attempt to orbit Moon/launch vehicle and spacecraft destroyed in explosion during static test before launch.
Luna 3	Oct. 4, 1959; USSR; 435 kg; photography of far side of Moon/successful—returned pictures of 70 percent of lunar far side.
Pioneer P-3	Nov. 26, 1959; USA; 169 kg; attempt to orbit Moon/failure—launch vehicle shroud tore away during ascent, payload impacted near Africa.
Pioneer P-30	Sept. 25, 1960; USA; 176 kg; attempt to orbit Moon/failure—launch vehicle malfunction, payload impacted in Africa.
Pioneer P-31	Dec. 15, 1960; USA; 176 kg; attempt to orbit Moon/failure—launch vehicle climbed to an altitude of 13 km, then exploded.
Ranger 1	Aug. 23, 1961; USA: 306 kg; high-Earth-orbit test of spacecraft/failure—intended to climb to 1,102,850 km but remained in low Earth orbit.
Ranger 2	Nov. 18, 1961; USA; 306 kg; high-Earth-orbit test of spacecraft/failure—intended to climb to 1,102,850 km but remained in low Earth orbit.
Ranger 3	Jan. 26, 1962; USA; 330 kg; attempted TV reconnaissance of Moon and hard landing/partial success—spacecraft missed Moon by 36,808 km, no TV pictures or landed instruments.
Ranger 4	April 23, 1962; USA; 331 kg; attempted TV reconnaissance of Moon and hard landing/partial success—timer failure, spacecraft fell on far side of Moon, no TV pictures.
Ranger 5	Oct. 18, 1962; USA; 342 kg; attempted TV reconnaissance of Moon and hard landing/partial success—power failure caused spacecraft to miss Moon by 725 km, entered solar orbit.
Luna (unannounced)	Jan. 4, 1963; USSR; 1,400 kg (?); attempted soft landing on Moon/failure—spacecraft achieved only Earth orbit.
Luna 4	April 2, 1963; USSR; 1,422 kg; attempted soft landing on Moon/partial success—lunar flyby at 8,500 km, entered solar orbit.
Ranger 6	Jan. 30, 1964; USA; 365 kg; attempted TV reconnaissance of Moon and hard landing/partial success—impacted on target but no TV pictures returned.
Ranger 7	July 28, 1964; USA: 366 kg; TV reconnaissance of Moon and hard landing/success—returned over 4,300 high-resolution images of Moon before impacting on target at Mare Nubium.
Ranger 8	Feb. 17, 1965; USA; 367 kg; TV reconnaissance of Moon and hard landing/success—returned 7,137 high-resolution images of Moon before impacting on target at Mare Tranquillitatis.
Kosmos 60	March 12, 1965; USSR; 1,470 kg; attempted soft landing on Moon/failure—spacecraft achieved only Earth orbit.
Ranger 9	March 21, 1965; USA; 366 kg; TV reconnaissance of Moon and hard landing/successful—returned 5,814 high-resolution images of Moon before impacting inside Crater Alphonsus.
Luna 5	May 9, 1965; USSR; 1,476 kg; attempted soft landing on Moon/partial success—retrofire failure caused spacecraft to crash-land on Mare Nubium.
Luna 6	June 8, 1965; USSR; 1,442 kg; attempted soft landing on Moon/partial success—missed Moon by 160,000 km, entered solar orbit.

[a]Soviet data are from TASS bulletins.

Table 1 A Summary of Moon Missions (1958–1989) (*continued*)

SPACECRAFT	LAUNCH DATES; NATIONALITY; WEIGHT; MISSION/RESULTS
Zond 3	July 18, 1965; USSR; 890 kg (?); photography of Moon's far side/success—returned 25 pictures of lunar far side, then entered solar orbit.
Luna 7	Oct. 4, 1965; USSR; 1,506 kg; attempted soft landing on Moon/partial success—retrofired early, spacecraft crashed in Oceanus Procellarum.
Luna 8	Dec. 3, 1965; USSR; 1,552 kg; attempted soft landing on Moon/partial success—retrofired late, crashed in Oceanus Procellarum.
Luna 9	Jan. 31, 1966; USSR; 1,583 kg; soft landing on Moon/success—first lunar soft landing, 100-kg capsule returned photographs from lunar surface at Oceanus Procellarum.
Kosmos 111	March 1, 1966; USSR; 1,600 kg (?); attempt to orbit Moon/failure—spacecraft only achieved Earth orbit.
Luna 10	March 31, 1966; USSR; 1,600 kg; to orbit Moon/successful—first spacecraft to achieve lunar orbit; returned physical data from lunar surface.
Surveyor 1	May 30, 1966; USA; 995 kg; soft landing on Moon/success—touchdown north of Flamsteed, returned 11,237 pictures from lunar surface.
Explorer 33	July 1, 1966; USA; 93 kg; attempt to orbit Moon/partial success—spacecraft failed to approach Moon at proper velocity, achieved Earth orbit instead.
Lunar Orbiter 1	Aug. 10, 1966; USA; 387 kg; to orbit Moon/successful—photographic mapping of lunar surface (Apollo landing sites).
Luna 11	Aug. 24, 1966; USSR; 1,604 kg (?); to orbit Moon/success—spacecraft achieved lunar orbit but did not return lunar surface pictures.
Surveyor 2	Sept. 20, 1966; USA; 1,000 kg; attempt to soft-land on the Moon/partial success—stabilization failure, spacecraft crashed southeast of Crater Copernicus.
Luna 12	Oct. 22, 1966; USSR; 1,625 kg (?); to orbit Moon/success—spacecraft returned images of lunar surface.
Lunar Orbiter 2	Nov. 6, 1966; USA; 390 kg; to orbit Moon/successful—photographic mapping of lunar surface (Apollo landing sites and far side).
Luna 13	Dec. 21, 1966; USSR; 1,595 kg (?); soft landing on Moon/successful—soft landing achieved on Oceanus Procellarum, returned pictures of lunar surface, studied lunar soil density.
Lunar Orbiter 3	Feb. 5, 1967; USA; 385 kg; to orbit Moon/successful—photographic mapping of lunar surface (Apollo landing sites).
Surveyor 3	April 17, 1967; USA; 1,035 kg; soft landing on Moon/success—touch down in Oceanus Procellarum, returned images of surface, dug lunar soil with shovel.
Lunar Orbiter 4	May 4, 1967; USA; 390 kg; to orbit Moon/successful—photographic mapping of large areas of Moon.
Surveyor 4	July 14, 1967; USA; 1,039 kg; attempted soft landing on Moon/partial success—signals ceased at lunar impact on Sinus Medii.
Explorer 35	July 19, 1967; USA; 104 kg; to orbit Moon/successful—returned physical data from lunar orbit.
Lunar Orbiter 5	Aug. 1, 1967; USA; 390 kg; to orbit Moon/successful—photographic mapping of Moon, including much of far side.
Surveyor 5	Sept. 8, 1967; USA; 1,005 kg; soft landing on Moon/success—soft-landed on Mare Tranquillitatis and returned 18,006 pictures of lunar surface, performed first chemical analysis of lunar soil.
Surveyor 6	Nov. 7, 1967; USA; 1,008 kg; soft landing on Moon/success—soft-landed on Sinus Medii, returned 30,065 pictures of lunar surface and performed chemical and mechanical studies of lunar soil.
Surveyor 7	Jan. 7, 1968; USA; 1,040 kg; soft landing on Moon/success—landed near north rim of crater Tycho, returned 21,274 images of lunar surface and performed chemical analysis of lunar soil from trench it dug.

Table 1 A Summary of Moon Missions (1958–1989) (*continued*)

SPACECRAFT	LAUNCH DATES; NATIONALITY; WEIGHT; MISSION/RESULTS
Zond 4	March 2, 1968; USSR; 5,800 kg (?); spacecraft test mission/partial success—flew to lunar distance but recovery in doubt.
Luna 14	April 7, 1968; USSR; 1,615 kg (?); to orbit Moon/success—achieved lunar orbit, returned data on lunar mass distribution.
Zond 5	Sept. 14, 1968; USSR; 5,800 kg (?); circumlunar flight/successful—ballistic reentry with biological specimens and pictures.
Zond 6	Nov. 10, 1968; USSR; 5,800 kg (?); circumlunar flight/successful—lifting reentry with biological specimens and pictures.
Luna 15	July 13, 1969; USSR; 5,800 kg (?); soft landing on Moon/partial success—lunar orbit achieved, but crash landing occurred on Mare Crisium.
Zond 7	Aug. 7, 1969; USSR; 5,800 kg (?); circumlunar flight/successful—photographs of lunar far side taken and lifting reentry accomplished.
Kosmos 300	Sept. 23, 1969; USSR; 5,800 kg (?); soft landing on Moon/failure—spacecraft only achieved Earth orbit.
Kosmos 305	Oct. 22, 1969; USSR; 5,800 kg (?); soft landing on Moon/failure—spacecraft only achieved Earth orbit.
Luna 16	Sept. 12, 1970; USSR; 5,800 (?); soft landing on Moon/successful—landed on Mare Faecunditatis, performed automated lunar soil sample collection and returned it to Earth.
Zond 8	Oct. 20, 1970; USSR; 5,800 kg (?); circumlunar flight/successful—took photographs and accomplished ballistic reentry.
Luna 17	Nov. 10, 1970; USSR; 5,800 (?); soft landing on Moon/successful—landed on Mare Imbrium, included *Lunokhod 1* automated rover vehicle, which conducted long-term exploration program.
Luna 18	Sept. 2, 1971; USSR; 5,800 kg (?); soft landing on Moon/partial success—spacecraft achieved lunar orbit but crashed on landing at Mare Faecunditatis.
Luna 19	Sept. 28, 1971; USSR; 5,800 kg (?); to orbit Moon/success—returned photographs and data.
Luna 20	Feb. 14, 1972; USSR; 5,800 kg (?), soft landing on Moon/successful—landed on Mare Crisium, made automated soil sample collection and returned it to Earth.
Luna 21	Jan. 8, 1973; USSR; 5,800 kg (?); soft landing on Moon/successful—landed near Le Monnier, included *Lunokhod 2* automated rover vehicle.
Explorer 49	June 10, 1973; USA; 328 kg; to orbit Moon/successful—performed radioastronomy experiments from far side of Moon.
Luna 22	May 29, 1974; USSR; 5,800 kg (?); to orbit Moon/success—achieved lunar orbit, took photographs and collected data.
Luna 23	Oct. 28, 1974; USSR; 5,800 kg (?); soft landing on Moon/partial success—soft landing in southern part of Mare Crisium but drill damaged so no soil sample returned to Earth.
Luna 24	Aug. 9, 1976; USSR; 5,800 kg (?); soft landing on Moon/successful—performed automated soil sample collection in Mare Crisium and returned to Earth.

SOURCE: NASA.

Table 2 Apollo Program Summary

Spacecraft Name	Crew	Date	Flight Time (Hrs., Min., Sec.)	Revolutions	Remarks
Apollo 7	Walter H. Schirra Donn Eisele Walter Cunningham	10/11–22/68	260:8:45	163	First manned Apollo flight demonstrated the spacecraft, crew and support elements. All performed as required.

Table 2 Apollo Program Summary

Spacecraft Name	Crew	Date	Flight Time (Hrs., Min., Sec.)	Resolutions	Remarks
Apollo 8	Frank Borman James A. Lovell, Jr. William Anders	12/21–27/68	147:00:41	10 rev. of Moon	History's first manned flight to the vicinity of another celestial body.
Apollo 9	James A. McDivitt David R. Scott Russell L. Schweickart	3/3–13/69	241:00:53	151	First all-up manned Apollo flight (with Saturn V and command, service and lunar modules). First Apollo EVA. First docking of CSM with LM.
Apollo 10	Thomas P. Stafford John W. Young Eugene A. Cernan	5/18–26/69	192:03:23	31 rev. of Moon	Apollo LM descended to within 14.5 km of Moon and later rejoined CSM. First rehearsal in lunar environment.
Apollo 11	Neil A. Armstrong Michael Collins Edwin E. Aldrin, Jr.	7/16–24/69	195:18:35	30 rev. of Moon	First landing of men on the Moon. Total stay time: 21 hrs., 36 min.
Apollo 12	Charles Conrad, Jr. Richard F. Gordon, Jr. Alan L. Bean	11/14–24/69	244:36:25	45 rev. of Moon	Second manned exploration of the Moon. Total stay time: 31 hrs., 31 min.
Apollo 13	James A. Lovell, Jr. John L. Swigert, Jr. Fred W. Haise, Jr.	4/11–17/70	142:54:41	—	Mission aborted because of service module oxygen tank failure.
Apollo 14	Alan B. Shepard, Jr. Stuart A. Roosa Edgar D. Mitchell	1/31–2/9/71	216:01:59	34 rev. of Moon	First manned landing in and exploration of lunar highlands. Total stay time: 33 hrs., 31 min.
Apollo 15	David R. Scott Alfred M. Worden James B. Irwin	6/26–7/7/71	295:11:53	74 rev. of Moon	First use of lunar roving vehicle. Total stay time: 66 hrs., 55 min.
Apollo 16	John W. Young Thomas K. Mattingly II Charles M. Duke, Jr.	3/16–27/72	265:51:05	64 rev. of Moon	First use of remote controlled television camera to record lift-off of the LM ascent stage from the lunar surface. Total stay time: 71 hrs., 2 min.
Apollo 17	Eugene A. Cernan Ronald E. Evans Harrison H. Schmitt	12/7–19/72	301:51:59	75 rev. of Moon	Last manned lunar landing and exploration of the Moon in the Apollo program returned 110 kg of lunar samples to Earth. Total stay time: 75 hrs.

SOURCE: NASA.

Table 3 Physical and Astrophysical Properties of the Moon

Diameter (equatorial)	3,476 km	Sidereal month (rotation period)	27.322 days
Mass	7.350×10^{22} kg		
Mass (Earth's mass = 1.0)	0.0123	Albedo (mean)	0.07
Average density	3.34 g/cm³	Mean visual magnitude (at full)	−12.7
Mean distance from Earth (center-to-center)	384,400 km	Surface area	37.9×10^6 km²
Surface gravity (equatorial)	1.62 m/sec²	Volume	2.20×10^{10} km³
Escape velocity	2.38 km/sec	Atmospheric density (at night on surface)	2×10^5 molecules/cm³
Orbital eccentricity (mean)	0.0549		
Inclination of orbital plane (to ecliptic)	5° 09′	Surface temperature	102 K–384 K

SOURCE: NASA.

Fig. 2 The heavily cratered far side of the Moon as seen from the *Apollo 16* Command Module, 1972. (Courtesy of NASA.)

scenario, the Moon may have either survived intact or else been fragmented by numerous collisions and then reaccumulated in Earth orbit. The third general lunar-origin theory suggests that the Moon separated from a partially differentiated Earth during a rapid rotational instability that occurred shortly after accretion.

Each of these general lunar origin theories suffers from several serious flaws. Recently, a new lunar origin theory has been suggested: a cataclysmic birth of the Moon. Scientists supporting this theory suggest that near the end of the Earth's accretion from the primordial solar nebula materials (i.e., after its core was formed, but while the Earth was still in a molten state), a Mars-size celestial

object hit the Earth at an oblique angle. This ancient explosive collision sent vaporized impactor and molten-Earth material into Earth orbit and the Moon then formed from these materials.

As previously mentioned, the surface of the Moon has two major regions with distinctive geologic features and evolutionary histories. First is the relatively smooth, dark areas that Galileo originally called "maria" (because he thought they were seas or oceans). Second is the densely cratered, rugged highlands (uplands), which Galileo called "terrae." The highlands occupy about 83 percent of the Moon's surface and generally have a higher elevation (as much as five kilometers above the Moon's mean radius). In other places the maria lie about five kilometers below the mean radius and are concentrated on the near side of the Moon.

The main external geologic process modifying the surface of the Moon is meteoroid impact. Craters range in size from very tiny pits only micrometers in diameter to gigantic basins hundreds of kilometers across (see fig. 2).

The lunar highlands, or terrae, are the bright (high-albedo) regions on the Moon's surface. They are primarily ancient surfaces that have become extensively cratered during the early history of the Moon, when the meteoroid flux was high. The anorthositic gabbros (consisting mainly of plagioclase, olivine and pyroxene) are the most abundant rocks in the lunar highlands. Almost all the samples returned by the Apollo expeditions have been extensively brecciated (fragmented), metamorphosed and chemically contaminated by repeated meteoroid impact.

The lunar maria are concentrated in the topographic basins on the near side of the Moon. The average normal albedo of the maria is a rather low 7 percent. These maria are vast plains of basaltic lava flows that were erupted after the highlands and the impact basins formed. The basalt in the middle of the basin may several kilometers thick but still represents only a small part of the crust, which is approximately 60 kilometers thick. Lunar basalts are chem-

Table 4   Major Chemical Elements Found on the Lunar Surface

	Highland Rocks		Mare Basalts	
	Anorthositic Gabbro	Gabbroic Anorthosite	Olivine Basalt (A12)	Green Glass (A15)
SiO_2	44.5	44.5	45.0	45.6
TiO_2	0.39	0.35	2.90	0.29
Al_2O_3	26.0	31.0	8.59	7.64
FeO	5.77	3.46	21.0	19.7
MnO	—	—	0.28	0.21
MgO	8.05	3.38	11.6	16.6
CaO	14.9	17.3	9.42	8.72
Na_2O	0.25	0.12	0.23	0.12
K_2O	—	—	0.064	0.02
P_2O_5	—	—	0.07	—
Cr_2O_3	0.06	0.04	0.55	0.41
Total	99.9	100.2	99.77	99.4

SOURCE: NASA.

ically and mineralogically similar to their terrestrial counterparts. They differ only in their high abundance of calcium-plagioclase and titanium and in their total absence of hydrous minerals.

The surface of the Moon is strongly brecciated, or fragmented. This mantle of weakly coherent debris is called regolith. It consists of shocked fragments of rocks, minerals and spherical pieces of glass formed by meteoroid impact. The thickness of the regolith is quite variable and depends on the age of the bedrock beneath and on the proximity of craters and their ejecta blankets. Generally, the maria are covered by 3 to 16 meters of regolith, while the older highlands have developed a "lunar soil" at least 10 meters thick. Table 4 provides a summary of the major chemical elements found on the lunar surface.

Because of its relatively close proximity to Earth and its mineral-resource potential, the Moon will play a very critical role in the development of our extraterrestrial civilization. To initiate the further exploration and eventual use of the Moon, we can first send sophisticated machines. For example, an unmanned *Lunar Observer* spacecraft could continue the scientific investigations started by the Apollo astronauts and would produce extensive maps of the entire lunar surface. Other robot space systems, similar to the Soviet *Luna 16* and *Luna 20* landers, could be used to return additional soil samples from previously unvisited regions such as the far side and the poles.

Then, when human beings return to the Moon, it will not be for another brief moment of scientific inquiry but rather to establish a permanent presence on another world. Within the next few decades, we will establish permanent bases and settlements on the surface of our celestial companion and perform the basic scientific and engineering experiments necessary to develop a complete industrial complex. The use of lunar materials has frequently been suggested by space technologists as an essential pathway in the industrialization of space. With its mineral wealth, strategic location and reduced surface gravity, the Moon could easily become our gateway to the entire Solar System. We can even speculate that the Moon settlements will serve as the technical and social "training ground" for the spacefaring portion of the human race.

See also: **lunar bases and settlements**

N

National Aeronautics and Space Administration (NASA) The U.S. National Aeronautics and Space Administration (NASA) conducts space and aeronautical activities for peaceful purposes. Since NASA's creation in Oc-

tober 1958, its network of centers and facilities has grown across the United States. It is at and through these field installations that NASA conducts many scientific and engineering programs, ranging from aerodynamic research to make civilian aviation safer here on Earth, to sending very sophisticated spacecraft throughout the Solar System and beyond to explore new worlds and search for extraterrestrial life.

The major NASA installations and facilities include the NASA Headquarters, Ames Research Center, Ames–Dryden Flight Research Facility, Goddard Space Flight Center, Jet Propulsion Laboratory, Johnson Space Center, Kennedy Space Center, Langley Research Center, Lewis Research Center, Marshall Space Flight Center, Michoud Assembly Facility, Stennis Space Center, Space Telescope Science Institute and Wallops Flight Facility. The overall functions and responsibilities of each of these facilities will be briefly described.

NASA HEADQUARTERS (WASHINGTON, DC 20546)

NASA headquarters, located in Washington, D.C., exercises management over the space flight centers, research centers and other installations that make up NASA. Responsibilities of headquarters include the determination of programs and projects, establishment of management policies, development of procedures and performance criteria, evaluation of progress within programs and projects and review of all phases of the NASA aerospace program.

AMES RESEARCH CENTER (MOFFET FIELD, CA 94035)

The NASA Ames Research Center (ARC) was founded in 1940 as an aircraft research laboratory by the National Advisory Committee for Aeronautics (NACA) and named for Dr. Joseph S. Ames, Chairman of NACA from 1927 to 1939. In 1958, Ames became part of NASA along with other NACA installations and certain Department of Defense facilities. In 1981, NASA merged Ames with the Dryden Flight Research Facility, and the two installations are currently referred to as "Ames–Moffett" and "Ames–Dryden" (which is described in the next section).

Ames–Moffett is located in Mountain View, California, in the heart of "Silicon Valley" at the southern end of San Francisco Bay. ARC is adjacent to the U.S. Naval Air Station, Moffett Field. The Ames Research Center specializes in scientific research, exploration and applications focused on the creation of new technologies for the nation.

This center's major program responsibilities are concentrated in computer science and applications, computational and experimental aerodynamics, flight simulation, flight research, hypersonic aircraft, rotorcraft and powered-life technology, aeronautical and space human factors, life sciences, space sciences, Solar System exploration, infrared astronomy and the search for extraterrestrial intelligence (SETI).

The laboratories at ARC are equipped to study solar and

geophysical phenomena, life evolution and life environmental factors, and the detection of life on other worlds.

AMES–DRYDEN FLIGHT RESEARCH FACILITY
(P.O. BOX 273, EDWARDS, CA 93523)

The Ames–Dryden Flight Research Facility (DFRF) is located at Edwards, California, in the Mojave Desert, some 80 miles (129 km) north of the Los Angeles metropolitan area. This facility enjoys almost ideal weather for flight testing and is situated at the southern end of a 500-mile (805-km) high-speed flight corridor. Situated adjacent to Rogers Dry Lake, a 65-square-mile (16,842-hectare) natural surface for landing, the facility is in an essentially isolated area.

DFRF's primary research tools are research aircraft, ranging from a B-52 carrier aircraft and high-performance jet fighters to the X-29 forward-swept-wing aircraft. Ground-based facilities include a high-temperature loads calibration laboratory, which permits testing of complete aircraft and structural components under the combined effects of loads and heat; a highly developed aircraft flight instrumentation capability; a flight systems laboratory with a diversified capability for avionics system fabrication, development and operations; a flow visualization facility that allows basic flow patterns and regimes to be observed on models or small components; and a remotely piloted vehicle (RPV) research facility.

This NASA facility was directly involved in the approach and landing tests of the Space Shuttle Orbiter *Enterprise* and continues to support Orbiter landings at the end of Space Shuttle missions, as well as Orbiter processing for transcontinental 747-aircraft "ferry flights" to the Kennedy Space Center. (See fig. 1). Facility researchers are also making preparations for the flight test program of the X-

Fig. 1 The Orbiter *Challenger* landing on runway 17 at Edwards Air Force Base, California, upon successful completion of Shuttle mission STS 61-A (November 6, 1985). (Courtesy of NASA.)

30, an experimental vehicle of the National Aerospace Plane program.

GODDARD SPACE FLIGHT CENTER
(GREENBELT, MD 20771)

The Goddard Space Flight Center (GSFC) is located 10 miles (16 km) northeast of Washington, D.C. The GSFC mission is being accomplished through scientific research centered in six space and Earth science laboratories and in the management, development and operation of several Earth-orbiting space systems.

One of these systems is the Hubble Space Telescope (HST). The movements of the HST are controlled from Goddard's Space Telescope Operations Control Center, as the robot observatory's five scientific instruments study the stars, planets and interstellar space. The Space Telescope Science Institute (STScI), where data from the HST is being analyzed, operates under a contract managed by GSFC. (The STScI will be discussed later in this entry).

Another recent GSFC Earth-orbiting spacecraft is the Cosmic Background Explorer (COBE). This spacecraft was launched in November 1989, and collected important data that tests the theory that the Universe began about 15 billion years ago with a Big Bang (a cataclysmic explosion) and then started expanding.

Also joining the Hubble Space Telescope and the Cosmic Background Explorer in the 1990s are the Gamma Ray Observatory (GRO) and the Upper Atmosphere Research Satellite (UARS). With four specially designed instruments, the GRO is now exploring the little understood processes that power and propel the energy-emitting objects of deep space. These objects include exploding galaxies, black holes and quasars. The UARS will soon look at the Earth's upper atmosphere and help scientists understand its composition and dynamic processes.

Once deployed in space, these and other GSFC scientific spacecraft fall under the 24-hour-a-day surveillance of a worldwide ground and space-based communication network, the nerve center of which is located at Goddard Space Flight Center. One of the major elements of this network is the Tracking and Data Relay Satellite System (TDRSS).

The GSFC role in the Space Station *Freedom* program is to develop the detailed design, construction, test and evaluation of the automated free-flying platform and to develop provisions for instruments and payloads to be attached externally to the Space Station. This center's responsibilities include the co-orbiting free-flying platform, external payload pointing system, the servicing facility and the Flight Telerobotic Servicer.

GSFC's tracking responsibility also extends to its Wallops Flight Facility (discussed later in this entry). In fact, the Wallops Flight Facility, located on Virginia's eastern shore, launches and tracks satellites and surborbital space vehicles and manages the National Scientific Balloon Facility in Palestine, Texas.

The GSFC suborbital vehicles include sounding rockets,

The Earth as viewed from space. This spectacular photograph was taken during the *Apollo 17* mission (1972) and shows Africa, the Indian Ocean, the Antarctic ice cap and the Arabian Peninsula. (Courtesy of NASA.)

The view from *Viking 2*, which landed on Mars on July 29, 1976. Scientists believe that the color of the Martian surface and that of the sky represent their true colors. The salmon color of the sky is caused by dust particles suspended in the atmosphere. (Courtesy of NASA.)

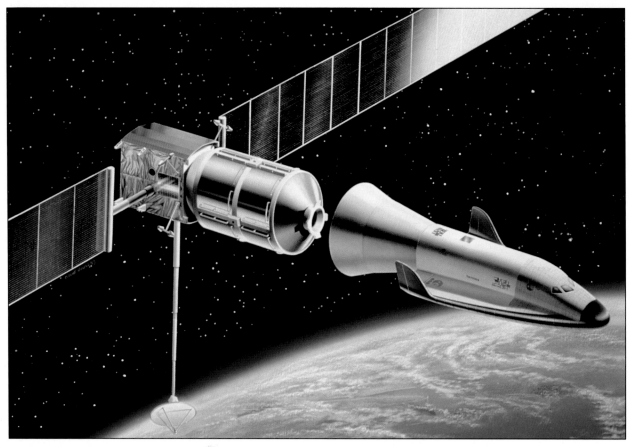

A drawing of the Hermes space vehicle docking with the *Columbus* Free-Flying Laboratory. (Courtesy of the European Space Agency [ESA].)

A Japanese communications satellite being prepared for launch. (Courtesy of NASDA.)

This montage of *Voyager II* images shows the blue-green Uranus overlaid with an artist's conception of the planet's dark rings as they might appear to a future space explorer. A high resolution image of Miranda is arranged in the foreground to show the view along one of the huge canyons on the tiny moon's surface. (Courtesy of NASA/JPL.)

The majestic blue gas giant Neptune and its most prominent feature, the Great Dark Spot (GDS), as imaged by the *Voyager 2* spacecraft during its 1989 encounter with the planet. (Courtesy of NASA/JPL.)

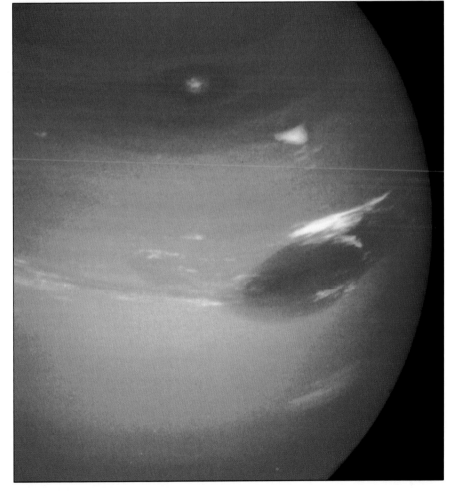

A 1989 NASA drawing of the baseline configuration of Space Station *Freedom* with the Shuttle Orbiter docked. Because of budget cuts, the station is currently being redesigned, but its basic components should remain the same. (Courtesy of NASA.)

Neptune's largest moon, Triton, as imaged on August 24, 1989 by the *Voyager 2* spacecraft at a range of 530,000 km (330,000 mi). Triton's highly reflective, frosty (frozen nitrogen) surface has a temparature of only about 38 K—making it the coldest celestial body yet observed in our Solar System. (Courtesy of NASA/JPL.)

Skylab astronaut Jack Lousma performing extravehicular activity during a mission to the orbiting *Skylab* in the summer of 1973. (Courtesy of NASA.)

The planet Jupiter as seen by the *Voyager* spacecraft.   (Courtesy of NASA.)

This photo aboard *Skylab 3* shows a huge gas cloud of helium erupting some 500,000 miles from the surface of the Sun. The magnitude of the eruption can be visualized by comparing it with the small white dot that represents the size of the Earth. (Courtesy of NASA/JSC.)

The *Ariane 4* expendable launch vehicle lifting off from its pad at the Guiana Space Center in Kourou, French Guiana. (Courtesy of the European Space Agency [ESA].)

An artist's concept of a cometary nucleus. Upon approaching the Sun, the comet's nucleus is heated up and releases enormous amounts of dust and gas (a few tons per second).   (Artwork courtesy of NASA/JPL. Color slide courtesy of ESA.)

A montage of Saturn and some of its major satellites from photos taken by *Voyager I* and *II*. The satellites shown are (clockwise from upper right) Titan, Iapetus, Tethys, Mimas, Dione in the foreground, Enceladus and Rhea. Not included are Hyperion and Phoebe and eight small satellites discovered in 1980.   (Courtesy of NASA.)

balloons and aircraft. These platforms are used to perform studies in galactic astronomy, high-energy astrophysics, solar physics and planetary astronomy, as well as performing atmospheric and oceanographic research and terrain mapping.

The scientific data from these and other spaceflight experiments are catalogued and archived at the National Space Science Data Center at GSFC in the form of magnetic tapes, microfilm and photographic images to satisfy numerous requests from the scientific community each year.

Finally, much of this center's theoretical research is performed at the Goddard Institute for Space Studies in New York City. Operated in close association with universities in the New York area, this institute provides supporting research to GSFC in geophysics, astrophysics, astronomy and meteorology.

JET PROPULSION LABORATORY (4800 OAK GROVE DRIVE, PASADENA, CA 91109)

NASA's Jet Propulsion Laboratory (JPL) is located near Pasadena, California, approximately 20 miles (32 km) northeast of Los Angeles. JPL is a government-owned facility operated by the California Institute of Technology under a NASA contract. In addition to the Pasadena site, JPL operates the worldwide Deep Space Network (DSN), including a DSN station at Goldstone, California.

This laboratory is engaged in activities associated with deep space automated scientific missions—engineering subsystem and instrument development and data reduction and analysis required by deep space flight. Current NASA projects under JPL management include Voyager, Galileo, Magellan and the Mars Observer. JPL also manages the NASA portion of the ESA/NASA Ulysses project. The laboratory designs and tests flight systems, including complete spacecraft, and provides technical direction to various aerospace contractor organizations.

LYNDON B. JOHNSON SPACE CENTER (HOUSTON, TX 77058)

NASA's Johnson Space Center (JSC) is located about 20 miles (32 km) southeast of downtown Houston. This center was established in September 1961, as NASA's primary center for (1) the design, development and testing of spacecraft and associated systems for manned flight; (2) the selection and training of astronauts; (3) the planning and supervision of manned missions; and (4) extensive participation in the medical, engineering and scientific experiments carried aboard human crew space flights.

JSC has program management responsibility for the Space Shuttle program, the nation's current manned space flight program. JSC also has a major responsibility for the development of Space Station *Freedom* (a permanently peopled, Earth-orbiting facility that will be operational in the late 1990s), as well as responsibility for interfaces between the Space Shuttle and the Space Station.

The Johnson Space Center also directs NASA operations

at the White Sands Test Facility (WSTF), which is located on the western edge of the U.S. Army White Sands Missile Range near Las Cruces, New Mexico. WSTF supports the Space Shuttle propulsion system, power system and materials testing.

Among the specialized training facilities at JSC are the Space Shuttle simulators, the Space Shuttle Orbiter Trainer, the Shuttle Manipulator Development Facility, Space Station mockups, and the Weightless Environment Training Facility. The Mission Control Center (MCC) at JSC is the central facility from which all U.S. manned space flights are monitored (see fig. 2).

JOHN F. KENNEDY SPACE CENTER (KENNEDY SPACE CENTER, FL 32899)

NASA's Kennedy Space Center (KSC) is located on the east coast of Florida, 150 miles (241 km) south of Jacksonville and approximately 50 miles (80 km) east of Orlando. KSC is immediately north and west of Cape Canaveral Air Force Station. This center is about 34 miles (55 km) long and varies in width from 5 miles (8 km) to 10 miles (16 km). The total land and water are occupied by the installation is 140,393 acres (56,817 hectares). Of this total area, 84,031 acres (34,007 hectares) are NASA-owned. The remainder is owned by the State of Florida. This large area, with adjoining bodies of water, provides the buffer space necessary to provide for the safety of the nearby civilian communities during space vehicle launches. Agreements have been made with the U.S. Department of the Interior supporting the use of nonoperational (buffer) areas as a wildlife refuge and national seashore.

KSC was established in the early 1960s to serve as the launch site for the Apollo–Saturn V lunar landing missions. After the Apollo program ended in 1972, Launch Complex 39 was used to support both the Skylab program (early nonpermanent U.S. space station) and the Apollo–Soyuz test project (an international rendezvous and docking demonstration involving the United States and the Soviet Union).

The Kennedy Space Center now serves as the primary center within NASA for the test, checkout and launch of space vehicles. KSC responsibility includes the launching of manned (Shuttle) space vehicles at Complex 39 and NASA unmanned (expendable) launch vehicles (e.g., a Delta II rocket) at both nearby Cape Canaveral Air Force Station and at Vandenberg Air Force Base in California.

The center is responsible for the assembly, checkout and launch of Space Shuttle vehicles and their payloads, Shuttle landing operations and "turnarounds" (Shuttle vehicle processing between flights), as well as for the preparation and launch of expendable NASA launch vehicles.

LANGLEY RESEARCH CENTER (HAMPTON, VA 23665-5225)

NASA's Langley Research Center (LaRC) in Hampton, Virginia, is approximately 100 miles (160 km) south of Washington, D.C. The center occupies 772 acres (312 hectares) of government-owned land divided into two areas

Fig. 2   Mission Control Center (MCC) at the Johnson Space Center during a Shuttle mission. (Courtesy of NASA.)

by the runway facilities of Langley Air Force Base. The runways, some utilities and certain facilities are used jointly by NASA and the Air Force.

The primary mission of LaRC is to conduct research and development concerning advanced concepts and technologies for future aircraft and spacecraft systems, with particular emphasis being given to environmental effects, performance, range, safety and economy.

The National Transonic Facility is a new cryogenic wind tunnel at LaRC that provides a unique capability for conducting high-Reynolds-number research (i.e., "turbulent-flow" research) at subsonic and transonic speeds. Major research disciplines include materials; flutter, aeroelasticity, dynamic loads and structural response; fatigue fracture; electronic and mechanical instrumentation; computer technology; flight dynamics and control; and communications technology.

One of LaRC's current major projects is developing technology for the National Aerospace Plane (NASP). This center also supports manned and unmanned space programs (including the Space Shuttle and the Space Station) through the development of experiments, sensors and data-handling systems.

LEWIS RESEARCH CENTER (21000 BROOKPARK
ROAD, CLEVELAND, OH 44135)

NASA's Lewis Research Center (LeRC) occupies 360 acres (146 hectares) of land adjacent to the Cleveland Hopkins International Airport, some 20 miles (32 km) southwest of Cleveland, Ohio. This center was established in 1941 by the National Advisory Committee for Aeronautics (NACA) and named for George W. Lewis, NACA's Director of Research from 1924 to 1947. LeRC is NASA's lead center for research, technology and development in aircraft propulsion, space propulsion, space power and satellite communications.

Aircraft propulsion activities in the early days of the jet age were to develop aircraft that would fly higher, faster and farther. Today's goals concentrate on fuel conservation, quieter flight and cleaner engine exhaust.

LeRC has responsibility for developing the largest space power system ever designed to provide the electric power needed to accommodate the life-support systems and research experiments on Space Station *Freedom*. In addition, this center is supporting the Space Station in other major technical areas, such as auxiliary propulsion systems and communications systems.

The Lewis Research Center is also the home of NASA's Microgravity Materials Science Laboratory, a unique facility that can qualify potential space experiments. Other LeRC facilities include a zero-gravity drop tower, wind tunnels, space environment chambers, chemical rocket thrust stands and chambers for testing jet engine efficiency and noise.

GEORGE C. MARSHALL SPACE FLIGHT CENTER (HUNTSVILLE, AL 35812)

The George C. Marshall Space Flight Center (MSFC) is located on 1,800 acres (728 hectares) inside the U.S. Army's Redstone Arsenal at Huntsville, Alabama. This center was formed on July 1, 1960, by transfer to NASA of part of the U.S. Army Ballistic Missile Agency, and was named for the famous soldier and statesman, General of the Army George C. Marshall. Two other NASA sites are also managed by MSFC: the Michoud Assembly Facility, New Orleans, Louisiana (discussed later in this entry), where the Space Shuttle external tanks are manufactured; and the Slidell Computer Complex, Slidell, Louisiana, which provides computer services to support the Michoud Assembly Facility.

In the past, MSFC has been most often identified as NASA's launch vehicle development center. Today, the center is a multiproject management, scientific and engineering establishment, with much emphasis placed on projects involving scientific investigation and the application of space technology to the solution of problems on Earth. The center has also a significant role in the development of the Space Shuttle. MSFC provides the Orbiter's main engines, the external tank that carries liquid hydrogen and liquid oxygen for those main engines, and the solid rocket boosters that assist in lifting the Shuttle vehicle from the launch pad.

MSFC also plays a key role in the development of major payloads to be flown onboard the Space Shuttle. One such payload is Spacelab, a reusable, modular scientific research facility carried inside the Orbiter's payload bay. This unique facility was designed, tested and provided to NASA by the European Space Agency. To support Spacelab missions, the center has built a state-of-the-art payload control center to support Spacelab missions from Huntsville.

The Marshall Space Flight Center has management responsibility for the Hubble Space Telescope, a large automated optical telescope deployed in space by the Shuttle. The Tethered Satellite System, often called "the satellite on a string," is also managed by MSFC. Expected to be flown in the early 1990s, this "satellite" will be suspended either upward or downward from the Orbiter's payload bay on a super-strong 0.254-cm (0.1-in) thick tether that is up to 97 km (60 mi) in length. When deployed upward from the Shuttle, the tethered satellite will examine electrodynamic and other space environment phenomena. When deployed downward (i.e., towards Earth), it will sample the Earth's upper atmosphere for magnetospheric, gravitational and atmospheric data.

The center also has major responsibilities in the Space Station *Freedom* program. This role includes the living quarters, U.S. laboratory module, logistics elements, and node structures for connecting the station's modules, fluids, environmental control, life-support and audio-video subsystems.

MSFC has been studying designs for an unmanned, cargo version of the Space Shuttle, called the Shuttle-C. This launch vehicle would use the existing Shuttle external tank, main engines and solid boosters, but would replace the Orbiter vehicle itself with a large cargo container.

MICHOUD ASSEMBLY FACILITY (P.O. BOX 29300, NEW ORLEANS, LA 70189)

The Michoud Assembly Facility (MAF) is located about 15 miles (24 km) east of downtown New Orleans. The primary mission of Michoud is the systems engineering, design, manufacture, fabrication and assembly of Space Shuttle external tanks. Marshall Space Flight Center exercises overall management control over this facility. A prime industrial contractor (Martin Marietta) provides Space Shuttle external tank production capability.

JOHN C. STENNIS SPACE CENTER (STENNIS SPACE CENTER, MS 39529)

The NASA John C. Stennis Space Center (SSC) is located near Bay St. Louis, Mississippi. This complex includes industrial, laboratory and specialized engineering facilities to support the testing of large rocket propulsion systems. The center has deep water access for transporting oversize cargo via the East Pearl River and Intercoastal Waterway.

The main mission of SSC is support of Space Shuttle main engine system testing. Static test firings are conducted on the same huge test towers that were used from 1966 to 1970 to captive-fire all first- and second-stage rocket engines of the huge Saturn V launch vehicle that supported the Apollo lunar landings and Skylab missions. Space Shuttle main engine testing has been underway at SSC since 1975.

SSC has also evolved into a center of excellence in the area of remote sensing applications. The center is involved in Earth sciences programs of national and international significance. SSC's Earth Resources Laboratory manages a balanced research and development program in Earth sciences, remote sensing technologies and applications and data systems development.

SPACE TELESCOPE SCIENCE INSTITUTE
(JOHNS HOPKINS HOMEWOOD CAMPUS,
BALTIMORE, MD 21218)

The Space Telescope Science Institute (STScI) is located in Baltimore, Maryland, on the campus of Johns Hopkins University. This institute is operated for NASA by the Association of Universities for Research in Astronomy. The STScI staff includes resident scientists and engineers from the European Space Agency. The facility consists of offices for scientists and administrative personnel and the computer and imaging systems needed to evaluate, prepare for and schedule the observations and science operations performed by the Hubble Space Telescope. Despite the initial problems with the HST, the institute is now receiving, analyzing, and displaying incoming data of value to space scientists.

WALLOPS FLIGHT FACILITY (WALLOPS
ISLAND, VA 23337)

NASA's Wallops Flight Facility (WFF), a part of the Goddard Space Flight Center (GSFC), is one of the oldest launch sites in the world. Established in 1945, this facility covers 6,166 acres (2,495 hectares), including about 1,100 acres (445 hectares) of marshland, in three separate areas of Virginia's eastern shore—the island, the main base and the mainland in back of the island. This facility manages and implements NASA's sounding rocket projects that use suborbital rocket vehicles to accommodate about 50 scientific missions annually.

WFF also manages and coordinates NASA's scientific balloon projects, involving some 45 high-altitude helium-filled scientific balloon missions each year. These balloon launches are conducted at Palestine, Texas, and at various sites throughout the world.

This NASA facility operates and maintains the Wallops launch range and data acquisition facilities. Approximately 100 to 150 rocket launches are conducted each year from the Wallops Island site. In addition, WFF mobile launch, tracking and data-acquisition systems are transported to and operated at various worldwide sites to accommodate sounding rocket, balloon and NASA data network mission requirements.

See also: **Civil Space Technology Initiative; Cosmic Background Explorer; Deep Space Network; Gamma Ray Observatory; Hubble Space Telescope; Magellan Mission; materials research and processing in space; National Aerospace Plane; space launch vehicles; space station; Space Transportation System**

National Aerospace Plane (NASP)   A research program funded by the U.S. Department of Defense, NASA and private industry to develop and demonstrate the technologies of hypersonic flight in a revolutionary piloted research vehicle called the X-30. *Hypersonic flight* is usually taken to mean flight through the Earth's atmosphere at speeds of at least Mach 5—that is, five times the speed of sound.

Fig. 1   One design concept for the National Aerospace Plane (NASP). (Courtesy of the United States Air Force.)

The speed of sound (Mach number = 1) in dry air is 331.4 meters per second (741.4 miles per hour) at a temperature of 0 degrees Celsius (273 degrees Kelvin).

If successful, the X-30 aerospace plan would demonstrate the capability to reach outer space using a single (propulsion) stage to orbit (SSTO) in a fully reusable flight vehicle—an accomplishment that would make unprecedented use of "air-breathing" engines. Air-breathing engines burn atmospheric oxygen during combustion instead of carrying an oxidant internally, as is done by rocket vehicles. As presently planned, by the year 2000 the X-30 vehicle will have successfully conducted a flight demonstration in which it takes off from a conventional 3,050-meter (10,000-ft) runway, accelerates to Mach 25 in the upper portions of the Earth's atmosphere, achieves Earth orbit and then returns to Earth, landing on a conventional runway.

In the next century, NASP-derived vehicles (NDVs) might offer a radically different approach to reaching space from the Earth's surface. These aerospace plane vehicles could become the foundational element of future space transportation systems, ferrying people and cargo into low Earth orbit on a dependable, low-cost basis. (See fig. 1.) In fact, some aerospace experts now suggest that a NASP-derived vehicle will supplement or entirely replace the Space Shuttle, when the current Shuttle fleet reaches the end of its useful lifetime in the early part of the next century.

See also: **space launch vehicles; Space Transportation System**

National Space Development Agency of Japan (NASDA)   Japan's space development activities are primarily implemented by the National Space Development Agency (NASDA) and the Institute of Space and Astronautical Science (ISAS), in cooperation with other related organizations, and in accordance with the space development program established by the Space Activities Commission, an advisory committee to the prime minister. NASDA was established on October 1, 1969.

NASDA's activities are limited to the peaceful uses of space. The agency is primarily engaged in research and development involving satellites and launch vehicles for practical uses, launch and tracking operations for Japanese satellites, promoting the development of remote-sensing technologies and promoting applications-oriented space experiments (e.g., materials processing in space).

In contrast, ISAS sponsors activities in the field of space science and is engaged in the research, development, launch and operation of scientific satellites, such as the *Sakigake* and *Suisei* space probes sent to Comet Halley in 1986. In January 1990, ISAS launched the *Muses-A* spacecraft to the Moon, making Japan the third nation (behind the United States and the Soviet Union) to visit our nearest celestial neighbor. The *Muses-A* spacecraft was placed in a large elliptical orbit around the Earth that will take the spacecraft within about 16,000 km (10,000 mi) of the Moon. When it is near the Moon, the *Muses-A* "mothership" will release a smaller satellite that will actually go in orbit around the Moon, sending data about the cislunar space environment back to the spacecraft.

NASDA's space development activities fall into four general categories: Earth observation, communication and broadcasting satellites; space transportation; space experiments; and basic space technology. The current NASDA space program can be placed in the following categories: (1) space development programs using the N-I N-II, H-I and H-II launch vehicles; (2) the GMS-CS-BS meteorological satellite, communications satellite and broadcasting satellite program using U.S. Delta launch vehicles; and (3) the Space Station Integrated Project, a group of programs focusing on the practical use of the space environment, including materials processing in space, using the U.S. Space Shuttle/*Spacelab* configuration and involving participation in the Space Station *Freedom* program with the Japanese Experiment Module.

NASDA is also working on the development of HOPE (H-II Orbiting Plane), a small spaceplane capable of being launched by the H-II launch vehicle and used to bring space experiment products back to Earth from orbiting facilities.

JAPANESE LAUNCH VEHICLES

The development of launch vehicles by NASDA for application satellites was initiated by technology transfer from the United States under a licensing agreement. The N-I launch vehicle is a three-stage vehicle capable of injecting a 130-kg (287-lbm) payload into geostationary orbit. The N-I's first- and third-stage and strap-on solid boosters used American Thor–Delta technology that was transferred under licensing agreement. However, NASDA developed the N-I's second-stage liquid propellant engine (called the LE-3 engine) and the reaction control system. Between

Table 1   Japanese Launch Vehicle Performance Characteristics

Launch Vehicle	Length (m)	Diameter (m)	Launch Performance Capability to Geostationary Orbit (Payload Mass, kg)
N-I	33	2.4	130
N-II	35	2.4	350
H-I	40	2.4	550
H-II	49	4.0	2,200

SOURCE: Data courtesy of NASDA.

1975 and 1982, seven satellites, including Japan's first geostationary satellite, were launched by the N-I vehicle.

The N-II launch vehicle is capable of placing about 350 kg (772 lbm) into geostationary orbit. The N-II's first stage was given enlarged propellant tanks. The first-stage engine and the strap-on solid rocket boosters were manufactured in Japan under the N-I licensing agreement with the United States. In the N-II's second stage, Thor–Delta rocket technology was also introduced from the United States as well as American inertial guidance system technology to improve flight precision. Between 1981 and 1986, eight satellites were launched successfully using the N-II vehicle.

In recent years, Japan's needs to launch larger, more massive satellites increased. The three-stage H-I launch vehicle was developed to satisfy these needs and to provide a current capability for placing satellites of about 550 kg (1,213 lbm) mass in geostationary orbit. The H-I's first-stage and strap-on boosters are the same as those used in the N-II vehicle. The H-I's second stage, however, uses a high-performance cryogenic propellant liquid engine (called the LE-5) developed by NASDA. The H-I's third stage employs a larger solid rocket motor.

The development of the H-II launch vehicle started in 1986. The H-II vehicle will provide Japan with the ability to launch large-capacity payloads in the mid-1990s, as for example approximately 2,200 kg (4,850 lbm) to geostationary orbit. The two-stage, liquid hydrogen/liquid oxygen fueled H-II rocket with its two large strap-on solid rockets will be the main Japanese launch vehicle of the 1990s and will give Japan the capability of sending significant payloads to low, medium and high Earth orbits, as well as to the Moon and to the planets. In addition, the H-II vehicle will also be capable of launching the HOPE spaceplane.

Table 1 provides a summary of the performance characteristics of these Japanese launch vehicles.

JAPANESE SATELLITES

Japanese satellites developed by NASDA include engineering test satellites (aimed at establishing and demonstrating basic space technologies), communications and broadcasting satellites, meteorological satellites and Earth observation satellites.

For example, NASDA's first satellite, Engineering Test Satellite-I (ETS-I), called *Kiku*, was launched in September

1975 from the Tanegashima Space Center. This spacecraft had a mass of approximately 82.5 kg (182 lbm) and was placed into a 1,000-km (620-mi) altitude orbit around the Earth. ETS-I helped demonstrate launch vehicle, tracking and control and antenna technologies.

NASDA satellites are also named after flowers popular in Japan. For example, broadcasting satellites are called *Yuri*, meaning lily in Japanese; engineering test satellites, *Kiku*, meaning chrysanthemum; meteorological satellites, *Himawari*, meaning sunflower; and communications satellites, *Sakura*, meaning cherry blossom.

The Marine Observation and Satellite-1 (MOS-1 or MOMO-1) is Japan's first marine observation satellite. This 740-kg (1,632-lbm) mass spacecraft was launched from the Tanegashima Space Center on February 19, 1987 and placed in a 909-km (564-mi) altitude, Sun-synchronous (polar) orbit around the Earth.

NASDA plans to launch the Earth Resources Satellite-1 (ERS-1) in 1992. ERS-1 will carry a synthetic aperture radar (SAR) and be able to observe the Earth's surface, regardless of weather conditions. This spacecraft will also have an optical sensor that will observe the Earth in the visible and near-infrared spectra regions. ERS-1 applications include resource identification, land-use planning, environmental monitoring, forestry, agriculture and coastal region monitoring.

JAPANESE MATERIALS PROCESSING IN SPACE (MPS) EXPERIMENTS

The First Material Processing Test (FMPT) program is now scheduled for the early 1990s and involves equipment flown onboard the U.S. Space Shuttle/Spacelab vehicle. Some 34 experiments (22 involving materials processing and 12 concerning life science) will be conducted by a Japanese astronaut/payload specialist using three-double experiment racks mounted inside the Spacelab module.

The Japanese Experiment Module (JEM) represents Japan's participation in the U.S. Space Station *Freedom* program. The Japanese experiment module will be attached to the space station and will consist of a pressurized module, an exposed facility and an experiment logistics module. The pressurized module is a multipurpose laboratory for performing astronaut-supervised materials processing experiments and life science experiments under microgravity conditions. The exposed facility consists of two pallets (exposed to the space environment) and a manipulator system. This facility will be primarily tended by remote control and will be used for science observations, communication system experiments, science and engineering experiments and materials processing research. The experiment logistics module is a container that will be used to store and resupply experimental specimens, various types of gases and other consumables, and to transport these materials between the JEM and the surface of the Earth.

NASDA SPACE FACILITIES

The main NASDA facilities include the Tsukuba Space Center, Tanegashima Space Center, Earth Observation Center and Kakuda Propulsion Center. Construction of the Tsukuba Space Center began in 1970 in Tsukuba, Ibaraki, and many new facilities have been added since. This center is engaged in research and development of space technologies and in engineering tests of spacecraft and launch vehicles. This center also plays an important role as the center of tracking and control of Japanese satellites. The center's large-capacity computing system supports real-time data processing during launch and initial orbital operations. The Tracking and Control Center at Tsukuba is supported by NASDA tracking stations at Katsuura, Okinawa and Masuda on Tanegashima Island. When needed, tracking support is also provided to NASDA by the U.S. National Aeronautics and Space Administration (NASA), the European Space Agency (ESA) and the French Space Agency (CNES).

Tanegashima Space Center is located on the southeastern portion of Tanegashima Island, Kagoshima. The center facilities include the Takesake range for small rockets, the Osaki range for the H-1 launch vehicle and the future H-II launch vehicle, and the Masuda tracking station. The Tanegashima Space Center is the largest launch site in Japan. Its major activities include launch vehicle assembly, checkout and flight and postlaunch tracking and control. This center also plays a major role in the launching of applications satellites and in combustion tests for both solid and liquid rocket engines.

The Earth Observation Center was established by NASDA in Hatoyama-machi, Saitama, in October 1978 as the receiving and processing station of remote-sensing data from a variety of Earth observation satellites, including the U.S. LANDSAT spacecraft, French SPOT spacecraft and the Japanese MOS-1. These remote-sensing data are recorded and processed at this center and then distributed to a variety of users.

Kakuda Propulsion Center was inaugurated in July 1980 and plays a major role in the development of new Japanese launch vehicle propulsion systems. The center is located north of the city of Kakuda, Miyagi, and contains a variety of sophisticated propulsion system test equipment, including a high-altitude simulation test stand for upper-stage engine evaluation.

See also: **European Space Agency; French Space Agency; National Aeronautics and Space Administration; space launch vehicles; space station**

nebula (plural: nebulae) A cloud of interstellar gas or dust. It can be seen either as a dark hole or band against a brighter background (this is called a dark nebula) or as a luminous patch of light (this is called a bright nebula).

Nemesis A postulated dark stellar companion to our Sun, whose close passage every 26 million years is thought to be responsible for the cycle of mass extinctions that seem to have also occurred on Earth at 26-million-year intervals. This "death star" companion has been named for the Greek goddess of retributive justice or vengeance. If it really does exist, it might be a white dwarf, a rouge star that was captured by the Sun or possibly a tiny but gravitationally influential neutron star.

The passage of such a death star through the Oort Cloud (a postulated swarm of comets surrounding our Solar System) could provoke a massive shower of comets into the Solar System. One or several of these comets impacting on Earth would then trigger massive extinctions and catastrophic environmental changes within a very short period of time.

A detailed astronomical search for Nemesis over the next few years should let us know whether our Solar System is really being stalked by a "dark," potentially deadly celestial companion to our Sun.

See also: **extraterrestrial catastrophe theory; rogue star; stars**

Neptune   The outermost of the Jovian planets and the first planet to be discovered using theoretical predictions. Neptune's discovery was made by J. G. Galle at the Berlin Observatory in 1846. This discovery was based on independent orbital perturbation (disturbance) analyses by the French astronomer Jean Joseph Urbain Le Verrier and the British scientist J. Adams. It is considered to be one of the triumphs of 19th-century theoretical astronomy.

Because of its great distance from Earth, little was known about this majestic blue giant planet until the *Voyager 2* spacecraft swept through the Neptunian system on August 25, 1989. Neptune's characteristic blue color comes from the selective absorption of red light by the methane (CH_4) found in its atmosphere—an atmosphere consisting primarily of hydrogen (over 75 percent) and helium (less than 25 percent). Neptune's most prominent "surface" feature

Fig. 1   The Neptunian ring system appears in these two long-exposure images taken by *Voyager 2* during its 1989 encounter with the planet. The two bright, thin outer rings are quite visible. The inner faint "fuzzy" ring can also be seen in this composite image, as well as the broad "dusty disk" that extends smoothly from about 53,000 km (33,000 mi) from the center of Neptune to roughly halfway between the two bright outer rings. (Courtesy of NASA/JPL.)

Table 1   Physical and Dynamic Properties of Neptune

Diameter (equatorial)	49,562 km
Mass	1.02×10^{26} kg
Density (mean)	1.64 g/cm³
Surface gravity	12 m/sec² (approx.)
Escape velocity	25 km/sec (approx.)
Albedo	0.5 (approx.)
Atmosphere	Hydrogen (>75%), helium (<25%), methane (<1%)
Effective temperature	59 Kelvin
Natural satellites	8
Rings	3 (two thin, one fuzzy) and a dusty disk
Period of rotation (a Neptunian day)	0.6715 day (16 hr 7 min)
Average distance from Sun	4.5×10^9 km (30.06 AU)
Eccentricity	0.009
Period of revolution around Sun (a Neptunian year)	165 yr
Mean orbital velocity	5.4 km/sec
Magnetic field	Yes (strong, complex; tilted 50° to planet's axis of rotation)
Radiation belts	Yes (complex structure)
Solar flux at planet	1.5 W/m² (at 30 AU)
(at top of atmosphere) Earth-to-Neptune distances	
Maximum	4.64×10^9 km (31 AU) [258 light-min]
Minimum	4.34×10^9 km (29 AU) [242 light-min]

SOURCE: NASA.

is called the Great Dark Spot (GDS), which is somewhat analogous in relative size and scale to Jupiter's Red Spot. On Neptune, the GDS is actually a large "hole in the clouds" that allows us to look deeper (and therefore darker) into Neptune's uniformly "blue" methane envelope.

The *Voyager 2* spacecraft discovered that, somewhat like Uranus, Neptune has an unusual magnetic field, which is tilted approximately 50 degrees to the planet's axis of rotation and is also offset from the center of the planet. Neptune has an effective temperature of about 59 degrees Kelvin. Considering its great distance from the Sun, this suggests to scientists that the planet actually emits about 2.7 times as much energy as it absorbs from the Sun. The Voyager encounter also revealed the existence of six additional satellites and an interesting ring system consisting of two thin, bright outer rings, a fuzzy inner ring and an extensive "dusty disk" (see fig. 1).

Table 1 provides contemporary physical and dynamic data for Neptune, and Table 2 describes Neptune's eight known satellites. Triton, Neptune's largest moon, is one of the most interesting and coldest objects (about 38 degrees Kelvin surface temperature) yet discovered in our Solar System. Because of its inclined retrograde orbit, density (2.0 g/cm³), rock and ice composition and frost-covered (frozen nitrogen) surface, scientist consider Triton to be a "first cousin" to the Planet Pluto. Both Pluto and Triton are believed to have formed independently in the outer regions of the Solar System, and then Triton (unlike Pluto) may have wandered too close to Neptune and was eventually captured as a permanent satellite.

See also: **Pluto; Uranus; Voyager**

neutrino [symbol: ν]   An electrically neutral elementary particle with a negligible mass (if any). It interacts very weakly with matter and is therefore very difficult to detect. The neutrino is produced in many nuclear reactions, such as beta decay, and has extremely high penetrating power. For example, neutrinos from the Sun, called solar neutrinos, usually pass right through the Earth without interacting.

See also: **Sun**

Table 2   Physical Data for Neptune's Moons

Name	Distance from Planet's Center (km)	Period (day)	Diameter (km)
Nereid	5,513,000	360.1	340
Triton	354,000	5.88 (retrograde)	2,720
1989N1	117,600	1.12	400
1989N2	73,600	0.55	200
1989N4[a]	62,000	0.43	180 (approx.)
1989N3[a]	52,500	0.33	150 (approx.)
1989N5	50,000	0.31	80 (approx.)
1989N6	48,000	0.30	50 (approx.)

[a] Ring shepherd satellite.

SOURCE: NASA.

neutron [symbol: n] An uncharged elementary particle with a mass slightly greater than that of the proton. It is found in the nucleus of every atom heavier than hydrogen. A free neutron is unstable, with a half-life of about 12 minutes, and decays into an electron, a proton and a neutrino. Neutrons sustain the fission chain reaction in a nuclear reactor.

Newton's laws of motion A set of three fundamental postulates that form the basis of the mechanics of rigid bodies. These were formulated in 1687 by the English scientist and mathematician Sir Isaac Newton (1652–1727).

Newton's first law is concerned with the principle of inertia and states that if a body in motion is not acted upon by an external force, its momentum remains constant. This can also be called the law of conservation of momentum.

The second law states that the rate of change of momentum of a body is proportional to the force acting upon the body and is in the direction of the applied force. A familiar statement of this law is the equation $\overrightarrow{F} = m\overrightarrow{a}$, where $\overrightarrow{F}$ is the vector sum of the applied forces, m is the mass and $\overrightarrow{a}$ the vector acceleration of the body.

Newton's third law is the principle of action and reaction. It states that for every force acting upon a body, there is a corresponding force of the same magnitude exerted by the body in the opposite direction.

nuclear–electric propulsion (NEP) system A propulsion system in which a nuclear reactor is used to produce the electricity needed to operate the electric propulsion engine(s). ANEP system provides shorter trip times and greater payload capacity than any of the advanced chemical propulsion technologies that might be used in the next two decades for detailed exploration of the outer planets, especially Saturn, Uranus, Neptune and their respective moons. An orbital NEP vehicle can also be effectively used in cislunar space to gently transport large cargoes and structures from low Earth orbit to geosynchronous orbit or to lunar orbit in support of expanding space-industrialization activities.

See also: **space nuclear power; space nuclear propulsion**

nucleosynthesis The production of heavier chemical elements from the fusion or joining of lighter chemical elements (such as hydrogen or helium nuclei) in thermonuclear reactions in stellar interiors.

See also: **astrophysics; fusion; stars**

O

occult, occulting The disappearance of one celestial object behind another. For example, a solar eclipse is an occulting of the Sun by the Moon; that is, the Moon comes between the Earth and the Sun, temporarily blocking the Sun's light and darkening regions of the Earth.

one g The downward acceleration of gravity at the Earth's surface.

open universe The open or unbounded universe model in cosmology assumes that there is not enough matter in the Universe to completely halt (by gravitational attraction) the currently observed expansion of the galaxies. Therefore, the galaxies will continue to move away from each other and the expansion of the Universe (which started with the "Big Bang") will continue forever. Compare with **closed universe**.

See also: **astrophysics; "Big Bang" theory; cosmology**

orbital-transfer vehicle (OTV) A propulsion system used to transfer a payload from one orbital location to another—as, for example, from low Earth orbit (LEO) to geostationary Earth orbit (GEO). Orbital-transfer vehicles can be expendable or reusable and may involve chemical, nuclear or electric propulsion systems. An expendable orbital-transfer vehicle is frequently referred to as an upper-stage unit, while a reusable OTV is often called a "space tug." OTVs can be designed to move people and cargo between different destinations in cislunar space.

See also: **space launch vehicles; space station; Space Transportation System**

orbits of objects in space We must know about the science and mechanics of orbits to launch, control and track spacecraft and to predict the motion of objects in space. An *orbit* is the path in space along which an object moves around a primary body. Common examples of orbits include the Earth's path around its celestial primary (the Sun) and the Moon's path around the Earth (its primary body). A single orbit is a complete path around a primary as viewed from space. It differs from a revolution. A single *revolution* is accomplished whenever an orbiting object passes over the primary's longitude or latitude from which it started. For example, the Space Shuttle *Columbia* completed a revolution whenever it passed over approximately 80 degrees west longitude on Earth. However, while *Columbia* was orbiting from west to east around the globe, the Earth itself was also rotating from west to east. Consequently, *Columbia* period of time for one revolution was actually longer than its orbital period (see fig. 1). If, on the other hand, *Columbia* were orbiting from east to west (not a practical flight path from a propulsion–economy standpoint), then because of the Earth's west-to-east spin, its period of revolution would be shorter than its orbital period. An east-to-west orbit is called a *retrograde orbit* around the Earth, while a west-to-east orbit is called a *posigrade orbit*. If *Columbia* were traveling in a north-south orbit, or *polar orbit*, it would complete a period of revolution whenever it passed over the latitude from which

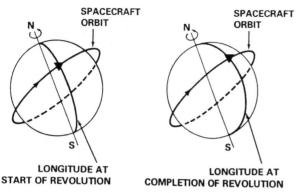

SPACECRAFT ORBIT

LONGITUDE AT START OF REVOLUTION

SPACECRAFT ORBIT

LONGITUDE AT COMPLETION OF REVOLUTION

Fig. 1 An illustration of a spacecraft's west-to-east orbit around the Earth and how the Earth's west-to-east rotation moves longitude ahead. As shown here, the period of one revolution can be longer than the orbital period. (Courtesy of NASA.)

it started. Its orbital period would be about the same as the revolution period, but not identical, because the Earth actually wobbles slightly north and south.

There are other terms used to describe orbital motion. The *apoapsis* is the farthest distance in an orbit from the primary; the *periapsis*, the shortest. For orbits around the planet Earth, the comparable terms are *apogee* and *perigee* (see fig. 2).

For objects orbiting the Sun, *aphelion* describes the point on an orbit farthest from the Sun; *perihelion*, the point nearest to the Sun.

Another term we frequently encounter is the *orbital plane*. An Earth satellite's orbital plane can be visualized by thinking of its orbit as the outer edge of a giant, flat plate that cuts the Earth in half. This imaginary plate is called the orbital plane.

Inclination is another orbital parameter. This term refers to the number of degrees the orbit is inclined away from the equator. The inclination also indicates how far north and south a spacecraft will travel in its orbit around the Earth. If, for example, a spacecraft has an inclination of 56

degrees, it will travel around the Earth as far north as 56 degrees north latitude and as far south as 56 degrees south latitude. Because of the Earth's rotation, it will not, however, pass over the same areas of Earth on each orbit. A spacecraft in a polar orbit has an inclination of about 90 degrees. As such, this spacecraft orbits the Earth traveling alternately in north and south directions. A polar-orbiting satellite eventually passes over the entire Earth because the Earth is rotating from west to east beneath it. The LANDSAT spacecraft is an example of a spacecraft whose cameras and multispectral sensors observe the entire Earth from a nearly polar orbit, providing valuable information about the terrestrial environment and resource base.

A satellite in an equatorial orbit around the Earth has zero inclination. The Intelsat communications satellites are examples of satellites in equatorial orbits. By placing such spacecraft into near-circular equatorial orbits at just the right distance above the Earth, these spacecraft can be made to essentially "stand still" over a point on the Earth's equator. Such satellites are called *geostationary*. They are in *synchronous orbits*, meaning they take as long to complete an orbit around the Earth as it takes for the Earth to complete one rotation about its axis (that is, approximately 24 hours). A satellite at the same "synchronous" altitude but in an inclined orbit may also be called synchronous. While this particular spacecraft would not move much east and west, it would move north and south over the Earth to the latitudes indicated by its inclination. The terrestrial ground track of such a spacecraft resembles an elongated figure eight, with the crossover point on the equator.

All orbits are elliptical, in accordance with Kepler's first law of planetary motion (described shortly). However, a spacecraft is generally considered to be in a circular orbit if it is in an orbit that is nearly circular. A spacecraft is taken to be in an elliptical orbit when its apogee and perigee differ substantially.

Two sets of scientific laws govern the motions of both celestial objects and human-made spacecraft. One is Newton's law of gravitation; the other, Kepler's laws of planetary motion.

The brilliant English scientist and mathematician Sir Isaac Newton observed the following physical principles:

1. All bodies attract each other with what we call gravitational attraction. This applies to the largest celestial objects and to the smallest particles of matter.
2. The strength of one object's gravitational pull upon another is a function of its mass—that is, the amount of matter present.
3. The closer two bodies are to each other, the greater their mutual attraction.

These observations can be stated mathematically as:

$$F = \frac{G\, m_1\, m_2}{r^2}$$

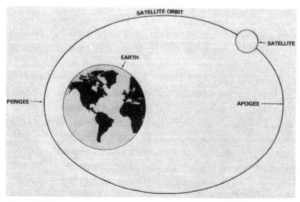

SATELLITE ORBIT

SATELLITE

EARTH

PERIGEE

APOGEE

Fig. 2 The terms *apogee* and *perigee* described in terms of a spacecraft's orbit around the Earth. (Drawing courtesy of NASA.)

where F is the gravitational force acting along the line joining the two bodies (N)

m_1, m_2 are the masses (in kilograms) of body one and body two, respectively

r is the distance between the two bodies (m)

and G is the universal gravitational constant ($6.6732 \times 1C^{-11}$ newton–meter^2/kilogram^2)

Specifically, Newton's law of gravitation states that two bodies attract each other in proportion to the product of their masses and inversely as the square of the distance between them. This physical principle is very important in launching spacecraft and guiding them to their operational locations in space and is frequently used by astronomers to estimate the masses of celestial objects. For example, Newton's law of gravitation tells us that for a spacecraft to stay in orbit, its velocity (and therefore its kinetic energy) must balance the gravitational attraction of the primary object being orbited. Consequently, a satellite needs more velocity in low than in high orbit. For example, a spacecraft with an orbital altitude of 250 kilometers (150 mi) will have an orbital speed of about 28,000 kilometers per hour (17,500 miles per hour). Our Moon, on the other hand, which is about 442,170 kilometers (238,857 mi) from Earth, has an orbital velocity of approximately 3,660 kilometers per hour (2,287 miles per hour). Of course, to boost a payload from the surface of the Earth to a high-altitude (versus low-attitude) orbit requires the expenditure of more energy, since we are in effect lifting the object further out of the Earth's "gravity well."

Any spacecraft launched into orbit moves in accordance with the same laws of motion that govern the motions of the planets around our Sun and the motion of the Moon around the Earth. The three laws that describe these planetary motions, first formulated by Johannes Kepler (1571–1630), may be stated as follows:

1. Each planet revolves around the Sun in an orbit that is an ellipse, with the Sun as it focus, or primary body.
2. The radius vector—such as the line from the center of the Sun to the center of a planet, from the center of the Earth to the center of the Moon or from the center of the Earth to the center (of gravity) of an orbiting spacecraft—sweeps out equal areas in equal periods of time.
3. The square of a planet's orbital period is equal to the cube of its mean distance from the Sun. We can generalize this last statement and extend it to spacecraft in orbit about the Earth by saying that a spacecraft's orbital period increases with its mean distance from the planet.

In formulating his first law of planetary motion, Kepler recognized that purely circular orbits did not really exist—rather, only elliptical ones were found in nature, being determined by gravitational pertubations (disturbances) and other factors. Gravitational attractions, according to Newton's law of gravitation, extend to infinity, although these forces weaken with distance and eventually become impossible to detect. However, spacecraft orbiting the Earth, while primarily influenced by the gravitational attraction of the Earth (and anomalies in the Earth's gravitational field), are also influenced by the Moon and the Sun and possibly other celestial objects, such as the planet Jupiter.

Kepler's third law of planetary motion states that the greater a body's mean orbital altitude, the longer it will take for it to go around its primary. Let's take this principle and apply it to a rendezvous maneuver between a Space Shuttle Orbiter and a satellite in low Earth orbit. To catch up with and retrieve an unmanned spacecraft in the same orbit, the Space Shuttle must first be decelerated. This action causes the Orbiter vehicle to descend to a lower orbit. In this lower orbit, the Shuttle's velocity would increase. When properly positioned in the vicinity of the target satellite, the Orbiter would then be accelerated, raising its orbit and matching orbital velocities for the rendezvous maneuver with the target spacecraft.

Another very interesting and useful orbital phenomenon is the Earth satellite that appears to "stand still" in space with respect to a point on the Earth's equator. Such satellites were first envisioned by the English scientist and writer Arthur C. Clarke, in a 1945 essay in *Wireless World*. Clarke described a system in which satellites carrying telephone and television would circle the Earth at an orbital altitude of approximately 35,580 kilometers (22,240 mi) above the equator. Such spacecraft move around the Earth at the same rate that the Earth rotates on its own axis. Therefore, they neither rise nor set in the sky like the planets and the Moon but rather always appear to be at the same longitude, synchronized with the Earth's motion. At the equator the Earth rotates about 1,600 kilometers per hour (1,000 miles per hour). Satellites placed in this type of orbit are called "geostationary" or "geosynchronous" spacecraft.

It is interesting to note here that the spectacular Voyager missions to Jupiter, Saturn and beyond used a "gravity-assist" technique to help speed them up and shorten their travel time. How is it, you may wonder, that a spacecraft can be speeded up while traveling past a planet? It probably seems obvious that a spacecraft will increase in speed as it approaches a planet (due to gravitational attraction), but the gravity of the planet should also slow it down as it begins to move away again. So where does this increase in speed really come from?

Let us first consider the three basic possibilities for a spacecraft trajectory when it encounters a planet (see fig. 3). The first possible trajectory involves a direct hit or hard landing. This is an *impact trajectory*. The second type of trajectory is an *orbital-capture trajectory*. The spacecraft is simply "captured" by the gravitational field of the planet and enters orbit around it (see trajectories b and c). Depending upon its precise speed and altitude (and other parameters), the spacecraft can enter this captured orbit from either the leading or trailing edge of the planet. In the third type of trajectory, a *flyby trajectory*, the spacecraft remains far enough away from the planet to avoid capture but passes close enough to be strongly affected by its gravity. In this case, the speed of the spacecraft will be increased if it approaches from the trailing side of the

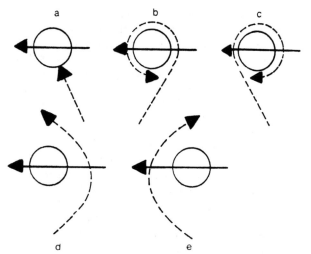

Fig. 3 Possible trajectories of a spacecraft encountering a planet. (NASA drawing, modified by the author.)

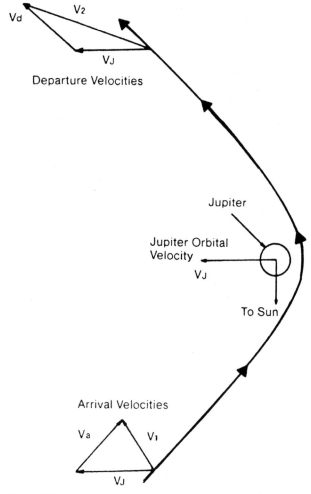

Fig. 4 Velocity changes during a Jupiter flyby. (Drawing courtesy of NASA.)

planet (see trajectory d) and diminished if it approaches from the leading side (see trajectory e). In addition to changes in speed, the direction of the spacecraft's motion also changes.

It may be obvious to you by now that the increase in speed of the spacecraft actually comes from a decrease in speed of the planet itself! In effect, the spacecraft is being "pulled" by the planet. Of course, this has been a greatly simplified discussion of complex encounter phenomena. A full account of spacecraft trajectories must consider the speed and actual trajectory of the spacecraft and planet, how close the spacecraft will come to the planet, and the size (mass) and speed of the planet in order to make even a simple calculation.

Perhaps an even better understanding of gravity assist can be obtained if we use vectors in a more mathematical explanation. The way in which speed is added to the flyby spacecraft during close encounters with the planet Jupiter is shown in figure 4. During the time that spacecraft, such as *Voyager 1* and *2*, were near Jupiter, the heliocentric (Sun-centered) path they followed in their motion with respect to Jupiter closely approximated a hyperbola.

The heliocentric velocity of the spacecraft is the vector sum of the orbital velocity of Jupiter (Vj) and the velocity of the spacecraft with respect to Jupiter (that is, tangent to its trajectory—the hyperbola). The spacecraft moves toward Jupiter along an asymptote, approaching from the approximate direction of the Sun and with asymptotic velocity (V_a). The heliocentric arrival velocity (V_1) is then computed by vector addition (see fig. 4):

$$V_1 = V_j + V_a$$

The spacecraft departs Jupiter in a new direction, determined by the amount of bending that is caused by the effects of the gravitational attraction of Jupiter's mass on the mass of the spacecraft. The asymptotic departure speed

(V_d) on the hyperbola is equal to the arrival speed. Thus, the length of V_a equals the length of V_d. For the heliocentric departure velocity, $V_2 = V_j + V_d$. This vector sum is also depicted in figure 4.

During the relatively short period of time that the spacecraft is near Jupiter, the orbital velocity of Jupiter (V_j) changes very little, and we assume that V_j is equal to a constant.

The vector sums in figure 4 illustrate that the deflection, or bending, of the spacecraft's trajectory caused by Jupiter's gravity results in an increase in the speed of the spacecraft along its hyperbolic path, as measured relative to the Sun. This increase in velocity reduces the total flight time necessary to reach Saturn and points beyond. This "indirect" type of deep-space mission to the outer planets saves two or three years of flight time when compared to "direct-trajectory" missions, which do not take advantage of gravity assist.

Of course, while the spacecraft gains speed during its Jovian encounter, Jupiter loses some of its speed. However, because of the extreme difference in their masses, the change in Jupiter's velocity is negligible.

other world enclosures (OWEs) Special facilities on large, orbiting space complexes used to simulate the conditions, especially gravity levels and atmospheres, encountered on other celestial bodies in our Solar System. These modular extraterrestrial laboratories would offer exobiologists, space scientists, interplanetary explorers and planetary settlers the unique opportunity of totally experiencing the alien-world conditions before actual expeditions or settlement activities are undertaken.

outer planets The planets in our Solar System with orbits larger than that of Mars. These are Jupiter, Saturn, Uranus, Neptune and Pluto. Of these outer planets, all except Pluto are also called the "giant planets."

P

panspermia The general hypothesis that microorganisms, spores or bacteria attached to tiny particles of matter have diffused through space, eventually encountering a suitable planet and initiating the rise of life there. The word itself means "all-seeding."

In the 19th century the British scientist Lord Kelvin (William Thompson) suggested that life may have arrived here on Earth from outer space, perhaps carried inside meteorites. Then, in 1903, the Nobel-Prize-winning Swedish chemist Svante Arrhenius put forward the idea that is now generally regarded as the panspermia hypothesis. Arrhenius said that life really didn't start here on Earth but rather was "seeded" by means of extraterrestrial spores (seed-like germs), bacteria or microorganisms. According to his hypothesis, these microorganisms, spores or bacteria originated elsewhere in the Galaxy (on a planet in another star system where conditions were more favorable for the chemical evolution of life) and then wandered through space attached to tiny bits of cosmic matter than moved under the influence of stellar radiation pressure.

The greatest difficulty most scientists today have with Arrhenius's original panspermia concept is simply the question of how these "life-seeds" can wander through interstellar space for up to several billion years, receive extremely severe radiation doses from cosmic rays and still be "vital" when they eventually encounter a solar system that contains suitable planets. Even on a solar-system scale, the survival of such microorganisms, spore or bacteria would be difficult. For example, "life seeds" wandering from the vicinity of the Earth to Mars would be exposed to both ultraviolet radiation from our Sun and ionizing radiation in the form or solar-flare particles and cosmic rays. This interplanetary spore migration might take several hundred thousand years in the airless, hostile environmental conditions of outer space.

Dr. Francis Crick and Dr. Leslie Orgel attempted to resolve this difficulty by proposing the directed-panspermia hypothesis. Feeling that the overall concept of panspermia was too interesting to abandon entirely, in the early 1970s they suggested that an ancient, intelligent alien race could have constructed suitable interstellar robot spacecraft; loaded these vehicles with an appropriate cargo of microorganisms, spores or bacteria; and then proceeded to "seed the Galaxy" with life, or at least the precursors of life. This "life-seed" cargo would have been protected during the long interstellar journey and then released into suitable planetary atmospheres or oceans when favorable planets were encountered by the robot starships.

Why would an extraterrestrial civilization undertake this type of project? Well, it might first have tried to communicate with other races across the interstellar void; then, when this failed, it could have convinced itself that it was *alone!* At this point in its civilization, driven by some form of "missionary zeal" to "green" (or perhaps "blue") the Galaxy with life as it knew it, the alien race might have initiated a sophisticated directed-panspermia program. Smart robot spacecraft containing well-protected spores, microorganisms or bacteria were launched into the interstellar void to seek new "life sites" in neighboring star systems. This effort might have been part of an advanced-technology demonstration program, a form of planetary engineering on an interstellar scale. These life-seeding robot spacecraft may also have been the precursors of an ambitious colonization wave that never came—or is just now on its way!

In their directed-panspermia discussions, Crick and Orgel identified what they called the theorem of detailed cosmic reversibility. This "theorem" suggests that if we can now contaminate other worlds in this Solar System with microorganisms hitchhiking on terrestrial spacecraft, then it is also reasonable to assume that an advanced, intelligent extraterrestrial civilization could have used its robot spacecraft to contaminate or seed our world with spores, microorganisms or bacteria sometime in the very distant past.

Others have been suggested that life on Earth might have evolved as a result of microorganisms inadvertently left here by ancient astronauts themselves. It is most amusing to speculate that we may be here today because ancient space travelers were "litterbugs," scattering their garbage on a then-lifeless planet. (This line of speculation is sometimes called "the extraterrestrial garbage theory of the origin of life.")

Sir Fred Hoyle and N. C. Wickramasinghe have also recently explored the issue of directed panspermia and the origin of life on Earth. In several publications they argue convincingly that the biological composition of living things on Earth has been and will continue to be radically influenced by the arrival of "pristine genes" from space. The further suggest that the arrival of these cosmic microorganisms, and the resultant complexity of terrestrial life, is not a random process, but one carried out under the influence of a greater cosmic intelligence.

This brings up another interesting point. As we here on Earth develop the technology necessary to send smart machines and humans to other worlds in our Solar System (and eventually even to other star systems), should we initiate a program of directed panspermia? If we became convinced that we might really be alone in the Galaxy, then strong intellectual and biological imperatives might urge us to "green the Galaxy," or to seed life as we know it where there is now none! Perhaps, late in the next century, robot interstellar explorers will be sent from our Solar System, not only to search for extraterrestrial life but also to plant life on potentially suitable extrasolar planets when none is found. This may be one of our higher cosmic callings, to be the first intelligent species to rise to a level of technology that permits the expansion of life itself within the Galaxy. Of course, our directed-panspermia effort might only be the next link in a cosmic chain of events that was started eons ago by a long-since-extinct alien civilization. Perhaps millions of years from now, on an Earthlike planet around a distant Sunlike star, other intelligent beings will start wondering whether life on their world started spontaneously or was seeded there by an ancient civilization (in this case, us) that has long since disappeared from view in the Galaxy!

While the panspermia or directed-panspermia hypotheses do not address how life originally started somewhere in the Galaxy, they certainly provide some intriguing concepts regarding how, once started, life might "get around."

See also: **ancient astronauts; extraterrestrial contamination; life in the Universe; robotics in space; search for extraterrestrial intelligence**

parsec (pa) A unit of distance frequently encountered in astronomical studies. The parsec is defined as a parallax shift of one second of arc. The term itself is a shortened form of "*par*allax *sec*ond." The parsec is the extraterrestrial distance at which the main radius of the Earth's orbit (one astronomical unit, by definition) subtends an angle of one arc-second. It is therefore also the distance at which a star would exhibit an annual parallax of one arc-second.

1 parsec = 3.26 light-years (or 206,265 AU)

The kiloparsec (kpa) represents a distance of 1,000 parsecs (or 3,260 light-years); and the megaparsec (Mpa), a distance of 1 million parsecs (or 3.26 million light-years).

See also: **Appendix A (table A–1)**

people in space In just a single, exciting decade, the 1960s, human beings extended their physical domain from the upper regions of the Earth's atmosphere to the mountains and valleys of the Moon. The first orbital flight by a human being occurred on April 12, 1961, when Soviet cosmonaut Yuri Gagarin flew around the Earth in the *Vostok 1* spacecraft. Then, on July 20, 1969, two American astronauts, Neil Armstrong and Buzz Aldrin, became the first human beings to walk on another world, when the *Apollo 11* mission took them successfully to the lunar surface.

Throughout the U.S. space program and especially since the tragic *Challenger* accident (January 28, 1986), there has been continued discussion concerning the relative value of "manned" versus "unmanned" missions. In a real sense, however, the question is essentially one involving the ultimate location of the human being (that is, in a command and control center on Earth or potentially at risk in space), because even sophisticated automated payloads are ultimately controlled by people. In the next century with permanent space stations, lunar bases and Mars settlements emerging as part of an overall space technology infrastructure, this apparent dichotomy of manned versus unmanned space mission should eventually vanish—both people and smart machines, working in a close, highly interactive partnership, will be needed to help us explore and settle the Solar System.

The human role in American space efforts was gradually expanded through five highly successful pre-Shuttle-era manned programs: Mercury, Gemini, Apollo, Skylab and the Apollo–Soyuz Test Project (ASTP).

Project Mercury, the first U.S. manned spaceflight program, was established on October 5, 1958, only five days after NASA was created. During Project Mercury, the United States accomplished its first human journey into the extraterrestrial environment. We learned that with proper equipment, people could indeed survive and function there.

The Gemini program extended U.S. manned spaceflight activities through the development of a two-person spacecraft that was designed for longer-duration flights (versus the Mercury program). From March 1965 to November 1966, ten manned Earth-orbital missions were flown. The Gemini program demonstrated that human beings could gainfully live, move about and work effectively in space. Techniques such as trajectory shaping and precise maneuvering for spacecraft rendezvous and docking were also developed. These sophisticated space operations techniques were then directly used in the Apollo program.

In the late 1960s and 1970s, the American manned spaceflight program was dominated by the Apollo program. With the flight of *Apollo 8* (December 1968), human beings first circled the Moon and returned safely to Earth. Starting with the *Apollo 11* mission (July 20, 1969), terrans walked on the lunar surface, and our extraterrestrial civilization was born! In all, 12 human beings had the opportunity to personally walk on an alien world in the Apollo program. In December 1972 humanity's first extraterrestrial-expedition program came to an end with the triumphant completion of the *Apollo 17* mission. Throughout the Apollo program, people played useful, often essential, roles in the operation of complex, multipurpose vehicles and in the conduct of sophisticated mission activities. In this program, humans continually exercised a high degree of judgment, selectivity and discrimination. Other situations (both planned and unplanned) took advantage of the astronauts' analytical capabilities, manual dexterity and ability to respond to the unexpected.

Skylab, the first American space station, was launched in 1973. This program expanded our knowledge of Earth-orbital operations and supported the performance of over 50 important scientific, technological and medical experiments. The 100-ton *Skylab* space station was placed in orbit by a Saturn V rocket; the three-person crews were carried into space in Apollo spacecraft lifted by Saturn 1B rockets. The first *Skylab* crew remained in orbit for 28 days; the second for 59 days; and the third for 84 days. The *Skylab* missions clearly showed that human beings can adapt well and function properly in the microgravity environment of low Earth orbit (LEO) for relatively long periods of time, provided they have a proper diet and adequate exercise, sleep, work and recreation. Soviet cosmonauts have stayed in space for even longer periods of time (6 to 12 months) during the Soviet *Salyut-6* and *Mir* space station missions. In *Skylab* it was clearly demonstrated that (1) people can function effectively in space for relatively long periods of time; (2) there are many worthwhile experiments, tasks and investigations that can *best* be accomplished through people in space; and (3) there are beneficial services (planned or emergency) that people can more advantageously perform in space.

The Apollo–Soyuz Test Project (ASTP) in July 1975 was a cooperative U.S.–U.S.S.R. space mission to test compatible rendezvous, docking and crew-transfer systems. Other ASTP goals included the performance of in-orbit experiments and Earth observations. Results from the Earth-observation and photography experiments confirmed the ability of a trained observer in orbit to greatly improve knowledge about the terrestrial environment and the major physical processes that help shape and drive it. ASTP results have clearly indicated that a trained astronaut-observer can expertly describe terrestrial features and phenomena (especially those that take place or change quickly), assimilate and interpret what has been seen (here the person exercises both judgement and recall), and rapidly select observational targets and modify observation-experiment protocols (procedures) as required. A person's unique ability to perform in this way directly complements the Earth-observation data obtained by automated remote-sensing spacecraft.

These five pre-Shuttle-era U.S. manned space programs have successfully demonstrated a variety of important roles for people in space. These roles include (1) space-vehicle pilot, (2) long-term inhabitant, (3) payload/experiment manager, (4) extravehicular activity (EVA), (5) mission planner and innovator, (6) in-orbit scientific investigator, (7) in-orbit equipment operator and experimenter and (8) in-orbit engineer and "troubleshooter." In all of these functions and roles, the performance of human beings is far superior to that of automated systems.

In the Space Shuttle era, the human role in space has not only taken advantage of these previously demonstrated capabilities but now places people (both male and female) in orbit with very flexible, highly sophisticated equipment that guarantees an expanding human role in extraterrestrial activities, including satellite retrieval and repair, advanced scientific investigations and on-orbit assembly and construction. Table 1 provides a brief summary of the American

Table 1 U.S. Manned Space-Flight Log (1961–1986)

Mission	Crew	Date	Mission Elapsed Time, (hr:min:sec)
Mercury-Redstone 3	Shepard	May 5, 1961	00:15:22
Mercury-Redstone 4	Grissom	July 21, 1961	00:15:37
Mercury-Atlas 6	Glenn	Feb. 20, 1962	04:55:23
Mercury-Atlas 7	Carpenter	May 24, 1962	04:56:05
Mercury-Atlas 8	Schirra	Oct. 3, 1962	09:13:11
Mercury-Atlas 9	Cooper	May 15 and 16, 1963	34:19:49
Gemini-Titan III	Grissom, Young	Mar. 23, 1965	04:53:00
Gemini-Titan IV	McDivitt, White	June 3 to 7, 1965	97:56:11
Gemini-Titan V	Cooper, Conrad	Aug. 21 to 29, 1965	190:55:14
Gemini-Titan VII	Borman, Lovell	Dec. 4 to 18, 1965	330:35:31
Gemini-Titan VI-A	Schirra, Stafford	Dec. 15 and 16, 1965	25:51:24
Gemini-Titan VIII	Armstrong, Scott	Mar. 16, 1966	10:41:26
Gemini-Titan IX-A	Stafford, Cernan	June 3 to 6, 1966	72:21:00
Gemini-Titan X	Young, Collins	July 18 to 21, 1966	70:46:39
Gemini-Titan XI	Conrad, Gordon	Sept. 12 to 15, 1966	71:17:08
Gemini-Titan XII	Lovell, Aldrin	Nov. 11 to 15, 1966	94:34:31
Apollo-Saturn 7	Schirra, Eisele, Cunningham	Oct. 11 to 22, 1968	260:09:03
Apollo-Saturn 8	Borman, Lovell, Anders	Dec. 21 to 27, 1968	147:00:42

Table 1   U.S. Manned Space-Flight Log (1961–1986) *(continued)*

Mission	Crew	Date	Mission Elapsed Time, (hr:min:sec)
Apollo-Saturn 9	McDivitt, Scott, Schweickart	Mar. 3 to 13, 1969	241:00:54
Apollo-Saturn 10	Stafford, Young, Cernan	May 18 to 26, 1969	192:03:23
Apollo-Saturn 11	Armstrong, Collins, Aldrin	July 16 to 24, 1969	195:18:35
Apollo-Saturn 12	Conrad, Gordon, Bean	Nov. 14 to 24, 1969	244:36:25
Apollo-Saturn 13	Lovell, Swigert, Haise	April 11 to 17, 1970	142:54:41
Apollo-Saturn 14	Shepard, Roosa, Mitchell	Jan. 31 to Feb. 9, 1971	216:01:57
Apollo-Saturn 15	Scott, Worden, Irwin	July 26 to Aug. 7, 1971	295:11:53
Apollo-Saturn 16	Young, Mattingly, Duke	April 16 to 27, 1972	265:51:05
Apollo-Saturn 17	Cernan, Evans, Schmitt	Dec. 7 to 19, 1972	301:51:59
Skylab SL-2	Conrad, Kerwin, Weitz	May 25 to June 22, 1973	672:49:49
Skylab SL-3	Bean, Garriott, Lousma	July 28 to Sept. 25, 1973	1427:09:04
Skylab SL-4	Carr, Gibson, Pogue	Nov. 16, 1973 to Feb. 8, 1974	2017:15:32
Apollo-Soyuz Test Program (ASTP)	Stafford, Brand, Slayton	July 15 to 24, 1975	217:28:23

Space Transportation System (Space Shuttle)

Mission	Crew	Date	Mission Elapsed Time, (hr:min:sec)
STS-1 (OFT)	Young, Crippen	April 12 to 14, 1981	54:20:53
STS-2 (OFT)	Engle, Truly	Nov. 12 to 14, 1981	54:13:12
STS-3 (OFT)	Lousma, Fullerton	March 22 to 30, 1982	192:04:49
STS-4 (OFT)	Mattingly, Hartsfield	June 27 to July 4, 1982	169:11:11
STS-5	Brand, Overmyer, Allen, Lenoir	Nov. 11 to 16, 1982	122:14:25
STS-6	Weitz, Bobko, Peterson, Musgrave	April 4 to 9, 1983	120:23:42
STS-7	Crippen, Hauck, Ride, Fabian, Thagard	June 18 to 24, 1983	146:23:59
STS-8	Truly, Brandenstein, D. Gardner, Bluford, W. Thornton	Aug. 30 to Sept. 5, 1983	145:08:40
STS-9	Young, Shaw, Garriott, Parker, Lichtenberg, Merbold	Nov. 28 to Dec. 8, 1983	247:47:24
41-B	Brand, Gibson, McCandless, McNair, Stewart	Feb. 3 to 11, 1984	191:15:55
41-C	Crippen, Scobee, van Hoften, G. Nelson, Hart	April 6 to 13, 1984	167:40:05
41-D	Hartsfield, Coats, Resnik, Hawley, Mullane, C. Walker	Aug. 30 to Sept. 5, 1984	144:57:00
41-G	Crippen, McBride, Ride, Sullivan, Leestma, Garneau, Scully-Power	Oct. 5 to Oct. 13, 1984	197:23:37
51-A	Hauck, D. Walker, D. Gardner, A. Fisher, Allen	Nov. 8 to Nov. 16, 1984	191:44:56
51-C	Mattingly, Shriver, Onizuka, Buchli, Payton	Jan. 24 to 27, 1985	73:33:27
51-D	Bobko, Williams, Seddon, Hoffman, Griggs, C. Walker, Garn	April 12 to 19, 1985	167:55:00
51-B	Overmyer, Gregory, Lind, Thagard, W. Thornton, van den Berg, Wang	April 29 to May 6, 1985	168:08:47
51-G	Brandenstein, Creighton, Lucid, Fabian, Nagel, Baudry, Al-Saud	June 17 to 24, 1985	169:39:00
51-F	Fullerton, Bridges, Musgrave, England, Henize, Acton, Bartoe	July 29 to Aug. 6, 1985	190:45:26
51-I	Engle, Covey, van Hoften, Lounge, W. Fisher	Aug. 27 to Sept. 3, 1985	170:18:00
51-J	Bobko, Grabe, Hilmers, Stewart, Pailes	Oct. 3 to 7, 1985	97:45:00
61-A	Hartsfield, Nagel, Buchli, Bluford, Dunbar, Furrer, Messerschmid, Ockels	Oct. 30 to Nov. 6, 1985	168:44:51
61-B	Shaw, O'Connor, Cleave, Spring, Ross, Neri-Vela, C. Walker	Nov. 26 to Dec. 3, 1985	165:04:49
61-C	Gibson, Bolden, Chang-Diaz, Hawley, G. Nelson, Cenker, B. Nelson	Jan. 12 to 18, 1986	146:03:51
51-L	Scobee, Smith, Resnik, Onizuka, McNair, Jarvis, McAuliffe	Jan. 28, 1986	00:01:13

SOURCE: NASA.

manned spaceflight experience from the first suborbital flight of Astronaut Shepard (May 5, 1961) to the loss of *Challenger* (January 28, 1986). On September 29, 1988 the Space Shuttle returned to flight with the successful liftoff of *Discovery* on the STS-26 mission. Many exciting missions are now planned for the 1990s and beyond. For example, as currently envisioned, the Space Shuttle crews will assemble Space Station *Freedom* in low Earth orbit. This permanent U.S. space station (and contemporary Soviet activities on the *Mir* and follow-on Soviet space stations) will help us learn more fully how to live, work and even play in the extraterrestrial environment.

The U.S. Space Station *Freedom* will be a highly structured, small-group environment where sustained productivity and safety over long periods of time in isolation will depend upon a number of sociopsychological, personality, task and environmental variables. However, several new mission characteristics (versus previous manned missions) could affect space station crew members and might affect overall group performance. These include longer-duration missions, crew heterogeneity, somewhat "routinized" missions, the need for high productivity and greater use of automation and robotics. The study and development of effective group performance processes on the space station will lead to a better definition and understanding of the optimum human role and initial population mix for permanent surface bases on the Moon and Mars.

In examining the human role in space, it is exciting to realize that some of the people born on Earth since Yuri Gagarin's first orbital flight will one day walk on the surface of Mars! And they and their children will live in space settlements orbiting the Earth, on the surface of the Moon and Mars and perhaps in hundreds of manmade planetoids sprinkled throughout heliocentric (Sun-centered) space.

See also: **extraterrestrial careers; hazards to space workers; space life sciences; space station; Space Transportation System**

Compare with: **robotics in space**

photon According to quantum theory, the elementary bundle or packet of electromagnetic radiation, such as a photon of light. Photons have no mass and travel at the speed of light. The energy of the photon is equal to the product of the frequency of the electromagnetic radiation (v) and Planck's constant (h);

$$E = hv$$

where h is equal to 6.626×10^{-34} joule-sec; v is the frequency (hertz).

photosynthesis The process by which photons (light energy) and chlorophyll manufacture carbohydrates out of carbon dioxide (CO_2) and (H_2O).

See also: **life in the Universe**

Pioneer plaque On June 13, 1983, the *Pioneer 10* spacecraft became the first human-made object to leave the Solar System. In an initial attempt at interstellar communication, the *Pioneer 10* spacecraft and its sistercraft (*Pioneer 11*) were equipped with identical special plaques (see fig. 1). The plaque is intended to show any intelligent alien civilization that might detect and intercept either spacecraft millions of years from now when the spacecraft was launched, from where it was launched and by what type of intelligent beings it was built. The plaque's design is engraved into a gold-anodized aluminum plate, 152 millimeters by 229 millimeters (or 6 by 9 inches). The plate is approximately 1.27 millimeters (0.05 in) thick. It is attached to the Pioneer spacecraft's antenna support struts in a position that helps shield it from erosion by interstellar dust.

Let's now review the message contained in the Pioneer plaque. Numbers have been superimposed on the plaque illustrated in figure 1 to assist in this discussion. At the far right, the bracketing bars (1) show the height of the woman compared to the Pioneer spacecraft. The drawing at the top left of the plaque (2) is a schematic of the hyperfine transition of neutral atomic hydrogen—a universal "yardstick" that provides a basic unit of both time and space (length) throughout the Galaxy. This figure illustrates a reverse in the direction of the spin of the electron in a hydrogen atom. The transition depicted emits a characteristic radio wave when an approximately 21-centimeter wavelength. Therefore, by providing this drawing, we are telling any technically knowledgeable alien civilization finding it that we have chosen 21 centimeters as a basic length in the message. While extraterrestrial civilizations will certainly have different names and defining dimensions for their basic system of physical units, the wavelength size associated with the hydrogen radio-wave emission will still be the same throughout the Galaxy. Science and commonly observable physical phenomena represent a general galactic language—at least for starters.

The horizontal and vertical ticks (3) represent the number 8 in binary form. Hopefully, the alien beings pondering

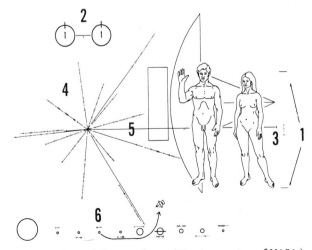

Fig. 1 Annotated Pioneer plaque. (Drawing courtesy of NASA.)

over this plaque will eventually realize that the hydrogen wavelength (21 centimeters) multiplied by the binary number representing 8 (indicated alongside the woman's silhouette) describes her overall height—namely, 8 × 21 centimeters = 168 centimeters, or approximately 5 feet by 5 inches tall. Both human figures are intended to represent the intelligent beings that built the Pioneer spacecraft. The man's hand is raised as a gesture of goodwill. These human silhouettes were carefully selected and drawn to maintain ethnic neutrality. Furthermore, no attempt was made to explain terrestrial "sex" to an alien culture—that is, the plaque makes no specific effort to explain the potentially "mysterious" differences between the man and woman depicted.

The radial pattern (4) should help alien scientists locate our Solar System within the Milky Way Galaxy. The solid bars indicate distance, with the long horizontal bar (5) with no binary notation on it representing the distance from the Sun to the galactic center, while the shorter solid bars denote directions and distances to 14 pulsars from our Sun. The binary digits following these pulsar lines represent the periods of the pulsars. From the basic time unit established by the use of the hydrogen-atom transition, an intelligent alien civilization should be able to deduce that all times indicated are about 0.1 second—the typical period of pulsars. Since pulsar periods appear to be slowing down at well-defined rates, the pulsars serve as a form of galactic clock. Alien scientists should be able to search their astrophysical records and identify the star system from which the Pioneer spacecraft originated and approximately when it was launched, even if each spacecraft isn't found for hundreds of millions of years. Consequently, through the use of this pulsar map, we have attempted to locate ourselves, both in galactic space and in time.

As a further aid to identifying the Pioneer's origin, a diagram of our Solar System (6) is also included on the plaque. The binary digits accompanying each planet indicate the relative distance of that planet from the Sun. The Pioneer's trajectory is shown starting from the third planet (Earth), which has been offset slightly above the others. As a final clue to the terrestrial origin of the Pioneer spacecraft, its antenna is depicted pointing back to Earth.

This message was designed by Drs. Frank Drake and Carl Sagan, and the artwork was prepared by Linda Salzman Sagan.

When the Pioneer 10 spacecraft sped past Jupiter in December 1973, it acquired sufficient kinetic energy (through the gravity-assist technique) to carry it completely out of the Solar System. Table 1 describes some of the "near" star encounters that Pioneer 10 will undergo in the next 860,000 years. Sometime, perhaps a billion years from now, it may pass through the planetary system of a distant stellar neighbor, one whose planets may have evolved intelligent life. If that intelligent life has also developed the technology capable of detecting the (by then derelict) Pioneer 10 spacecraft, it may also possess the curiosity and technical systems needed to intercept it and eventually decipher the message from Earth!

The Pioneer 11 spacecraft carries an identical message. After that spacecraft's encounter with Saturn, it also acquired sufficient velocity to escape the Solar System, but in almost the opposite direction to Pioneer 10. In fact, Pioneer 11 is departing in the same general direction in which our Solar System is moving through Space.

Table 1　Near Star Encounters Predicted for the *Pioneer 10* Spacecraft

Star No.	Name	Information
1	Proxima Centauri	Red dwarf "flare" star. Closest approach is 6.38 light-years after 26,135 years.
2	Ross 248	Red dwarf star. Closest approach is 3.27 light-years after 32,610 years.
3	Lamda Serpens	Sun-type star. Closest approach is 6.9 light-years after 173,227 years.
4	G 96	Red dwarf star. Closest approach is 6.3 light-years after 219,532 years.
5	Altair	Fast-rotating white star (1.5 times the size of the Sun and 9 times brighter). Closest approach is 6.38 light-years after 227,075 years.
6	G 181	Red dwarf star. Closest approach is 5.5 light-years after 292,472 years.
7	G 638	Red dwarf star. Closest approach is 9.13 light-years after 351,333 years.
8	D + 19 5036	Sun-type star. Closest approach is 4.9 light-years after 423,291 years.
9	G 172.1	Sun-type star. Closest approach is 7.8 light-years after 847,919 years.
10	D + 25 1496	Sun-type star. Closest approach is 4.1 light-years after 862,075 years.

[a]SOURCE: NASA/Kennedy Space Center.

As some scientists have philosophically noted, the Pioneer plaque represents "at least one intellectual cave painting, a mark of humanity, which might survive not only all the caves on Earth, but also the Solar System itself!"

See also: **interstellar communication; *Pioneer 10, 11***

Pioneer 10, 11   The *Pioneer 10* and *11* spacecraft, as their names imply, have been true extraterrestrial explorers—the first human-made objects to navigate the main asteroid belt, the first to encounter Jupiter and its fierce radiation belts, the first to encounter Saturn and its magnificent ring system and the first spacecraft to leave the Solar System.

The Pioneer spacecraft investigated magnetic fields, cosmic rays, the solar wind and the interplanetary dust concentrations as they flew through interplanetary space. At Jupiter and Saturn, scientists used the spacecraft to investigate the giant planets and their interesting complement of moons in four main ways: (1) by measuring particles, fields and radiation; (2) by spin-scan imaging the planets and some of their moons; (3) by accurately observing the paths of the spacecraft and measuring the gravitational forces of the planets and their major satellites acting on them; and (4) by observing changes in the frequency of the S-band radio signal before and after occultation (the temporary "disappearance" of the spacecraft caused by their passage behind these celestial bodies) to study the structures of their ionospheres and atmospheres.

The *Pioneer 10* spacecraft was launched from Cape Canaveral by an Atlas–Centaur rocket on March 2, 1972. It became the first spacecraft to cross the main asteroid belt and the first to make close-range observations of the Jovian system. Sweeping nearby Jupiter on December 3, 1973, (its closest approach to the giant planet), it discovered no solid surface under the thick layer of clouds enveloping the giant planet—an indication that Jupiter was a liquid hydrogen planet. *Pioneer 10* also explored the giant Jovian magnetosphere, made close-up pictures of the intriguing Red Spot and observed at relatively close range the Galilean satellites Io, Europa, Ganymede and Callisto. When *Pioneer 10* flew past Jupiter, it acquired sufficient kinetic energy to carry it completely out of the Solar System.

Departing Jupiter, *Pioneer 10* continued to map the heliosphere (the Sun's giant magnetic bubble, or field, drawn out from it by the action of the solar wind). Then, on June 13, 1983, *Pioneer 10* crossed the orbit of Neptune, which at the time was (and until 1999 will be) the planet farthest out from the Sun, due to the eccentricity in Pluto's orbit, which now takes it inside that of Neptune. This historic date marked the first passage of a human-made object beyond the known boundary of the Solar System—an event that can only occur once in all human history! Beyond the boundary of the Solar System, *Pioneer 10* continues to measure the extent of the heliosphere as it flies through interstellar space. Along with its sistership, *Pioneer 11*, this spacecraft is also helping scientists search

for a postulated massive, dark companion to our Sun, called Nemesis.

The *Pioneer 11* spacecraft was launched on April 5, 1973, and swept by Jupiter at an encounter distance of only 43,000 kilometers on December 2, 1974. It provided additional detailed data and pictures on Jupiter and its moons, including the first views of Jupiter's polar regions. Then on September 1, 1979, *Pioneer 11* swept by Saturn, demonstrating a safe flight path for the more sophisticated Voyager spacecraft to follow through the rings. *Pioneer 11* (officially renamed *Pioneer Saturn*) provided the first close-up observations of Saturn, its rings, satellites, magnetic field, radiation belts and atmosphere. It found no solid surface on Saturn but discovered at least one additional satellite and ring. After rushing past Saturn, *Pioneer 11* also headed out of the Solar System toward the distant stars.

Today, both *Pioneer 10* and *11* are still performing missions, exploring the outer extent of the heliosphere and searching for gravitational effects of a possible Tenth Planet (Planet X) or a dark stellar companion (Nemesis) to our Sun. The data now being collected by these two spacecraft will be compared with similar data gathered by the Voyager spacecraft as they also pass beyond the outer regions of the Solar System in different interstellar directions (see fig. 1). Scientists have found that the magnetic field in the outer Solar System has corresponded reasonably well with a spiral field model developed in the late 1950s. The effects of solar maximum and solar minimum activities are also being observed. In fact, most cosmic ray particles from the Galaxy are excluded from the Solar System at periods of solar maximum activity.

Space scientists currently speculate that the region of the heliosphere where modulation of the incoming galactic cosmic rays occurs will be found at a distance of between 40 and 90 astronomical units (AU) from the Sun. It is highly likely, therefore, that sometime in the early 1990s, one of these Pioneer or Voyager spacecraft will become the first spacecraft to reach and record the boundary of cosmic ray modulation. Their cosmic ray experiments should be able to tell scientists back on Earth whether there is a "clean shock" between the heliosphere and interstellar space or only a turbulent transition region.

As both Pioneer spacecraft pass beyond the previously known limits of the Solar System, scientists are also using Doppler tracking data to check for gravity-induced trajectory perturbations that might indicate the presence of a trans-Plutonian planet or a dark star companion to our Sun. At present, the results of these efforts have been negative, indicating that if there really is a planet beyond those presently known or if the Sun has a dark star companion, either must now be extremely distant, perhaps far out along a highly elliptical orbit. Depending on the actual operating times remaining in the *Pioneer 10* and *11* spacecraft nuclear (radioisotope thermoelectric generator) power supplies, these interstellar mission activities are expected to continue well beyond the year 2000.

Finally, both Pioneer spacecraft carry a special message

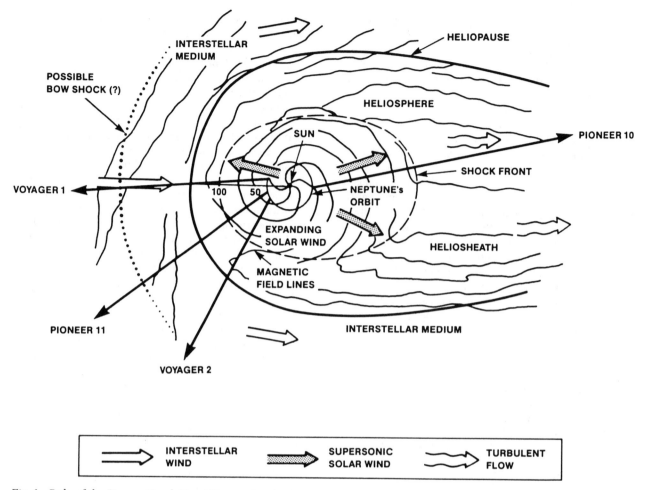

Fig. 1 Paths of the *Pioneer 10* and *11* and *Voyager 1* and *2* spacecraft through the heliosphere and into the interstellar medium. (Courtesy of NASA.)

for any intelligent alien civilization that might find them wandering through the interstellar void millions of years from now. This message is a drawing-map, engraved on an anodized aluminum plaque. The plaque depicts the location of Earth and the Solar System, a man and a woman, and other points of science and astrophysics that should be decipherable by a technically intelligent extraterrestrial civilization.

See also: **cosmic rays; Jupiter; Nemesis; Pioneer plaque; Saturn; Tenth Planet**

Pioneer Venus The *Pioneer Venus* mission consisted of two separate spacecraft launched by the United States to the planet Venus in 1978. The *Pioneer Venus* Orbiter (also called *Pioneer 12*) was a 553-kilogram spacecraft that contained a 45-kilogram payload of scientific instruments. It was launched on May 20, 1978, and placed into a highly eccentric orbit around Venus on December 4, 1978. The 875-kilogram *Pioneer Venus* Multiprobe spacecraft (also called *Pioneer 13*) consisted of a basic bus spacecraft, a large probe and three identical small probes. The Multi-

probe spacecraft was launched on August 8, 1978, and separated about three weeks before entry into the Venusian atmosphere into four probes and the bus. These four probes and the bus successfully entered the Venusian atmosphere at widely separated locations on December 9, 1978, and returned valuable scientific data as they plunged toward the planet's surface. Although the probes were not designed to survive landing, one hardy probe did and transmitted data for about one hour after impact.

The Orbiter spacecraft has made an extensive radar map, covering about 90 percent of the Venusian surface. Using its radar to look through the dense Venusian clouds, the spacecraft revealed that the planet's surface was mostly gentle, rolling planes with two prominent plateaus: Ishtar Terra and Aphrodite Terra. Venus was also found to possess a volcanic structure larger than the Earth's Hawaii-Midway chain and a mountain, called Maxwell Montes, that is larger than Mount Everest.

Today, the *Pioneer Venus* Orbiter *(Pioneer 12)* continues to orbit Venus, providing important information about the planet and about nearby heliocentric space, including how the solar wind interacts with a planet lacking an intrinsic magnetic field. Sometime in 1992, the Orbiter will enter a final phase of mission operations. With its supply of orbital trajectory control propellant exhausted, the spacecraft will eventually enter the atmosphere of Venus and be destroyed. However, important data will be collected until the very end, including in-situ measurements as the spacecraft's decaying orbit takes it lower and lower into the atmosphere.

The Orbiter and probe data have provided more refined information about the Venusian atmosphere. For example, the lower atmosphere of Venus consists of approximately 96 percent carbon dioxide (CO_2), 3 percent nitrogen (N_2) and about 1 percent other gases, including trace amounts of water vapor and sulfur dioxide (SO_2). The Venusian atmosphere is now believed to have at least four distinct cloud and haze layers at different altitudes above the surface. The haze layers appear to contain small aerosol particles, possibly droplets of sulfuric acid.

Finally, the *Pioneer Venus* data have confirmed that a runaway greenhouse effect is indeed responsible for Venus's inferno-like surface temperatures—an issue of great concern to scientists here on Earth who are now studying global change phenomena.

See also: **global change, Magellan Mission, Venus**

planet   A nonluminous celestial body that orbits around the Sun or some other star. There are nine such large objects, or "major planets," in our Solar System and numerous "minor planets," or asteroids. The distinction between a planet and a satellite may not always be clear-cut, except for the fact that a satellite orbits around a planet. For example, our Moon is nearly the size of the planet Mercury and is very large in comparison to its parent planet, Earth. In some cases, the Earth and the Moon can almost be treated as a "double-planet system," with the same being true for icy Pluto and its large satellite, or moon, Charon.

The largest planet is Jupiter, which has more mass than all the other planets combined. Mercury is the planet nearest the Sun, while (on the average) Pluto is farthest away. At perihelion (the point in an orbit at which a celestial body is nearest the Sun), Pluto is actually closer to the Sun than Neptune. Saturn is the least dense planet in our Solar System. If we could find some giant cosmic swimming pool, Saturn would float, since it is less dense than water! Seven of the nine planets have satellites, or moons, some of which are larger than the planet Mercury.

See also: **asteroid; Earth; Jupiter; Mars; Mercury; Neptune; Pluto; Saturn; Uranus; Venus**

planetary engineering   Planetary engineering or terraforming, as it is sometimes called, is the large-scale modification or manipulation of the environment of a planet to make it more suitable for human habitation. In the case of Mars, for example, human settlers would probably seek to make its atmosphere more dense and breathable by adding more oxygen. Early "Martians" would most likely also attempt to alter the planet's harsh temperatures and modify them to fit a more terrestrial thermal pattern. Venus represents an even larger challenge to the planetary engineer. Its current atmospheric pressure would have to be significantly reduced, its inferno-like surface temperatures greatly diminished, the excessive amounts of carbon dioxide in its atmosphere reduced, and perhaps the biggest task of all, its rotation rate would have to be increased to shorten the length of the solar day.

It should now be obvious that when we discuss planetary engineering projects, we are speaking of truly large, long-term projects. Typical time estimates for the total terraforming of a planet like Mars or Venus range from centuries to a few millennia. However, we can also develop ecologically suitable enclaves or niches, especially on the Moon or Mars. Such localized planetary modification efforts could probably be accomplished within a few decades of project initiation.

Just what are the "tools" of planetary engineering? The planetary pioneers in the next century will need at least the following if they are to convert presently inhospitable worlds into new ecospheres that permit human habitation with little or no personal life-support equipment: first, and perhaps the most often overlooked, human ingenuity; second, a thorough knowledge of the physical processes of the particular planet or moon undergoing terraforming (especially the existence and location of environmental pressure points at which small modifications of the local energy or material balance can cause global environmental effects); third, the ability to manipulate large quantities of energy; fourth, the ability to manipulate the surface or material composition of the planet; and fifth, the ability to move large quantities of extraterrestrial materials (for example, small asteroids, comets or water-ice shipments from the Saturnian rings) to any desired locations within heliocentric space.

One frequently suggested approach to planetary engineering is the use of biological techniques and agents to manipulate alien worlds into more desirable ecospheres. For example, scientists have proposed seeding the Venusian atmosphere with special microorganisms (such as genetically engineered algae) capable of converting excess carbon dioxide into free oxygen and combined carbon. This biological technique would not only provide a potentially more breathable Venusian atmosphere, it would also help to lower the currently intolerable surface temperatures by reducing the runaway greenhouse effect.

Other individuals have suggested the use of special vegetation (such as genetically engineered lichen, small plants or scrubs) to help modify the polar regions on Mars. The use of specially engineered, survivable plants would reduce the albedo of these frigid regions by darkening the surface, thereby allowing more incident sunlight to be captured. In time, an increased amount of solar energy absorption would elevate global temperatures and cause melting of the long frozen volatiles, including water. This would raise the atmospheric pressure on Mars and possibly cause a green-house effect. With the polar caps melted, large quantities of liquid water would be available for transport to other regions of the planet. Perhaps one of the more interesting Martian projects late in the next century will be to construct a series of large irrigation canals.

Of course, there are other alternatives to help melt the Martian polar caps. The Martian settlers could decide to construct giant mirrors in orbit above the Red Planet. These mirrors would be used to concentrate and focus raw sunlight directly on the polar regions. Other scientists have suggested dismantling one of the Martian moons (Phobos or Deimos) or perhaps a small dark asteroid, and then using its dust to physically darken the polar regions. This action would again lower the albedo and increase the absorption of incident sunlight.

Another approach to terraforming Mars is to use non-biological replicating systems (that is, self-replicating robot systems). These self-replicating machines will probably be able to survive more hostile environmental conditions than genetically engineered microorganisms or plants. To examine the scope and magnitude of this type of planetary engineering effort, we first assume that the Martian crust is mainly silicone dioxide (SiO_2) and then that a general purpose 100-ton, self-replicating system (SRS) "seed machine" can make a replica of itself on Mars in just one year. This SRS unit would initially make other units like itself, using native Martian raw materials. In the next phase of the planetary engineering project, these SRS units would be used to reduce SiO_2 into oxygen that is then released into the Martian atmosphere. In just 36 years from the arrival of the "seed machine," a silicon dioxide reduction capability would be available that could release up to 220,000 tons per second of pure oxygen into the thin atmosphere of the Red Planet. In only 60 years of operation, this array of SRS units would have produced and liberated 4×10^{17} kilograms of oxygen into the Martian environment. Assuming negligible leakage through the Martian exosphere, this is enough "free" oxygen to create a 0.1 bar pressure breathable atmosphere across the entire planet. This pressure level is roughly equivalent to the terrestrial atmosphere at an altitude of 3,000 meters (16,000 feet).

What would be the environmental impact of all these mining operations on Mars? Scientists estimate that the total amount of material that must be excavated to terraform Mars is on the order of 10^{18} kilograms of silicon dioxide. This is enough soil to fill a surface depression 1 kilometer deep and about 600 kilometers in diameter. This is approximately the size of the crater Edom near the Martian equator. The Martians might easily rationalize: "just one small hole for Mars, but a new ecosphere for humankind!"

Asteroids have also played an interesting role in planetary engineering scenarios. People have suggested crashing one or two "small" asteroids into depressed areas on Mars (such as the Hellas Basin) to instantly deepen and enlarge the depression. The goal would be individual or multiple (con-nected) instant depressions about 10 kilometers deep and 100 kilometers across. These manmade impact craters would be deep enough to trap a more dense atmosphere— allowing a small ecological enclave or niche to develop. Environmental conditions in such enclaves could range from typical polar conditions to perhaps something almost balmy.

Others have suggested crashing selected asteroids into Venus to help increase its spin. If the asteroid hits the Venusian surface at just the right angle and speed, it could conceivably help speed up the planet's rotation rate— greatly assisting any overall planetary engineering project. Unfortunately, if the asteroid is too small or too slow, it will have little or no effect; while if it is too large or hits too fast, it could possibly shatter the planet! This would be a truly cosmic "oops"!

Another scientist has proposed that several large-yield nuclear devices be used to disintegrate one or more small asteroids that had previously been maneuvered into orbits around Venus. This would create a giant dust and debris cloud that would encircle the planet and reduce the amount of incoming sunlight. This, in turn, would lower surface temperatures on Venus and allow the rocks to cool sufficiently to start absorbing carbon dioxide from the dense Venusian atmosphere.

Finally, other scientists have suggested mining the rings of Saturn for frozen volatiles, especially water-ice and then transporting these back for use on Mars, the Moon or Venus for large-scale planetary engineering projects.

Can you think of anything else planetary engineers might do in the next few centuries to make Mars, Venus, the Moon or other celestial objects potential "garden spots" in heliocentric space?

planetesimals   Small celestial objects in the Solar System, such as asteroids, comets and moons.

planet fall   The landing of a spacecraft or space vehicle on a planet.

planetoid   An asteroid, or minor planet.

Pluto   Even though it was discovered more than six decades ago, in February 1930, by the Kansan farm boy-astronomer Clyde W. Tombaugh, Pluto still remains somewhat of an astronomical mystery. Orbiting the Sun at the frigid extremes of our Solar System, this planet lies at the resolution limit of the largest optical telescopes on Earth. It is also the only one of the nine major planets not yet encountered by flyby scientific spacecraft.

Because of Pluto's highly eccentric orbit, its distance from the Sun varies from about 4.4 to 7.4 billion kilometers, or some 29.5 to 49.2 astronomical units. For most of its orbit, Pluto is the outermost of the planets. However, since 1979 it has actually been orbiting closer to the Sun than Neptune. This condition will remain until 1999, when Pluto becomes the outermost planet in the Solar System.

Despite this obviously chaotic, highly inclined (more

than 17 degrees) orbit, which takes Pluto around the Sun once every 248 years, the planet appears to have settled into a stable 3:2 orbital resonance with its planetary neighbor, Neptune. This means that Neptune experiences three complete orbits around the Sun in the same amount of time it takes Pluto to complete just two orbits around the Sun. As a result of this orbital mechanics arrangement, Pluto never actually gets within 2.7 billion kilometers of Neptune and has so far avoided a cosmic encounter of the most disruptive kind.

Pluto's physical discovery in 1930 came as no real astronomical surprise. Its existence had already been predicted by astronomers at the turn of the century as a result of perturbations (disturbances) observed in the orbits of both Uranus and Neptune. Pluto was found after a dedicated, deliberate search in just about the anticipated location. This discovery, however, was somewhat fortuitous, because Pluto with its tiny mass could not be the source of these outer planet orbital perturbations. In fact, the planetary perturbation data that stimulated the original search for a planet beyond Neptune were mainly erroneous, the result of inadequate information about Neptune's orbit. However, recent and more precise orbit data for Neptune and Uranus have suggested other unexplained perturbations. Planetary scientists and astronomers are not currently sure what the source of these unexplained perturbations is. Pluto, because of its tiny mass, is not considered solely responsible. The speculation that there is yet another undiscovered planet—a "Tenth Planet," or "Planet X"—located somewhere beyond Pluto, does not appear to match the perturbation data either. At present, these unexplained perturbations remain an open astronomical question.

In 1978, James W. Christy of the U.S. Naval Observatory triggered a revolution in our understanding of Pluto with the discovery of its large moon, Charon. As you might remember, Charon was the boatman in Greek and Roman mythology who ferried the dead across the River Styx to the Underworld, ruled by Hades (Pluto in Roman mythology).

Recent observations of Pluto–Charon eclipses and stellar occultations have provided new details about the physical characteristics of this distant planet and its celestial companion. Scientists now believe that Pluto and perhaps Charon have a density of about 2.0 grams per cubic centimeter, a value that suggests an overall planetary composition of rocky materials as well as water and methane ices. Pluto is the smallest of the major planets, with a diameter of about 2,300 km; while its large moon, Charon, has a diameter of approximately 1,190 km—forming what some scientists regard as the only true double-planet system in our Solar System. Charon circles Pluto at an orbital radius of 19,640 km. Its orbit is gravitationally synchronized with Pluto's rotational period (approximately 6.4 days) so that both the planet and its moon keep the same hemisphere facing each other.

If we assume a density of 2.0 grams per cubic centimeter for both Pluto and Charon, then the mass of Pluto is approximately 1.3×10^{22} kg and the mass of Charon about 1.7×10^{21} kg. Tables 1 and 2 summarize the most recent physical data about Pluto and its moon, Charon. Of course, these data will remain quite speculative until an instrumented robot spacecraft successfully encounters the distant

Table 1   Dynamic Properties and Physical Data for Pluto

Diameter	2,300 km
Mass	1.3×10^{22} kg
Mean density	2.0 g/cm^3
Albedo (overall)	0.5
Surface temperature	58 Kelvin
Atmosphere (a transient phenomenon)	Methane (CH$_4$) possibly mixed with argon (Ar), nitrogen (N$_2$), oxygen (O$_2$) and carbon monoxide (CO)
Period of rotation	6.387 days
Inclination of axis (of rotation)	122°
Orbital period (around Sun)	247.69 years
Distance from the Sun	
Aphelion	7.375×10^9 km (49.2 AU) [409.2 light-min]
Perihelion	4.43×10^9 km (29.5 AU) [245.3 light-min]
Semimajor axis	5.94×10^9 km (39.7 AU)
Orbit inclination	17.139°
Eccentricity of orbit	0.25235
Mean solar flux (at 30 AU)	1.5 W/m^2
Number of natural satellites	1

Note: These data are speculative.
SOURCE: NASA.

Table 2   Dynamic Properties and Physical Data for Charon

Diameter	1,186 km
Mass	1.7×10^{21} kg
Mean density	2.0 g/cm^3 (?)
Albedo (overall)	<0.4
Semimajor axis (of orbit around Pluto)	19,640 km
Period (gravitationally synchronized orbit)	6.387 days
Inclination to Pluto's equator	94° ± 3°

Note: These data are speculative.
SOURCE: NASA.

double planet system sometime in the first quarter of the next century.

Scientists currently believe that Pluto possesses a very thin atmosphere that contains methane (CH_4) as well as possibly argon (Ar), nitrogen (N_2), carbon monoxide (CO) and oxygen (O_2). Pluto's atmosphere is unique in the Solar System in that it undergoes a formation and decay cycle each orbit around the Sun. The atmosphere begins to form several decades before perihelion (the planet's closest approach to the Sun) and then slowly collapses and freezes out decades later, as the planet's orbit takes it farther and farther away from the Sun to the frigid outer extremes of the Solar System. In September 1989, Pluto experienced perihelion. Several decades from now its thin atmosphere will freeze out and collapse, leaving a fresh layer of methane snow on the planet's surface.

The surface of Pluto appears covered with methane ice, which would appear reddish to human eyes. This frozen world reflects about half of the light striking it. Charon, on the other hand, would appear gray or bland to the human eye and reflects less than 40 percent of incoming light. Scientists believe that Charon may be covered by a large expanse of water ice.

Data collected by the Infrared Astronomical Satellite indicate that Pluto's equatorial surface temperature is about 58 degrees Kelvin, a planetary surface temperature (during the planet's perihelion approach) that is physically consistent with the presence of methane in Pluto's thin atmosphere.

NASA scientists are currently considering a flyby mission to Pluto. One mission scenario involves a pair of identical spacecraft, each launched by an expendable Delta-class rocket vehicle. Each of these robot explorers would carry imaging sensors, spectrometers and particle-plasma sensor package. The spacecraft would be launched approximately one year apart early in the next century and sent on a 14-year deep space trajectory to the Pluto-Charon system. The proposed trajectories would take advantage of gravity assist maneuvers at Earth and Jupiter. These flyby missions would provide scientists an opportunity to obtain high resolution images of both Pluto and Charon, characterize Pluto's atmosphere, investigate the internal structure and composition of both celestial objects and even search for other moons. Two separate spacecraft would greatly increase the overall chances of a successful deep space mission to the very outer regions of our Solar System.

Plutonian   Of or relating to the planet Pluto; (in science fiction) a native of the planet Pluto.

pressurized habitable environment   Any module or enclosure in space in which an astronaut may perform activities in a "shirt-sleeve" environment.

primary body   The celestial body about which a satellite, moon or other object orbits or from which it is escaping or toward which it is falling. For example, the primary body

of the Earth is the Sun; the primary body of the Moon is the Earth.

primitive atmosphere   The atmosphere of a planet or other celestial object as it existed in the early stages of its formation. For example, the primitive atmosphere of the Earth some 3 billion years ago was thought to consist of water vapor, carbon dioxide, methane and ammonia.

principle of mediocrity   A general assumption or speculation often used in discussions concerning the nature and probability of extraterrestrial life. It assumes that things are pretty much the same all over—that is, it assumes that there is nothing special about the Earth or our Solar System. By invoking this hypothesis, we are guessing that other parts of the Universe are pretty much as they are here. This philosophical position allows us to then take the things we know about the Earth, the chemical evolution of life that occurred here and the facts we are learning about other objects in our Solar System and extrapolate these to develop concepts of what may be occurring on alien worlds around distant suns.

The simple premise of the principle of mediocrity is very often employed as the fundamental starting point for contemporary speculations about the cosmic prevalence of life. If the Earth is indeed *nothing special*, then perhaps a million worlds in our own Galaxy (which is one of billions of galaxies) not only are suitable for the origin of life but have witnessed its chemical evolution in their primeval oceans and are now (or at least were) habitats for a myriad of interesting living creatures. Some of these living systems may also have arisen to a level of intelligence where they are at this very moment gazing up into the heavens of their own world and wondering if they, too, are alone!

If, on the other hand, the Earth and its delicate biosphere really are something special, then life—especially intelligent life capable of comprehending its own existence and contemplating its role in the cosmic scheme of things—may be a rare, very precious jewel in a vast, lifeless cosmos. In this latter case, the principle of mediocrity would be most inappropriate to use in estimating the probability that extraterrestrial life exists elsewhere in the Universe.

Today, we cannot pass final judgment on the validity of the principle of mediocrity. We must, at an absolute minimum, wait until human and robot explorers have made more detailed investigations of the interesting objects in our own Solar System. Celestial objects of particular interest to exobiologists include the planet Mars and certain moons of the giant outer planets Jupiter and Saturn. Once we have explored these alien worlds in depth, scientists will have a much more accurate technical basis for suggesting that we are either "something special" or "nothing special"—as the principle of mediocrity implies.

See also: **search for extraterrestrial intelligence**

Project Cyclops   A very large array of dish antennas proposed for use in a detailed search of the radio-frequency spectrum (especially the 18- to 21-centimeter wavelength "waterhole") for interstellar signals from intelligent alien civilizations. The engineering details of this SETI (search for extraterrestrial intelligence) configuration were derived in a special summer institute design study sponsored by NASA at Stanford University in 1971. The stated object of the Project Cyclops study was to assess what would be required in hardware, manpower, time and funding to mount a realistic effort, using present (or near-term future) state-of-the-art techniques, aimed at detecting the existence of extraterrestrial (extrasolar system) intelligent life.

Named for the one-eyed giants found in Greek mythology, the proposed Cyclops Project would use as its "eye" a large array of individually steerable 100-meter-diameter parabolic dish antennas. These Cyclops antennas would be arranged in a hexagonal matrix, so that each antenna unit was equidistant from all its neighbors. A 292-meter separation distance between antenna dish centers would help avoid shadowing. In the Project Cyclops concept, an array of about 1,000 of these antennas would be used to simultaneously collect and evaluate radio signals falling on them from a target star system. The entire Cyclops array would function like a single giant radio antenna, some 30 to 60 square kilometers in size.

Project Cyclops can be regarded as one of the foundational studies in our contemporary search for extraterrestrial intelligence. Its results—based on the pioneering efforts of such individuals as Dr. Frank Drake, Dr. Philip Morrison, Dr. John Billingham and Dr. Bernard Oliver—have established the technical framework for subsequent SETI activities. Project Cyclops also reaffirmed the interstellar microwave window, the 18- to 21-centimeter-wavelength "waterhole," as perhaps the most suitable part of the electromagnetic spectrum for interstellar civilizations to communicate with each other and for us here on Earth as we begin our cosmic search for signals from intelligent alien beings.

See also: **Drake equation; extraterrestrial civilizations; Fermi paradox; search of extraterrestrial intelligence**

Project Daedalus   The name given to the most extensive study of interstellar space exploration yet undertaken. From 1973 to 1978 a team of scientists and engineers under the auspices of the British Interplanetary Society studied the feasibility of performing a simple interstellar mission using only current technology and reasonable extrapolations of imaginable near-term capabilities.

In Roman mythology Daedalus was the grand architect of King Minos's labyrinth for the Minotaur on the island of Crete. But Daedalus also showed the Greek hero Theseus, who slew the Minotaur, how to escape from the labyrinth. An enraged King Minos imprisoned both Daedalus and his son Icarus. Undaunted, Daedalus (a brilliant engineer) fashioned two pairs of wings out of wax, wood and leather. Before their aerial escape from a prison tower, Daedalus cautioned his son not to fly too high, so that the Sun would

not melt the wax and cause the wings to disassemble. They made good their escape from King Minos's Crete, but while over the sea, Icarus, an impetuous teen-ager, ignored his father's warnings and soared high into the air. Daedalus (who reached Sicily safely) watched his young son, wings collapsed, tumble to his death in the sea below.

The proposed Daedalus spaceship structure, communications systems and much of the payload were designed entirely within the parameters of 20th-century technology. Other components, such as the advanced machine intelligence flight controller and on-board computers for inflight repair, required artificial-intelligence capabilities expected to be available in the mid-21st century. The propulsion system, perhaps the most challenging aspect of any interstellar mission, was designed as a nuclear-powered, pulsed-fusion rocket engine that burned an exotic thermonuclear fuel mixture of deuterium and helium 3 (a rare isotope of helium). This pulsed-fusion system was believed capable of propelling the robot interstellar probe to velocities in excess of 12 percent of the speed of light (that is, above $0.12\ c$). The best source of helium 3 was considered to be the planet Jupiter, and one of the major technologies that had to be developed for Project Daedalus was an ability to mine the Jovian atmosphere for helium 3. This mining operation might be achieved by using "aerostat" extraction facilities (floating balloon-type factories).

The Project Daedalus team suggested that this ambitious interstellar flyby (one-way) mission might possibly be undertaken at the end of the 21st century—when the successful development of humankind's extraterrestrial civilization had generated the necessary wealth, technology base and exploratory zeal. The target selected for this first interstellar probe was Barnard's star, a red dwarf (spectral type M 5) about 5.9 light-years away in the constellation Ophiuchus.

The Daedalus spaceship would be assembled in cislunar space (partially fueled with deuterium from Earth) and then ferried to an orbit around Jupiter, where it could be fully fueled with the helium 3 propellant that had been mined out of the Jovian atmosphere. These thermonuclear fuels would then be prepared as pellets, or "targets," for use in the ship's two-stage pulsed-fusion power plant. Once fueled and readied for its epic interstellar voyage, somewhere around the orbit of Callisto, the ship's mighty pulsed-fusion first-stage engine would come alive. This first-stage pulsed-fusion unit would continue to operate for about two years. At first-stage shutdown, the vessel would be traveling at about 7 percent of the speed of light ($0.07c$).

The expended first-stage engine and fuel tanks would be jettisoned in interstellar space, and the second-stage pulsed-fusion engine would ignite. The second stage would also operate in the pulsed-fusion mode for about 2 years. Then, it, too, would fall silent, and the giant robot spacecraft, with its cargo of sophisticated remote sensing equipment and nuclear (fission)-powered probe ships, would be traveling at about 12 percent of the speed of light ($0.12c$). It would take the Daedalus spaceship about 47 years of coasting (after second-stage shutdown) to encounter Barnard's star.

When the Daedalus interstellar probe was about 3 light-years away from its objective (about 25 years mission elapsed time), smart computers on board would initiate long-range optical and radio astronomy observations. A special effort would be made to locate and identify any extrasolar planets that might exist in the Barnardian system.

Of course, traveling at 12 percent of the speed of light, Daedalus would only have a very brief passage through the target star system. This would amount to a few days of "close-range" observation of Barnard's star itself and only "minutes" of observation of any planets or other interesting objects by the robot mothership.

However, several years before the Daedalus mothership passed through the Barnardian system, it would launch its complement of nuclear-powered probes (also traveling at 12 percent of the speed of light initially). These probe ships, individually targeted to objects of potential interest by computers on board the robot mothership, would "fly ahead" and act as data-gathering scouts. A complement of 18 of these scout craft or small robotic probes was considered appropriate in the Project Daedalus study.

Then, as the main Daedalus spaceship flashed through the Barnardian system, it would gather data from its own on-board instruments as well as information telemetered to it by the numerous probes. Then, over the next day or so, it would transmit all these mission data (billions and billions of data bits) back toward our Solar System, where team scientists would patiently wait the approximately six years it takes for these information-laden electromagnetic waves, traveling at light speed, to cross the interstellar void.

Its mission completed, the Daedalus mothership—without its probes—would continue on a one-way journey into the darkness of the interstellar void, to be discovered perhaps millennia from then by an advanced alien race, which might puzzle over humankind's first attempt at the direct exploration of another star system.

The main conclusions that can be drawn from the Project Daedalus study might be summarized as follows: (1) exploration missions to other star systems are, in principle, technically feasible; (2) a great deal could be learned about the origin, extent and physics of the Galaxy, as well as the formation and evolution of stellar and planetary systems, by missions of this type; (3) the prerequisite interplanetary and initial interstellar space system technologies necessary to successfully conduct this class of mission also contribute significantly to humankind's search for extraterrestrial intelligence (for example, smart robot probes and interstellar communications); (4) a long-range societal commitment on the order of a century would be required to achieve such a project; and (5) the prospects for interstellar flight by human beings do *not* appear very promising using current or foreseeable 21st-century technologies.

The Project Daedalus study also identified three key technology advances that would be needed to make even a robot interstellar mission possible. These are (1) the development of controlled nuclear fusion, especially the use of the deuterium/helium 3 thermonuclear reaction; (2) advanced machine intelligence; and (3) the ability to extract helium 3 in large quantities from the Jovian atmosphere.

Although the choice of Barnard's star as the target for the first interstellar mission was somewhat arbitrary, if future human generations can build such an interstellar robot spaceship and successfully explore the Barnardian system, then with modest technology improvements, all star systems within 10 to 12 light-years of Earth become potential targets for a more ambitious program of (robotic) interstellar exploration.

See also: **Barnard's star; fusion; interstellar contact; robotics in space; search for extraterrestrial intelligence; space nuclear propulsion; starship**

Project Galileo   A NASA planetary exploration mission launched October 18, 1989, to orbit Jupiter and to send an instrumented probe into the giant planet's dense atmosphere. The project is named after the famous Italian astronomer, Galileo Galilei (1564–1642), who in 1610 discovered the four major moons of Jupiter: Io, Europa, Ganymede and Callisto. These four Jovian moons are called the "Galilean satellites" in his honor. Galileo helped confirm the theory that the Sun indeed was the center of our Solar System—a scientific view that almost cost him his life in 1633. Because of his pioneering astronomical work, Galileo became engaged in a bitter conflict with religious officials of the day, who supported the Ptolemaic (Earth-centered) concept of the Universe. Forced to renounce his belief in the Copernican (Sun-centered, or heliocentric) model of the Solar System, legend has it that the 70-year-old Galileo defiantly mumbled, while rising from his knees before his persecutors, "And yet it [the Earth] moves!"

Project Galileo will enable planetary scientists to study, at close range and for almost two years, Jupiter, several of its satellites and its magnetospheric domain. Today, scientists believe that Jupiter is composed of the original material from which the Solar System evolved some 4.6 billion years ago. This material, mainly hydrogen and helium, is largely unmodified by thermonuclear processes, because Jupiter—an "almost-star"—isn't supporting the nuclear-fusion burning found in stellar interiors. Close-up studies of the giant planet should provide interesting information about the beginning and development of the Solar System as well as new insights into phenomena that directly relate to our understanding of all the planets.

The mission has as its main scientific objectives the study of the chemical composition and physical state of Jupiter's atmosphere, a close-up examination of the physical state and chemical composition of selected Jovian moons and a detailed investigation of the structure and dynamics of Jupiter's magnetosphere. On the way to Jupiter, Project Galileo will observe Venus, the Earth–Moon system, one or two asteroids and various phenomena in interplanetary space.

The *Galileo* spacecraft has two major components: a planetary orbiter and an atmospheric entry probe. Both the orbiter and the probe carry a set of instruments. The orbiter has a dual-spin design, with a spinning or "spun" section and a fixed or "despun" section. The main spacecraft section—containing the fields and particles sensors, main antenna, propulsion module, and power supply—can be spun from 3 to 10 revolutions per minute (rpm). The instruments located on the spun section of the spacecraft enjoy a wide viewing range and support very comprehensive observations.

The camera and other remote-sensing equipment are located on a three-axis stabilized platform in the despun (aft) section of the *Galileo* spacecraft. These particular instruments require a fixed orientation in space and will be pointed at targets selected for observation. Instead of the vidicon television tubes used on the Voyager spacecraft, the *Galileo* has a new solid-state imaging system with a charge-coupled device. Because the Galileo orbiter spacecraft will fly closer to the Galilean moons and because of the new imaging system, scientists will enjoy images of these interesting Jovian satellites at resolutions of 20 to 1,000 times better than acquired during the highly successful Voyager encounters.

The orbiter spacecraft has an overall mass of about 2,370 kilograms (5,230 lb), including instruments and about 907 kg (2,000 lb) of propellant. The smaller atmospheric probe has a mass of 338 kg (744 lb), including instruments.

Communication with the Earth is accomplished through one 4.88-meter (16-foot) diameter high-gain antenna and two low-gain antennas. The two low-gain antennas assist with communications when the main (high-gain) antenna is not activated or not pointed toward the Earth. The spacecraft uses S-band and X-band (2,295 megahertz and 8,415 megahertz frequencies, respectively) to transmit to and receive signals from Earth. The orbiter spacecraft's maximum communication rate is 134 kilobits per second. The probe can transmit data at a rate of 128 bits per second to the orbiter spacecraft, which then relays these data back to Earth. When the spacecraft arrives at Jupiter, it will take just under an hour for its signals to reach Earth.

The spacecraft is powered by two plutonium-238 fueled radioisotope thermoelectric generators (RTGs), which will provide approximately 500 watts of electric power for the entire eight-year mission. Propulsion for the orbiter is provided by a single 400-newton (89 pound-force) retroengine and two clusters of six 10-newton (2.2 pound-force) thrusters, which use monomethylhydrazine (fuel) and nitrogen tetroxide (oxidizer). The probe has no propulsion system and is powered by long-life batteries.

The spinning section of *Galileo* contains the following instruments and experiments: (1) a magnetometer to measure magnetic fields and their dynamic behavior; (2) a plasma instrument to provide information on low-energy particles and clouds of ionized gases in the Jovian magnetosphere; (3) a plasma wave instrument to investigate plasma waves generated inside Jupiter's magnetosphere and plasma waves generated by possible lightning discharges in the Jovian atmosphere; (4) energetic particle detectors to measure the composition, distribution and energy spectra of high-energy particles trapped in Jupiter's magnetic field; (5) dust detection instrumentation to determine the size, speed and charge of small particles such as micrometeorites; (6) a celestial mechanics/radio science experiment to measure the gravity fields of Jupiter and its moons, and to search for gravity waves propagating through space; and (7) a radio-wave propagation experiment that uses radio signals from the orbiter spacecraft and probe to study the structure of the atmospheres and ionospheres of Jupiter and its major satellites.

The instruments and experiments on the despun portion of the spacecraft include (1) a solid-state imaging camera (using a charge-coupled device) to provide color images of the Jovian system, (2) a near-infrared mapping spectrometer to study the surface-composition and properties of selected Jovian moons and the asteroids encountered enroute, (3) an ultraviolet spectrometer to examine the composition and structure of the upper atmosphere of Jupiter and its satellites and (4) a photopolarimeter radiometer to measure the temperature profile and energy balance of the Jovian atmosphere.

Finally, the probe into the Jovian atmosphere contains the following instruments and experiments: (1) an instrument to provide information about the temperature, density and pressure of Jupiter's atmosphere; (2) a neutral mass spectrometer to measure the composition of the gases in Jupiter's atmosphere and compositional variations at different levels in the atmosphere; (3) a helium abundance interferometer to measure the ratio of hydrogen to helium in the Jovian atmosphere; (4) a nephelometer to determine the properties of cloud particles and the location of cloud layers; (5) a net flux radiometer to measure the difference in the energy radiated from Jupiter and the energy received from the Sun at different levels in the upper atmosphere; (6) a lightning and radio emission instrument to measure the electromagnetic waves generated by any lightning occurring in the Jovian atmosphere, including a measurement of optical and radio wave emissions from these flashes; and (7) an energetic particle detector to measure electrons and protons in the inner regions of the Jovian radiation belts.

Galileo spacecraft began its approximately 4.0 billion kilometer (2.5 billion mi) journey aboard the Space Shuttle *Atlantis,* which carried it to low Earth orbit. A two-stage inertial upper stage (IUS) propulsion system then successfully placed the spacecraft on a complicated interplanetary trajectory, involving a triple-gravity-assist maneuver, to give it the total velocity needed to reach Jupiter. (See fig. 1.) The Galileo mission was previously designed for a direct flight of about two and one half years from the Earth to Jupiter. However, new safety restrictions involving the use of upper-stage launch systems following the Space Shuttle *Challenger* accident (January 28, 1986) precluded this short flight when the Centaur liquid-fueled upper stage rocket was replaced by a two-stage (solid-fueled) inertial upper

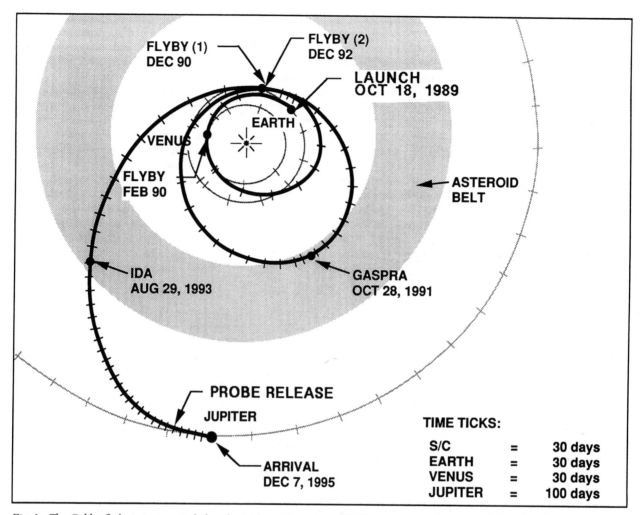

Fig. 1 The *Galileo* flight trajectory, including the triple-gravity-assist or VEEGA maneuvers. (Courtesy of NASA/JPL.)

stage system. Trajectory engineers designed a new inter-planetary flight path using gravity-assist maneuvers with Venus and Earth (twice) to build up the spacecraft's speed to reach Jupiter. This six-year flight path is called the Venus-Earth-Earth-gravity-assist (VEEGA) trajectory. During its journey to the planet Jupiter, *Galileo* will accomplish an historic first in October 1991, when it flies by the asteroid Gaspra at a range of about 1,000 kilometers (620 mi) and a relative speed of 29,000 kilometers per hour (18,000 miles per hour). Instruments onboard the spacecraft will take pictures and examine Gaspra's surface composition, roughness, optical and thermal properties, rotation rate and mass. A second asteroid flyby, this time involving the larger asteroid Ida, will occur in August 1993. The planned encounter range is also 1,000 kilometers (620 mi) although the relative speed of this encounter will be higher at approximately 48,300 kilometers per hour (30,000 miles per hour). Data similar to that taken during the Gaspra encounter will be collected.

Then, in early July, 1995, about five months before

arrival at the planet Jupiter, *Galileo* will make precise maneuvers to point its probe in the right direction for entry into the Jovian atmosphere. The probe, encased in its heat shield, will be spun up to ten revolutions per minute and released. The orbiter spacecraft and probe will then continue on independent trajectories toward the Jovian encounter.

In December 1995, *Galileo* will come within 1,000 kilometers (620 mi) of the innermost Galilean moon, Io. This close encounter will help slow the spacecraft and reduce the propulsive requirements of the retrofire maneuver needed to achieve orbit around Jupiter. During this sequence of events, the spacecraft will also have passed within 35,400 kilometers (22,000 miles) of the Jovian moon Europa.

On December 7, 1995, a few hours after the orbiter spacecraft's encounter with Io, the Galileo probe will enter the Jovian atmosphere at about 48 kilometers per second (more than 100,000 miles per hour) and slow down by aerodynamic braking in about 2 minutes before dropping its heat shields. This will allow the probe to drop through the Jovian clouds about 200 kilometers (125 mi), passing from a pressure of one-tenth that on the Earth's surface to about 25 Earth atmospheres in 75 minutes. The probe, sending its data back to the orbiter for relay to Earth, is not expected to survive beyond this point.

Then, for the next 22 months, the orbiter spacecraft will follow a series of elliptical paths through a wide region of the Jovian environment. Observations will include the Giant Red Spot, a long-persisting storm system in Jupiter's atmosphere. The orbiter will make at least ten gravity-assist orbits of Jupiter, flying as close as 200 kilometers (125 miles) above the surface of a Jovian moon each time. During at least one orbit, the spacecraft will fly deep into Jupiter's magnetotail—the area of the Jovian magnetic environment directly opposite the Sun. Following this 22-month orbital observation phase, *Galileo* will orbit around Jupiter indefinitely, and will continue to gather and transmit data until its propellant supply (for maneuver and attitude control) is exhausted.

See also: **Jupiter; orbits of objects in space** (gravity assist); **space nuclear power; Voyager**

Project Orion (I)   A design study sponsored by NASA's Office of Space Science in 1976 to examine techniques and instruments that could be used in a search for extrasolar planets. Are planets common throughout our Galaxy? Are they formed regularly from a nebula of gas and dust surrounding protostars (developing stars)? (This is called the nebula hypothesis of planetary formation.) Or are they created only rarely, through such unusual cosmic events as the close passage of a rogue star, a stellar collision or a supernova event? The answers to these and many other intriguing questions about the cosmic prevalence of extrasolar planets will help us better understand the origin of our own Solar System and will assist us in planning the search for extraterrestrial intelligence.

At present, most astronomers support the nebula hypothesis of planetary formation. They assume that planetary systems are the rule rather than the exception. This implies that planets occur as a natural, if not inevitable, part of the stellar-formation process. A careful, systematic investigation of how frequently such extrasolar planets occur would provide observational support for modern theories of stellar formation.

In addition, a precise hunt for extrasolar planets helps solve a most puzzling cosmic mystery: "Are we alone?" If planets are necessary for the origin of life and the eventual rise of intelligent creatures (as occurred here on Earth), then we really must know how often extrasolar planets occur in order to develop credible estimates of the probability of encountering intelligent alien beings. For example, if planets—especially Earthlike planets around Sunlike stars—are really rare, then the chances for finding intelligent alien creatures around distant suns would indeed be very remote. If, on the other hand, suitable planets occur as part of the formation process of most stars, then there may

indeed be many life-bearing worlds in our Galaxy—some spawning intelligent planetary civilizations like our own!

Unfortunately, the detection of extrasolar planets is a difficult task, because obtaining the extremely precise astrometric data that would be required is beyond the capability of today's technology. Present observational techniques and ground-based instruments are only marginal in their ability to conduct such an effort. NASA's Project Orion examined design concepts for an advanced ground-based astrometric system that could support a technically realistic search effort.

Scientific and technical advancement at the turn of the century made it possible to measure the separation between two photographically recorded star images with accuracies of a few thousandths of a millimeter. As a result, astronomers were able to measure stellar angular separations as small as a few hundredths of an arc-second. (An arc-second is an angle equal to 1/3,600 degree.)

Astrometry is that branch of astronomy that deals with the precise measurement of the position and motion of stars. Massive, but unseen or "dark," stellar companions will cause a star to wobble from its predicted course. The traditional technique in searching for extrasolar planets is consequently to perform precise astrometric studies. If astronomers detect a slight wobble in a star's predicted motion, the presence of planetary companions could be implied.

However, the first unseen object to be detected outside the Solar System was not a planet, but rather a new type of star called a white dwarf. For a long time, astronomers had been puzzled by the tiny variations they observed in the motion of the star Sirius. Instead of traveling in a straight line, it was seen to wobble from one side to the other of its predicted trajectory. Its motion was much like the wobble pattern described above. Astronomers have now established that this tiny wobble (about 4 arc-seconds) in Sirius's projected trajectory is being caused by a dark stellar companion (a white dwarf star) that has about the same mass as our Sun.

Although severely limited by the measurement capabilities of existing ground-based astrometric telescopes, astronomers over the last few decades have attempted to search for evidence of extrasolar planets by carefully looking for a telltale wobble or perturbation (disturbance) in a particular star's motion. Unfortunately, these efforts currently remain scientifically inconclusive. For example, a great deal of interest has recently centered around Barnard's star, a red dwarf about 5.9 light-years away. Certain wobble data have suggested that this star could possess one or more Jovian-mass planets. But because these data lie at the very threshold of current measurement accuracies, such conclusions are not regarded as "technically firm." More accurate wobble data are needed before scientists can confidently accept the possible fact that one of our stellar neighbors is accompanied by massive planetary companions.

Three techniques for detecting extrasolar planets were

considered in the Project Orion study: (1) very accurate astrometric searches for perturbations or wobbles in a star's motion, (2) infrared searches for intrinsic thermal radiation from extrasolar planets and (3) optical searches for reflected stellar (visible) radiation.

The first technique, using precise astrometric observations, is often referred to as "indirect" detection of extrasolar planets. This is because the presence of a planetary companion around an alien sun is deduced from observations of the parent star, and not from observations of the extrasolar planet itself.

The techniques for detecting extrasolar planets by looking for telltale planetary radiations are called "direct" detection techniques, because detection arises from a direct observation of the planet. Planetary companions to stars can be considered as sources of electromagnetic radiation. We can characterize such planetary radiation as either thermal or reflected optical. A planet will radiate at some temperature that is determined by its overall energy balance. A planetary energy balance is simply the sum of the incoming energy (radiant energy from the parent star) plus internal planetary energy sources (such as radioactive decay) minus outgoing energy (energy radiated away by the planet). This planetary radiation is generally most pronounced in the infrared portion of the electromagnetic spectrum and is often called "intrinsic thermal radiation."

A planet might also reflect radiation from its parent star. This reflected radiation is usually strongest at those wavelengths where the star emits the majority of its radiant energy—typically, in the visible portion of the spectrum. The amount of reflected visible radiation is a function of both the size of the extrasolar planet and the nature of the reflecting medium (that is, the composition of the planet's atmosphere, its surface conditions and so on).

The main effort of Project Orion was to develop a design concept for an improved ground-based astrometric telescope. This telescope needed the following features: (1) reflecting optics, (2) large aperture and (3) "electric" (automated) detection of interference fringes from stars. Project Orion's imaging stellar interferometer (ISI) design incorporated as many technical advantages as possible, while avoiding most of the problems found in other astrometric systems. It is essentially a long-baseline interferometer that simultaneously "images" the white-light fringes of many stars.

An interferometer splits an incoming beam of electromagnetic radiation (for example, a beam of light) into two or more components that are then reunited after each has traveled a different pathlength. As a result, an interference pattern is formed. If the incoming radiation is in the visible portion of the spectrum, light and dark rings or bands are formed. Stellar interferometers are used by optical astronomers to measure very small angles, such as the "apparent" diameter of a single star or the angular separation of a binary star system. Astronomers use the word *apparent* to describe a measured or observed property. Apparent values must often be corrected to obtain the actual, or true, value.

For example, astronomers must compensate for the effects of the Earth's atmosphere.

Each of the three extrasolar-planet-detection techniques considered during Project Orion appeared to be feasible. In the next few years, improvements in both ground-based astrometric telescopes and space-based astronomy should greatly assist astronomers in their search for positive evidence of planetary companions around alien suns.

The search for extrasolar planets involves many different scientific disciplines. Knowledge about how frequently planets occur in our Galaxy and their distribution as a function of star type would help scientists confirm current theories about stellar evolution, including hypotheses about the origin of our Solar System. Data on the frequency of extrasolar planets will also help us in our efforts to search for extraterrestrial intelligence.

See also: **Drake equation; electromagnetic spectrum; extrasolar planets; Hubble Space Telescope; search for extraterrestrial intelligence; stars**

Project Orion (II)   *Project Orion* was also the name given to a nuclear-fission pulsed rocket concept studied in the early 1960s. A manned, interplanetary spaceship would be propelled by exploding a series of nuclear-fission devices behind it. A giant pusher plate, mounted on large shock absorbers, would receive the energy pulse from each successive nuclear detonation, and the spaceship configuration would be propelled forward by Newton's action–reaction principle.

In theory, this concept is capable of achieving specific-impulse values ranging from 2,000 to 6,000 seconds, depending on the size of the pusher plate. Specific impulse is a performance index for rocket propellants. It is defined as the thrust produced by the propellant divided by the mass flow rate. As a point of comparison, the best chemical rockets have specific-impulse values ranging from 450 to 500 seconds.

A manned Orion spaceship would move rapidly throughout interplanetary space at a steady acceleration of perhaps 0.5 g (one-half the acceleration of gravity on the Earth's surface). Typically, a 1- to 10-kiloton fission device would be exploded every second or so close behind the giant pusher plate. A kiloton is the energy of a nuclear explosion that is equivalent to the detonation of 1,000 tons of TNT (trinitrotoluene). A few thousand such detonations would be needed to propel a crew of 20 astronauts to Mars or the moons of Jupiter.

Work by the United States on this nuclear-fission pulse rocket concept came to an end in the mid-1960s, as a result of the Limited Test Ban treaty of 1963. This treaty prohibited the signatory nations from testing nuclear devices in the Earth's atmosphere, underwater or in outer space.

Advanced versions of the original Orion concept have also emerged. In these new spaceship concepts, externally detonated nuclear-fission devices have been replaced by many small, controlled, thermonuclear-fusion explosions taking place inside a specially constructed thrust chamber. These mini-thermonuclear explosions would occur in an inertial confinement fusion (ICF) process in which many powerful laser, electron or ion beams simultaneously impinge on a tiny fusion pellet. Each miniature thermonuclear explosion would have an explosive yield equivalent to a few tons of TNT. The expanding shell of very hot, ionized gas from the thermonuclear explosion would be directed into a thrust-producing exhaust stream. Such pulsed nuclear-fusion spaceships, when developed, open up our entire Solar System to human visitation. For example, Earth-to-Neptune travel would take less than 15 days at a steady, comfortable acceleration of one g. Pulsed-fusion systems also represent a possible propulsion system for interstellar travel.

See also: **fission (nuclear); fusion; Project Daedalus; space nuclear propulsion**

Project Ozma The first attempt to detect interstellar radio signals from an intelligent extraterrestrial civilization. It was conducted by Dr. Frank Drake at the National Radio Astronomy Observatory (NRAO) in Green Bank, West Virginia, in 1960. Drake derived the name for this effort from the queen of the imaginary land of Oz, since in his own words Oz was "a place very far away, difficult to reach, and populated by exotic beings."

A frequency of 1,420 megahertz (MHz) was selected for this initial search—the frequency of the 21-centimeter interstellar hydrogen line. Since this is a radio frequency at which most emerging technical civilizations would first use narrow-bandwidth, high-sensitivity radio telescopes, scientists reasoned that this would also be the frequency that more advanced alien civilizations would use in trying to signal emerging civilizations across the vast interstellar void.

In 1960 the 29.5-meter (85-ft) Green Bank radio telescope was aimed at two sunlike stars about 11 light-years away, Tau Ceti and Epsilon Eridani. Patiently, Frank Drake and his Project Ozma team listened for intelligent signals. But after over 150 hours of listening, no evidence of strong radio signals from intelligent extraterrestrial civilizations was obtained. Project Ozma is generally considered the first serious attempt to listen for intelligent interstellar radio signals in our search for extraterrestrial intelligence—the birth of modern SETI!

See also: **Drake equation; extraterrestrial civilizations; search for extraterrestrial intelligence**

Project Pathfinder As an augmentation to the Civil Space Technology Initiative (CSTI), NASA's Project Pathfinder will provide the advanced space technologies necessary to enable missions beyond Earth orbit, such as a permanent lunar base and human expeditions to Mars. Currently, Project Pathfinder is focused on four major technology thrusts: (1) planetary surface exploration, (2) in-space operations, (3) the long-term support of humans in space and (4) advanced space transfer vehicles. (See table 1.)

Table 1 Project Pathfinder Technology Thrusts

Exploration technology

- Planetary rover
- Sample acquisition, analysis and preservation
- Surface power
- Optical communication

Operations technology

- Autonomous rendezvous and docking
- Resource processing pilot plant
- In-space assembly and construction
- Cryogenic fluid depot
- Space nuclear power (SP-100)

Humans-in-space technology

- Extravehicular activity (EVA) suit
- Human performance
- Closed-loop life support

Transfer vehicle technology

- Chemical transfer propulsion
- Cargo vehicle propulsion
- High-energy aerobraking
- Autonomous lander
- Fault-tolerant systems

SOURCE: Author (based on NASA data).

See also: **Civil Space Technology Initiative**

protogalaxy A galaxy at the early stages of its evolution.

protoplanet Any of a star's planets as it emerges during the process of accretion (accumulation), in which planetesimals collide and coalesce into larger objects.

protostar A star in the making. Specifically, the stage in a young star's evolution after it has separated from a gas cloud but prior to its collapsing sufficiently to support thermonuclear reactions.

protosun The Sun as it emerged in the formation of the Solar System.

quasars Mysterious objects that appear almost like stars but are far more distant than any individual star we can now observe. These unusual objects were first discovered in the 1960s with radio telescopes, and they were called "quasi-stellar radio sources"—or "quasars" for short. Quasars emit tremendous quantities of energy from very small volumes. The most distant quasars observed to date are so far away that they are receding at more than ninety percent of the speed of light.

As bright, concentrated radiation sources, quasars are now thought to be the nuclei of active galaxies. The optical brightness of some quasars has been observed to change by a factor of two in about a week, with detectable changes occurring in just one day. Therefore, astrophysicists now speculate that such quasars cannot be much larger than about one light-day across (a distance about twice the dimension of our Solar System), because a light source cannot change brightness significantly in less time than it takes for light itself to travel. The problem facing scientists today is to explain how quasars can generate more energy than is possessed by an entire galaxy and generate this energy in so small a region of space! As we place more sophisticated observatories in space, we will learn more about these very unusual extraterrestrial objects.

See also: **active galaxies; astrophysics**

quiet sun A term used by solar physicists to describe the minimum portion of the sunspot cycle for our Sun.

See also: **Sun**

R

radar astronomy The use of radar by astronomers to study objects in our Solar System, such as the Moon, the planets, asteroids and even planetary ring systems. For example, a powerful radar telescope, like the Arecibo Observatory, can hurl a radar signal through the "opaque" Venusian clouds (some 80 kilometers thick) and then analyze the faint return signal to obtain detailed information for the preparation of high-resolution surface maps. Radar astronomers can precisely measure distances to celestial objects, estimate rotation rates and also develop unique maps of surface features, even when the actual physical surface is obscured from view by thick layers of clouds.

See also: **Arecibo Observatory; Magellan Mission**

radio astronomy Branch of astronomy that collects and evaluates radio signals from extraterrestrial sources. Radio astronomy is a relatively young branch of astronomy. It was started in the early 1930s, when the American radio engineer Karl Jansky detected the first extraterrestrial radio signals. Until Jansky's discovery, astronomers had only used the visible portion of the electromagnetic spectrum to view the Universe. The detailed observation of cosmic-radio

sources, however, is very difficult, because these sources shed so little energy on the Earth.

The radio telescope has been used to discover some extraterrestrial radio sources so unusual that their very existence had not even been imagined or predicted by scientists. One of the strangest of these cosmic radio sources is the pulsar—a collapsed giant star that emits pulsating radio signals as it spins. When the first pulsar was detected in 1967, it created quite a stir in the scientific community. Because of the regularity of its signal, scientists thought they had just detected the first interstellar signals from an intelligent alien civilization!

Another interesting celestial object is the quasar, or quasi-stellar radio source. Discovered in 1964, quasars are now considered to be entire galaxies in which a very small part (perhaps only a few light-days across) releases enormous amounts of energy—amounts equivalent to the total annihilation of millions of stars. Quasars are the most distant known objects in the Universe, some of which are receding from us at over ninety percent of the speed of light.

Scattered throughout interstellar space are atoms and molecules of materials such as hydrogen, formaldehyde and methyl alcohol. Some of these interstellar molecules represent the basic ingredients of life as we know it. Radio telescopes are being used to collect valuable information about the abundance and nature of these interstellar molecules, as well as to search for other freely floating chemicals in the interstellar medium. If these potential "seeds of life" are found fairly commonly throughout interstellar space, then their existence might be considered as substantiating the principle of mediocrity; that is, the chemistry needed to initiate life on Earth may be typical of what occurs elsewhere.

See also: **Arecibo Observatory; interstellar medium; life in the Universe; quasars; stars; Very Large Array**

radio frequency (RF) In general, a frequency at which electromagnetic radiation is useful for communication purposes; specifically, a frequency above 10,000 hertz (cycles per second) and below 3×10^{11} hertz.

radio galaxy A galaxy (often exhibiting a dumbbell-shaped structure) that produces very strong signals at radio wavelengths.

See also: **Arecibo Observatory; radio astronomy; Very Large Array**

radio telescope A large, metallic, parabolic (dish-shaped) device that collects and focuses radio waves onto a sensitive radio-frequency (RF) receiver.

See also: **Arecibo Observatory; astrophysics; radio astronomy; search for extraterrestrial intelligence; Very Large Array**

relativity The theory of space and time developed by Albert Einstein, which has become one of the foundations of 20th-century physics. Einstein's theory of relativity is often discussed in two general categories: the special theory of relativity, which he first proposed in 1905, and the general theory of relativity, which he presented in 1915.

The special theory of relativity is concerned with the laws of physics as seen by observers moving relative to one another at constant velocity—that is, by observers in non-accelerating or inertial reference frames. Special relativity has been well demonstrated and verified by many types of experiments and observations.

Einstein proposed two fundamental postulates in formulating special relativity:

• *First postulate of special relativity:* The speed of light (*c*) has the same value for all (inertial-reference-frame) observers, regardless and independent of the motion of the light source or the observers.

• *Second postulate of special relativity:* All physical laws are the same for all observers moving at constant velocity with respect to each other.

The first postulate appears contrary to our everyday "Newtonian mechanics" experience. Yet the principles of special relativity have been more than adequately validated in experiments. Using special relativity, scientists can now predict the space–time behavior of objects traveling at speeds from essentially zero up to those approaching that of light itself. At lower velocities the predictions of special relativity become identical with classical Newtonian mechanics. However, when we deal with objects moving close to the speed of light, we must use relativistic mechanics.

What are some of the consequences of the theory of special relativity?

The first interesting relativistic effect is called *time dilation.* Simply stated—with respect to a stationary observer/clock—time moves more slowly on a moving clock/system. This unusual relationship is described by the equation:

$$(1) \qquad \Delta t = (1/\beta)\, \Delta T_p$$

where Δt is called the time dilation (the apparent slowing down of time on a moving clock relative to a stationary clock/observer)

and ΔT_p is the "proper time" interval as measured by an observer/clock on the moving system.

$$\beta \equiv \sqrt{1 - (v^2/c^2)} \qquad (2)$$

where v is the velocity of the object

and c is the velocity of light.

Let's now explore the time-dilation effect with respect to a postulated starship flight from our Solar System. We start with twin brothers, Astro and Cosmo, who are both astronauts and are currently 25 years of age. Astro is selected for a special 40-year-duration starship mission, while Cosmo is selected for the ground control team. This particular starship, the latest in the fleet, is capable of cruising at 99 percent of the speed of light (0.99 *c*) and can quickly reach this cruising speed. During this mission,

Cosmo, the twin who stayed behind on Earth, has aged 40 years. (We are taking the Earth as our fixed or stationary reference frame "relative" to the starship.) But Astro, who has been on board the starship cruising the Galaxy at 99 percent of the speed of light for the last 40 Earth-years, has aged just 5.64 Earth-years! Therefore, when he returns to Earth from the starship mission, he is a little over 30 years old, while his twin brother, Cosmo, is now 65 and retired in Florida. Obviously, starship travel (if we can overcome some extremely challenging technical barriers) also presents some very interesting social problems.

The time-dilation effects associated with near-light speed travel are real, and they have been observed and measured in a variety of modern experiments. All physical processes (chemical reactions, biological processes, nuclear-decay phenomena and so on) appear to slow down when in motion relative to a "fixed" or stationary observer/clock.

Another interesting effect of relativistic travel is *length contraction.* We first define an object's proper length (L_p) as its length measured in a reference frame in which the object is at rest. Then, the length of the object when it is moving (*L*)—as measured by a stationary observer—is always smaller, or contracted. The relativistic length contraction is given by:

$$L = \beta\, (L_p) \qquad (3)$$

This apparent shortening, or contraction, of a rapidly moving object is seen by an external observer (in a different inertial reference frame) only in the object's direction of motion. In the case of a starship traveling at near-light speeds, to observers on Earth this vessel would appear to shorten, or contract, in the direction of flight. If an alien starship was 1 kilometer long (at rest) and entered our Solar System at an encounter velocity of 90 percent of the speed of light (0.9 *c*), then a terrestrial observer would see a starship that appeared to be about 435 meters long. The aliens on board and all their instruments (including tape measures) would look contracted to external observers but would not appear any shorter to those on board the ship (that is, to observers within the moving reference frame). If this alien starship was "burning rubber" at a velocity of 99 percent of the speed of light (0.99 *c*), then its apparent contracted length to an observer on Earth would be about 141 meters. If, however, this vessel was a "slow" interstellar freighter that was limping along at only 10 percent of the speed of light (0.1 *c*), then it would appear about 995 meters long to an observer on Earth.

Special relativity also influences the field of dynamics. Although the rest mass (m_o) of a body is invariant (does not change), its "relative" mass increases as the speed of the object increases with respect to an observer in another fixed or inertial reference frame. An object's relative mass is given by:

$$m = (1/\beta)\, m_o \qquad (4)$$

This simple equation has far-reaching consequences. As an object approaches the speed of light, its mass becomes infinite! Since things can't have infinite masses, physicists conclude that material objects cannot reach the speed of light. This is basically the *speed-of-light barrier,* which appears to limit the speed at which interstellar travel can occur. From the theory of special relativity, scientists now conclude that only a "zero-rest-mass" particle, like a photon, can travel at the speed of light. There is one other major consequence of special relativity that has greatly affected our daily lives—the equivalence of mass and energy from Einstein's very famous formula:

$$E = \Delta m \; c^2 \qquad (5)$$

where E is the energy equivalent of an amount of matter (Δm) that is annihilated or converted completely into pure energy

and c is the speed of light.

This simple yet powerful equation explains where all the energy in nuclear fission or nuclear fusion comes from. The complete annihilation of just one gram of matter releases about 9×10^{13} joules of energy.

In 1915 Einstein introduced his general theory of relativity. He used this development to describe the space–time relationships developed in special relativity for cases where there was a strong gravitational influence—such as white dwarf stars, neutron stars and black holes. One of Einstein's conclusions was that gravitation is not really a force between two masses (as Newtonian mechanics indicates) but rather arises as a consequence of the curvature of space and time. In our four-dimensional Universe (x, y, z and time), space–time becomes curved in the presence of matter.

The fundamental postulate of general relativity is also called *Einstein's principle of equivalence:* The physical behavior inside a system in free-fall is indistinguishable from the physical behavior inside a system far removed from any gravitating matter (that is, the complete absence of a gravitational field).

Several experiments have been performed to confirm the general theory of relativity. These experiments included observation of the bending of electromagnetic radiation (starlight and radio-wave transmissions from Project Viking on Mars) by the Sun's immense gravitational field and recognizing the subtle perturbations (disturbances) in the orbit (at perihelion—that is, at the point of closest approach to the Sun) of the planet Mercury as caused by the curvature of space–time in the vicinity of the Sun. While some scientists do not think that these experiments have conclusively demonstrated the validity of general relativity, additional experimental evidence is anticipated in the upcoming years, when we continue investigating phenomena such as neutron stars and black holes with more powerful space-based observatories.

remote sensing   The sensing of an object, event or phenomenon without having the sensor in direct contact with the object being studied. Information transfer from the object to the sensor is accomplished through the use of the electromagnetic spectrum. Modern remote-sensing technology uses many different portions of the electromagnetic spectrum, not just the visible portion we see with our eyes. As a result, very different and very interesting "images" are often created by modern remote sensing instruments. For example, figure 1 is a radar image of the Los Angeles, California, region taken from an altitude of 800 kilometers by NASA's *Seasat* satellite in July 1978. This radar image shows an area of approximately 100 by 100 kilometers. Downtown Los Angeles is located slightly lower and left of center in this figure. *Seasat's* radar instrument was called a synthetic aperture radar, or "SAR" for short. This device measured the intensity (brightness) of its radar waves as they were reflected back from the Earth's surface. The bright patch in the city of Burbank (upper left portion of fig. 1) and other bright patches, were caused by reflections from building walls and hillsides that were at or near a 90-degree angle to the radar beams from *Seasat.* The spacecraft was located over the Pacific Ocean to the west, when the image shown here was made. In this type of remote sensing imagery, streets, highways, airports, flood channels and calm water appear dark. This is because few or no radar beams were reflected back to the spacecraft.

Since 1966 NASA has conducted an Earth Observations Program. This program explores the use of multispectral remote sensing in many application areas, including the measurement of environmental pressures on the terrestrial biosphere, the search for mineral and petroleum resources, the state of health and yield estimations for agricultural crops, and the monitoring of the marine environment. These studies began with Earth surveys from conventional aircraft. Then, in 1972 NASA's *Landsat 1,* previously called the Earth Resources Technology Satellite or ERTS, provided scientists with a new experimental tool for collecting valuable data about the terrestrial environment from space. The initial effect of *Landsat 1* was the availability of large quantities of remotely sensed data about the Earth, with regional coverage repeated at 18 day intervals.

For example, figure 2 shows an important Landsat capability—to detect crop disease. The Landsat image on the right shows healthy crops, while the Landsat image on the left shows severely "stressed" crops. Similarly, figure 3 is a Landsat image of Travis County, Texas, that illustrates the application of modern remote sensing platforms in land management. In a false-color, multispectral image, each shade represents a particular land-use category, such as urban, forest and farm land. Landsat imagery has also been used in water resource surveys, coastal zone management and in the identification of wildlife areas.

Just before the launch of *Landsat 5* in March 1984, the U.S. Department of Commerce solicited proposals for Landsat commercialization (in response to the Land Remote-Sensing Act of 1984). In September 1985, the federal government awarded a contract to the Earth Observation Satellite Company (EOSAT) to distribute commercial Landsat

Fig. 1 A synthetic aperture radar (SAR) view of Los Angeles taken at an altitude of 800 km (500 mi) by NASA's SEASAT (July 28, 1978). (Courtesy of NASA/Jet Propulsion Laboratory.)

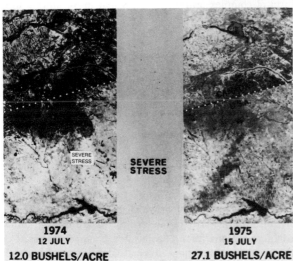

Fig. 2 LANDSAT remote sensing images showing healthy and stressed crops. (Courtesy of NASA.)

Fig. 3 The use of LANDSAT imagery for land classification. (Courtesy of NASA.)

data and to design, construct and launch the next generations of Landsat spacecraft *(Landsat* 6 and 7). For example, *Landsat 6,* based on the Advanced TIROS-N series of polar-orbiting weather satellites, will be the largest and most sophisticated of the Landsat series of spacecraft flown to date. Its advanced sensor package will provide coregistered

15-meter (49-foot) resolution panchromatic imagery and 30-meter (98-foot) resolution multispectral (7-band) imagery data. The numerous applications of these high-resolution imagery data include geologic thermal inertia modeling, improved plant species discrimination, detailed regional planning surveys and estuarine pollution monitoring.

Remote sensing of the Earth from outer space is one of the major contributions of modern space technology toward improving our overall quality of life. Future remote sensing systems, like the Earth Observing System (EOS), will provide even more detailed information to assist us in the careful, enlightened stewardship of our home planet in the next century.

See also: **Earth Observing System; electromagnetic spectrum; global change; Mission to Planet Earth**

ringed world A planet with a ring or series of rings encircling it. In our Solar System, Jupiter, Saturn, Uranus and Neptune possess ring systems. This indicates that ring systems may be a common feature of Jovian type planets.

Astronomers speculate that there are three general ways such ring systems are formed around a planet. The first is meteoroidal bombardment of a large body, so that the fragments inside the planet's Roche limits form a ring. The second is condensation of material inside the Roche limit when the planet was forming. This trapped nebular material can neither join the parent planet nor form a moon or satellite. Finally, ring systems can also form when a satellite or other celestial object (for example, comet) comes within the planet's Roche limit and gets torn apart by tidal forces.

The Roche limit, named after the French mathematician Edouard Roche who formulated it in the 19th century, is the critical distance from the center of a massive celestial object or planet within which tidal forces will tear apart a moon or satellite. If we assume that (1) both celestial objects have the same density and (2) that the moon in question is held together only by gravitational attraction of its matter, then the Roche limit is typically about 1.2 times the diameter of the parent planet or primary celestial body.

See also: **Jupiter; Neptune; Saturn; Uranus**

robotics in space Robotics is basically the science and technology of designing, building and programming robots. It is a rapidly growing field that is playing an increasingly important role in humanity's exploration and exploitation of the extraterrestrial environment. For example, robotic devices have been used on lunar and planetary missions and on the Space Shuttle. Advanced robotic devices are now being considered for servicing spacecraft, for assembling large space structures and for conducting sophisticated exploration of both the lunar and Martian surfaces.

Robotic devices, or robots as they are usually called, are primarily "smart machines" with manipulators that can be programmed to do a variety of manual or human labor tasks automatically. A robot therefore is simply a machine that does mechanical, routine tasks on human command. The expression "robot" is attributed to the Czechoslovakian writer, Karel Capek, who wrote the play *R.U.R. (Rossum's Universal Robots.)* This play first appeared in English in 1923 and is a satire on the mechanization of civilization. The word "robot" is derived from *robata,* a Czechoslovakian word meaning compulsory labor or servitude.

Since then, robots have become the essential elements found in many exciting science fiction stories. These mechanical "actors" have played the part of both villain and hero. The well-known writer, Isaac Asimov, formulated his three classic "Laws of Robotics" in the 1940s to describe a "machine code of ethics" by which robots could be built and programmed so as not to harm their human masters. Asimov's three laws of robotics are as follows:

Law 1: A robot may not injure a human being, or, through inaction, allow a human being to come to harm.

Law 2: A robot must obey the orders given it by human beings except where such orders would conflict with the First Law.

Law 3: A robot must protect its own existence as long as such protection does not conflict with the First and Second Law.

If we try to build robots that obey such laws, the following design conditions should certainly be satisfied: (a) our robot must not be capable of deliberately injuring a human being, (b) our robot should be able to exercise common sense, (c) our robot must be intelligent and (d) our robot must also be conscious. For a robot or any smart machine to exercise common sense, intelligence and consciousness, scientists must make great advances in the current field of artificial intelligence.

Actually, today's robots bear very little resemblance to the delightful robots found in such science fiction adventure stories as George Lucas' *Star Wars.* Instead of a beeping R2D2 or an apologetic C3PO, Space-Age robots take the form of complex arm-like devices that move objects from one place to another; or they may be very sophisticated, automated surface vehicles that wander across alien worlds in search of resources.

Before we begin a detailed discussion about the use of robots in space, it will be helpful to review some of the basic concepts and terminology related to robots here on Earth.

A PRIMER ON ROBOTS

A typical robot consists of one or more manipulators (arms), end effectors (hands), a controller, a power supply and possibly an array of sensors to provide information about the environment in which the robot must operate. Because the majority of modern robots are used in industrial applications, their classification is currently based on these industrial functions. Terrestrial robots are frequently divided into the following classes: non-servo (or pick-and-place) servo, programmable, computerized, sensory and assembly robots.

The non-servo robot is the simplest type. It picks up an object and places it at another location. The robot's freedom of movement is usually limited to two or three directions.

The servo robot represents several categories of industrial robots. This type of robot has servo-mechanisms for the manipulator and end effector to enable it to change direction in midair without having to trip or trigger a mechanical limit switch. Five to seven directions of motion are common, depending on the number of joints in the manipulator.

The programmable robot is essentially a servo robot that is driven by a programmable controller. This controller memorizes (stores) a sequence of movements and then repeats these movements and actions continuously. This type of robot is programmed by "walking" the manipulator and end effector through the desired movement.

The computerized robot is simply a servo robot run by a computer. This kind of robot is programmed by instructions fed into the controller electronically. These "smart robots" may even have the ability to improve upon their basic work instructions.

The sensory robot is a computerized robot with one or more artificial senses to observe and record its environment and to feed information back to the controller. The artificial senses most frequently employed are sight (robot or computer vision) and touch.

Finally, the assembly robot is a computerized robot, generally with sensors, that is designed for assembly line and manufacturing tasks, both on Earth and eventually in space.

In industry, robots are mainly designed for manipulation purposes. The actions that can be produced by the end effector or hand include (1) motion (from point to point, along a desired trajectory or along a contoured surface); (2) a change of orientation; and (3) rotation.

Non-servo robots are capable of point-to-point motions. For each desired motion, the manipulator moves at full speed until the limits of its travel are reached. As a result, non-servo robots are often called "limit sequence," "bang-bang," or "pick-and-place" robots. When non-servo robots reach the end of a particular motion, a mechanical stop or limit switch is tripped, stopping the particular movement.

Servo robots are also capable of point-to-point motions; but their manipulators move with controlled variable velocities and trajectories. Servo robot motions are controlled without the use of stop or limit switches.

Four different types of manipulator arms have been developed to accomplish robot motions. These are the rectangular, cylindrical, spherical and anthropomorphic (articulated or jointed) arm. Each of these manipulator arm designs features two or more degrees of freedom (DOF), a term that refers to the directions a robot's manipulator arm is able to move. For example, a simple straight line or linear movement represents one DOF. If the manipulator arm is to follow a two-dimensional curved path, it needs two degrees of freedom: up and down and right and left. Of course, more complicated motions will require many degrees of freedom. To locate an end effector at any point and to orient this effector in a particular work volume requires six DOF. If the manipulator arm needs to avoid obstacles or other equipment, even more degrees of freedom will be required. For each DOF, one linear or rotary joint is needed. Robot designers sometimes combine two or more of these four basic manipulator arm configurations

to increase the versatility of a particular robot's manipulator.

Actuators are used to move a robot's manipulator joints. There are three basic types of actuators currently used in contemporary robots: pneumatic, hydraulic and electrical actuators. Pneumatic actuators employ a pressurized gas to move the manipulator joint. When the gas is propelled by a pump through a tube to a particular joint, it triggers or actuates movement. Pneumatic actuators are inexpensive and simple, but their movement is not precise or "squooshy." Therefore, this kind of actuator is usually found in non-servo or pick-and-place robots. Hydraulic actuators are quite common and capable of producing a large amount of power. The main disadvantages of hydraulic actuators are their accompanying apparatus (pumps and storage tanks) and problems with fluid leaks. Electrical actuators provide smoother movements, can be very accurately controlled and are very reliable. However, these actuators cannot deliver as much power as comparable mass hydraulic actuators. Nevertheless, for modest power actuator functions, electrical actuators are often preferred.

Many industrial robots are fixed in place or move along rails or guideways. Some terrestrial robots are built into wheeled carts, while others use their end effectors to grasp handholds and pull themselves along. Advanced robots use articulated manipulators as legs to achieve a walking motion.

A robot's end effector (hand or gripping device) is generally attached to the end of the manipulator arm. Typical functions of this end effector include grasping, pushing and pulling, twisting, using tools, performing insertions and various types of assembly activities. End effectors can be mechanical, vacuum or magnetically operated, can use a snare device or have some other unusual design feature. The final design of an end effector is determined by the shapes of the objects that the robot must grasp. Most end effectors are usually some type of gripping or clamping device.

Robots can be controlled in a wide variety of ways, from simple limit switches tripped by the manipulator arm to sophisticated computerized remote sensing systems that provide machine vision, touch and hearing. In the case of computer-controlled robots, the motions of the manipulator and end effector are programmed; that is, the robot "memorizes" what it's supposed to do. Sensor devices on the manipulator help to establish the proximity of the end effector to the object to be manipulated and feed information back to the computer controller concerning any modifications needed in the manipulators trajectory.

CONTEMPORARY SPACE ROBOTS

Robotic systems will play a major role in mankind's exploration and exploitation of the Solar System. Some contemporary applications include the Space Shuttle's versatile remote manipulator system (RMS); space station assembly and maintenance; automated servicing of orbiting spacecraft; the construction of very large space structures;

and sophisticated interplanetary spacecraft, probes, landers and surface rovers. On the more distant extraterrestrial horizon, one might envision highly automated interstellar probes and perhaps one of the most interesting smart devices of all, the self-replicating system (SRS).

The Space Shuttle has been designed primarily to transport a variety of payloads between the Earth and low Earth orbit. The Shuttle Orbiter contains a cargo bay large enough to carry one and one-half school buses! Once the Orbiter vehicle has achieved low Earth orbit, payloads in the cargo bay need to be deployed. For example, spacecraft are removed from the cargo bay and employed to their higher-altitude, operational orbits. In addition, defective spacecraft that are already on orbit must be brought into the cargo bay for on-orbit repair work or for a return to Earth. To handle these spacecraft deployment and retrieval operations as well as to permit the assembly of large structures (such as a permanent space station in orbit), a very versatile robot arm has been installed in the Space Shuttle Orbiter along the port (left) side cargo bay door hinges. This robot arm is called the remote manipulator system (RMS).

The RMS was designed and built by the National Research Council of Canada. It is a highly sophisticated robotic device that is similar to a human arm. The 15-meter-long RMS features a shoulder, wrist and hand—although its hand does not at all look like a human hand. The skeleton of this mechanical arm is made of lightweight graphite composite materials. Covering the skeleton are skin layers consisting of thermal blankets. The muscles driving the joints are electric actuators (motors). Built-in sensors act like nerves and sense joint positions and rotation rates.

Beyond the initial similarities to the human arm, the RMS features additional capabilities. To provide all the needed degrees of freedom, the RMS has six independently operating joints. There is a pitch joint (up and down) and a yaw joint (right and left) in the RMS shoulder. The mechanical elbow has a pitch joint, while the RMS wrist has pitch, yaw and roll (rotation) joints—a definite improvement over the human arm. The shoulder joint allows the RMS end effector to move to any point along the surface of a sphere. The elbow joint gives the RMS freedom to reach any point within that sphere, while the wrist joints provide any desired orientations of the end effector.

The RMS also includes two closed-circuit television cameras, one at the wrist and one at the elbow. These cameras allow an astronaut, who is operating the RMS from the Orbiter's aft flight deck, to see critical points along the arm and the target toward which the arm is moving. The television camera at the RMS wrist is also linked to a computer, producing a machine vision that enables the arm to operate in a fully automatic mode. To permit this automatic operation, the target or grapple fixture to be grasped by the RMS must have four white dots. The computer in the machine vision system will sense the physical separation of these white dots. From that it will determine the distance and direction of the RMS end effector from the target, as well as the target's rate of motion.

The RMS end effector represents a radical departure from the human arm analogy we have used so far. Instead of finger-like clamps, the end effector has a three-wire snare. Each payload to be grasped by the RMS must have a special grapple fixture. This fixture is a small knobby projection that is mounted to a disc. To grasp the payload, the end effector slides over the fixture and the snare wires are pulled together in the center. Once tightly held by the wires, the disc is pulled into the effector and the payload is ready for deployment or retrieval.

The versatile RMS can be operated in several modes, ranging from fully manual to fully automatic. In the manual mode, an astronaut controls the robot arm with rotational and translational hand controllers. These devices look similar to the controllers on home video games. The astronaut, watching the arm through windows or on closed circuit television, can move the RMS end effector with ease. In the automatic mode, the payload manipulator task to be performed is programmed into the computer and, like a dutiful servant, the RMS carries it out.

Robotics and machine intelligence have also been used in the exploration of the Solar System. The development of advanced integrated circuits, microprocessors and silicon chip technology has made it possible to build systems with compact memories and processors that provide robot explorers with a form of portable machine intelligence. These space exploration robots are able to sense their environments, plan and execute actions, and perform complicated manipulations, frequently with a degree of dexterity normally associated with human beings.

By means of telecommunication, human operators can activate and control instruments stationed on alien worlds or flying through the deep regions of interplanetary space. During the 1960s such procedures were developed as part of the planetary exploration program and became known as "teleoperation." Teleoperators are simply machine systems that augment and extend human sight, touch and even thinking abilities to remote places. In this context, the term "robot" is often applied to the remote portion of a teleoperator system if this part of the machine has at least some degree of independent sensing, decision-making or movement capabilities.

System autonomy is necessary in space exploration, because the distances between the planets are so great. Even communicating at the speed of light, it still takes many minutes for instructions from human operators on Earth to reach these robot explorers. This makes direct control of space exploration instruments either a very slow and tedious process, or even technically impossible.

However, a remote machine with independent decision-making can accomplish a great deal on its own without step-by-step instructions from its human supervisors. In fact, less "talking" between human and smart robot can often result in the successful accomplishment of more technical and scientific tasks by the robot—thereby enhancing the overall results of a particular exploration mission.

In such extraterrestrial applications, these robots will

take on many forms, although few will look like the popular concepts of androids and robots that are found in science fiction. Their appearance will follow strictly functional lines and will change with the specific requirements of a particular mission. For example, contemporary robot space explorers may take the form of a large, automated telescope in Earth-orbit, or they may be sophisticated interplanetary flyby and orbiter spacecraft. They may be stationary landers, like the *Viking 1* and *2 Landers* placed on Mars in 1976, or they may take the form of wheeled-surface rovers that travel across the plains of an alien world.

During future planetary missions, the robot rover would move itself to a new position on the planet's surface every day. Each day the vehicle would observe the alien environment and relay appropriate terrain imagery and scientific data back to Earth. Based on the mission's overall technical objectives and on the topographical features encountered on Mars or on one of the major moons of the outer planets, scientists on Earth would issue general commands to the robot explorer vehicle. The vehicle or vehicles would then proceed autonomously throughout the next day performing automated tasks and scientific investigations with no or minimal interaction from humans on Earth.

Space industrialization will also depend extensively on automated and robotic machine technology. Industrial activities in space will include the construction of very large antennas and solar arrays, and the construction of permanent habitats, such as space stations and space bases. Robotic cranes, manipulators and teleoperator systems will make these construction jobs practical. In addition, orbital manufacturing and processing stations will most likely be operated and maintained through highly automated control systems and robotic devices.

Future automated space systems now under consideration include space-based manufacturing modules to produce biological, metallic and fluid products for scientific and commercial applications on Earth; systems to provide on board health care to space workers, astronaut crews and even space tourists; lunar rovers to collect and stockpile materials for the construction of lunar bases; and complete lunar base systems to perform mining, processing and manufacturing operations.

FLIGHT TELEROBOTIC SERVICER

NASA's Flight Telerobotic Servicer (FTS) program was initiated to provide new capabilities for the accomplishment of tasks in the space environment. The need for these capabilities was identified recently in both the Space Station *Freedom* definition studies and NASA strategic planning activities for servicing spacecraft in orbits beyond the reach of current manned systems. The Flight Telerobotic Servicer is being designed to use evolving technologies for telerobotic space systems that result in increased capability and autonomy.

Currently, the FTS program is composed of four major elements: (1) the FTS operational flight system for Space Station *Freedom,* (2) a series of test flights, (3) a ground-based support infrastructure and (4) a technology utilization and transfer process. Many contemporary Space Station *Freedom* definition and development studies have identified a need to perform sophisticated mechanical manipulations in space and also the need to augment the astronaut crew's extravehicular activity (EVA) capabilities. Analysis of an operational Space Station *Freedom* further indicates that many external (i.e., in-space) tasks would be required for routine maintenance and servicing of the station and its payloads. These activities include battery replacement, fuel transfer and refurbishment of thermal control systems. As a matter of fact, astronaut time on-orbit is one of the most valuable assets on the Space Station—far too valuable a commodity to be expended performing routine and repetitious EVA tasks. Instead, astronauts should perform creative and supervisory activities, while various automated systems and semiautonomous manipulators monitor the status of various systems and subsystems and inspect for trouble. A robotic system like NASA's Flight Telerobotic Servicer is being designed to fill the gap between the large-scale (and somewhat "coarse") Remote Manipulator System (RMS) positioning capabilities and the potentially dextrous astronaut EVA capabilities that can consume large quantities of precious crew time. The FTS represents a baseline approach to accomplish hazardous tasks in space and to provide at least a "first generation" of automated manipulation capability that frees the Space Station crew for more productive and creative activities.

The FTS might also be joined with an orbital transfer vehicle. This configuration would then allow the sophisticated robot servicer to reach more distant orbiting platforms, instead of having to bring these spacecraft back to the Space Station for servicing and repair.

The major objectives of NASA's Flight Telerobotic Servicer Program are to (1) reduce Space Station *Freedom* dependence on crew EVA, (2) improve overall Space Station crew safety, (3) enhance crew utilization and (4) provide remote servicing capabilities for space platforms both at the Space Station and eventually at locations outside the reach of manned systems.

THE NEED FOR MACHINE INTELLIGENCE

Tomorrow's advanced robotic space systems promise to take over much of the data processing and information sorting activities that are now performed by human mission controllers here on Earth. In the past, the amount of data made available by space missions has been considerably larger than scientists could comfortably sift through. For example, the Viking missions to Mars returned image data of the Red Planet that were transferred onto approximately 75,000 reels of magnetic tape. Smart robot spacecraft, with onboard computers capable of deciding what information gathered by a spacecraft or surface rover is worth relaying back to Earth and what information should be stored or discarded, would greatly relieve the current extraterrestrial data glut.

Robots with advanced machine intelligence, capable of

making these kinds of decisions, would have a large number of pattern classification templates or "world models" stored in their computer memories. These templates would represent the characteristics of objects or features of interest in a particular mission. The robot explorers would compare the patterns or objects they see with those stored in their memories and discard any unnecessary or unusable data. As soon as something unusual appeared, the smart machine explorer would examine this object or event more closely. The robot explorer would then dutifully alert its human controllers on Earth and report the unusual findings. Through these automated selection and data filtering operations, the smart system would free human experts for more demanding and judgmental intellectual tasks.

The advanced machine intelligence (or artificial intelligence) requirements for general purpose robotic space exploration systems can be summarized mainly in terms of two fundamental tasks: (1) The smart robot must be capable of learning about new environments and (2) it must be able to formulate hypotheses about these new environments. Hypothesis formation and learning represent the key problems in the successful development of machine intelligence. Deep interplanetary and interstellar robotic space systems will need a machine intelligence system capable of autonomously conducting intense studies of alien objects. The machine intelligence levels supporting these missions must be capable of producing scientific knowledge concerning previously unknown objects. Since the production of scientific knowledge is a high-level intelligence capability, the machine intelligence requirements for "smart" autonomous space missions are often called "advanced-intelligence machine intelligence" or simply "advanced machine intelligence."

For a really autonomous deep-space exploration system to undertake knowing and learning tasks, it must have the ability to mechanically or artificially formulate hypotheses, using all three of the logical patterns of inference: analytic, inductive and abductive. Analytic inference is needed by the robot explorer system to process raw data and to identify, describe, predict and explain extraterrestrial events and processes in terms of existing knowledge structures. Inductive inference is needed so that the robot explorer can formulate quantitative generalizations and abstract the common features of events and processes occurring on alien worlds. Such logic activities amount to the creation of new knowledge structures. Finally, abductive inference is needed by the smart robotic explorer system to formulate hypotheses about new scientific laws, theories, concepts, models and so forth. The formulation of this type of hypothesis is really the key to the ability to create a full range of new knowledge structures. These new knowledge structures, in turn, are needed if we are to successfully explore and investigate alien worlds.

Although the three patterns of inference just described are distinct and independent, they can be ranked by order of difficulty or complexity. Analytic inference is at the low end of the new-knowledge-creation scale. An automated

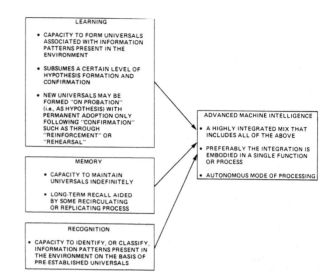

Fig. 1   The adaptive machine intelligence needed for advanced interplanetary and interstellar space exploration. (Drawing courtesy of NASA.)

system that performs only this type of logic could probably successfully undertake only extraterrestrial reconnaissance missions. A machine capable of performing both analytic and inductive inference could most likely successfully perform space missions combining reconnaissance and exploration. This assumes, however, that the celestial object being visited is represented well enough by the world models with which the smart robot has been preprogrammed. However, if the target alien world cannot be well represented by such fundamental world models, then automated exploration missions will also require an ability to perform abductive inference. This logic pattern is the most difficult to perform and lies at the heart of knowledge creation. An automated space system capable of abductive reasoning could successfully undertake missions combining reconnaissance, exploration and intensive study. Figure 1 summarizes the adaptive machine intelligence required for advance robotic space exploration systems.

TITAN DEMONSTRATION MISSION

Scientists and engineers have examined a concept for an advanced robotic space mission that could eventually lead to a deep space exploration system. This autonomous robot explorer would incorporate advanced machine intelligence technology capable of performing NASA's three general phases of extraterrestrial object investigation: reconnaissance, exploration and intensive study.

A general purpose robotic explorer spacecraft to Saturn's interesting moon, Titan, was proposed as a possible mission to demonstrate this advanced technology. Titan was chosen as the target because it lies far enough from the Earth to prevent effective ground-based investigation or even simple teleoperator control of robot explorer vehicles. Yet it still lies close enough for effective system monitoring and even

Table 1 The General Features of a Proposed Titan Demonstration Mission

Status: Opportunity Mission (not in current NASA plans)
Lifetime: 10 years; includes 5 years at Titan
Launch/transfer vehicle: Shuttle/400 kW Nuclear Electric Propulsion (NEP)
Operational location: Titan, Saturn's largest satellite
Total mass: 13,000–17,000 kg
Total power: About 400 kW

SOURCE: NASA.

human intervention during the operation of this advanced robot system. The knowledge gained from this initial automated mission would be applied to other deep space exploration missions within our Solar System and would also serve as the basis for the design and development of the first generation of interstellar robot probes.

Tables 1 through 3 provide some of the technical features of this conceptual Titan demonstration mission. As described in these tables, the mission would involve a full complement of sophisticated robotic devices. Perhaps the

Table 2 Candidate Space Systems for the Proposed Titan Demonstration Mission

Spacecraft type	Typical number	Operational location	Mass, kg	Power, kW
Nuclear electric propulsion	1	Earth to Titan orbit	10,000[a]	400
Main orbiting spacecraft	1	Circular polar Titan orbit at 600 km altitude	1,200	—[b]
Lander/Rover	2	Surface	1,800	1
Subsatellites	~3	One at a Lagrange point; others on 100 km tethers from NEP	300	0.3
Atmospheric probe	~6	Through Titan atmosphere to surface	200	0.1
Powered air vehicle	1	Atmosphere	1,000	10
Emplaced science	~6	Surface	50	0.1

[a]Does not include propellant.
[b]Uses NEP power.
SOURCE: NASA.

Table 3 Possible Accomplishments of Titan Demonstration Mission

Spacecraft type	Possible accomplishments
Nuclear electric propulsion	Spiral escape from low Earth orbit; interplanetary transfer to Saturn; rendezvous with Titan; and spiral capture into 600 km circular polar orbit.
Main orbiting spacecraft	Automated mission operations during interplanetary and Titan phases: this includes interfacing with one supporting other spacecraft before deployments; deploying other spacecraft; communicating with other spacecraft and with Earth; studying Titan's atmosphere and surface using remote sensing techniques at both global characterization and intensive study levels; and selecting landing sites.
Lander/Rover	Lands at preselected site, avoids hazards; intensive study of Titan's surface; selects, collects and analyzes samples for composition, life, etc., explores several geologic regions.
Subsatellite	Lagrange point satellite monitors environment near Titan and is continuous communications relay; tethered satellite measure magnetosphere and upper atmosphere properties.
Atmospheric probe	Determines surface engineering properties and atmospheric structure at several locations/times.
Powered air vehicle	Intensive study of Titan's atmosphere; aerial surveys of surface; transport of surface samples or surface systems.
Emplaced science package	Deployed by long-range rover to form meteorological and seismological network. (Alternatives are penetrators or extended lifetime probes.)

SOURCE: NASA.

single most important technology driver for this type of automated mission is advanced machine intelligence, especially a sophisticated machine intelligence (MI) system capable of learning about new environments and generating scientific hypotheses using analytic, inductive and abductive reasoning. Numerous other supporting technologies are also needed to perform such automated space exploration missions in the next century. These technologies include low-thrust propulsion systems; general-purpose surface exploration vehicles that can function on both solid and liquid surfaces; smart sensors and sensor networks that can reconfigure themselves; flexible, highly adaptive, general-purpose robotic manipulators; and distributed intelligence/database systems.

Once we have developed the sophisticated robotic devices needed for the detailed investigation of the outer regions of our own Solar System, the next step becomes quite obvious. Sometime in the 21st century, humankind will build and launch its first automated explorer to a nearby star system. This interstellar probe will be capable of searching for extrasolar planets around alien suns, targeting any suitable planets for detailed investigation, and then initiating the search for extraterrestrial life. Light-years away, terrestrial scientists will patiently wait for its faint, distant radio signals by which the robot starship describes any new worlds it has encountered and sheds light on the greatest cosmic mystery of all: Does life exist elsewhere in the depths of space?

ROLE OF ROBOTICS IN ADVANCED SPACE MANUFACTURING ACTIVITIES

Two of the most important products that will be manufactured in space in the next century are robots and teleoperator systems. The ultimate goals for advanced space manufacturing facilities cannot be achieved without a large expansion of the automation equipment initially provided from Earth. Eventually, space robots and teleoperators must be manufactured in space, drawing from the working experience gathered during the use of the first generation of space industrial robots. These second- and third-generation robotic devices must be far more versatile and fault-tolerant than the first-generation devices created on Earth and shipped to extraterrestrial locations as "seed" or starter machines. The most critical technologies needed for the manufacture of second- and third-generation robots and teleoperator systems appear to be space-adaptive sensors and computer vision. Enhanced decision-making capabilities and self-preservation features must also be provided for use in space robots and teleoperators.

Once we develop the ability to make robots in space, another step in robotics becomes possible—that of the self-replicating system (SRS). A single SRS unit is a machine system that contains all the elements required to maintain itself, to manufacture desired products, and even (as the name implies) to reproduce itself! In fact, an SRS unit would behave much like a biological cell.

THE THEORY AND OPERATION OF SELF-REPLICATING SYSTEMS

The brilliant Hungarian–American mathematician, John von Neumann, was the first person to seriously consider the problem of self-replicating systems. He became interested in the study of automatic replication as part of his wide-ranging interests in complicated machines. His work during the World War II Manhattan Project led him into automatic computing and he became fascinated with the idea of large, complex electronic computing machines. In fact, he invented the scheme used today in the great majority of general-purpose computers—the von Neumann concept of serial processing stored-program—which is also referred to as the "von Neumann machine." Following this pioneering work, von Neumann decided to tackle the larger problem of developing a self-replicating machine. Von Neumann actually conceived of several types of self-replicating systems, which he called the "kinematic machine," the "cellular machine," the "neuron-type machine," the "continuous machine" and the "probabilistic machine." Unfortunately, before his death in 1957, he was only able to develop a very informal description of the kinematic machine.

The kinematic machine is the one we hear about most often in discussing von Neumann type self-replicating systems. For this type of SRS, the brilliant mathematician had envisioned a machine residing in a "sea of spare parts." Von Neumann's kinematic machine would have a memory tape that instructed it to go through certain mechanical procedures. Using manipulator arms and its ability to move around, this SRS would gather and assemble parts. The stored computer program would instruct the machine to reach out and pick up a certain part, and then go through an identification and evaluation routine to determine whether the part selected was or was not the one called for by the master tape. (Note: in von Neumann's day microcomputers, floppy disks and bubble memory devices did not exist.) If the component picked up by the manipulator arm did not meet the selection criteria, it was tossed back into the "sea of parts." The process would continue until the required part was found and then an assembly operation would be performed. In this way, von Neumann's kinematic SRS would eventually make a complete replica of itself—without, however, really understanding what it was doing. When the duplicate was physically completed, the parent machine would make a copy of its own memory tape on the (initially) blank tape in its offspring. The last instruction on the parent machine's tape would be to activate the tape of its mechanical progeny. The offspring kinematic SRS could then start searching the "sea of parts" for components to build yet another generation of SRS units.

In dealing with his self-replicating machine concepts, von Neumann concluded that they should include the following characteristics and capabilities: (1) logical universality, (2) construction capability, (3) constructional universality and (4) self-reproduction. Logical universality is simply the device's ability to function as a general purpose computer. To be able to self-replicate, a machine must be capable of manipulating information, energy and materials. This is what is meant by "construction capability." The closely related term "constructional universality" is a characteristic that implies the machine's ability to manufacture any of the finite-sized machines that can be built from a finite number of different parts that are available from an indefinitely large supply. The characteristic of self-reproduction means that the original machine, given a sufficient number of component parts (of which it is made) and sufficient instructions, can make additional replicas or copies of itself. One characteristic of SRS devices that von Neumann did not address, but that has been addressed by subsequent investigators, is the concept of evolution. If we have a series of machines making machines, making machines, and so forth, can successive generations of machines learn to make themselves better?

One conceptual problem associated with the kinematic SRS concept is that the device lives in a "sea of parts," that is, the SRS unit inhabits a universe that provides it precisely what it needs to duplicate itself. This, of course, brings up the issue of closure, a major problem in thinking about self-replicating machines that will be discussed shortly.

From von Neumann's work and the more recent work of other investigators, we arrive at five broad classes of SRS behavior:

(1) *Production*—The generation of useful output from useful input. In the production process, the unit machine remains unchanged. Production is a simple behavior demonstrated by all working machines, including SRS devices.

(2) *Replication*—The complete manufacture of a physical copy of the original machine unit by the machine unit itself.

(3) *Growth*—An increase in the mass of the original machine unit by its own actions, while still retaining the integrity of its original design. For example, the machine might add an additional set of storage compartments in which to keep a larger supply of parts or constituent materials.

(4) *Evolution*—An increase in the complexity of the unit machine's function or structure. This is accomplished by additions or deletions to existing subsystems, or by changing the characteristics of these subsystems.

(5) *Repair*—Any operation performed by a unit machine on itself that helps reconstruct, reconfigure or replace existing subsystems, but does not change the SRS unit population, the original unit mass or its functional complexity.

In theory, replicating systems can be designed to exhibit any or all of these machine behaviors. When such machines are actually built, however, a particular SRS unit will most likely emphasize just one or several kinds of machine behavior, even if it were capable of exhibiting all of them. For example, the fully autonomous, general-purpose self-replicating lunar factory, proposed by Georg von Tiesenhausen and Wesley A. Darbo of the Marshall Space Flight Center in 1980, is an SRS design that is intended for unit replication. There are four major subsystems that make up this proposed SRS unit. First, a materials-processing subsystem gathers raw materials from its extraterrestrial environment and prepares industrial feedstock. Next, a parts production subsystem uses this feedstock to manufacture other parts or entire machines. At this point, the conceptual SRS unit has two basic outputs. Parts may flow to the universal constructor subsystem, where they are used to make a new SRS unit (this is replication); or else parts may flow to a production facility subsystem, where they are made into commercially useful products. This self-replicating lunar factory has other secondary subsystems, such as a materials depot, parts depot, product depot, power supply and command and control center. The universal constructor (UC) manufactures complete SRS units that are exact replicas of the original SRS unit. Each replica can then make more replicas of itself until a preselected SRS unit population is achieved. The universal constructor would retain overall command and control (C&C) responsibilities for its own SRS unit as well as for its mechanical progeny—until at least the C&C functions themselves have been duplicated and transferred in the new units. To avoid cases of uncontrollable exponential growth of such SRS units in some planetary resource environment, the human masters of these devices may reserve the final step of C&C function transfer to themselves or so design the SRS units such that the final C&C transfer function from machine to machine can be overridden by external human commands.

EXTRATERRESTRIAL IMPACT OF SELF-REPLICATING SYSTEMS

The issue of closure (total self-sufficiency) is one of the fundamental problems in designing self-replicating systems. In an arbitrary SRS unit there are three basic requirements necessary to achieve closure: (1) matter closure, (2) energy closure and (3) information closure. In the case of matter closure we ask: Can the SRS unit manipulate matter in all the ways needed for complete self-construction? If not, the SRS unit has not achieved matter or material closure. Similarly, we ask whether the SRS unit can generate a sufficient amount of energy and in the proper form to power the processes needed for self-construction. Again, if the answer is no, then the SRS unit has not achieved energy closure. Finally, we must ask: Does the SRS unit successfully command and control all the processes necessary for complete self-construction? If not, information closure has not been achieved.

If the machine device is only partially self-replicating, then we say that only partial closure of the system has occurred. In this case, some essential matter, energy or information must be provided from external sources, or else the machine system would fail to reproduce itself.

Just what are the applications of self-replicating systems? The early development of self-replicating system technology for use on Earth and in space will trigger an era of superautomation that will transform most terrestrial industries and lay the foundation for efficient space-based industries. One interesting machine proposed by physicist Theodore Taylor is called the "Santa Claus" machine. In his concept of an SRS unit, a fully automatic mining, refining and manufacturing facility gathers scoopfuls of terrestrial or extraterrestrial materials. It then processes these raw materials by means of a giant mass spectrograph that has huge superconducting magnets. The material is converted into an ionized atomic beam and sorted into stockpiles of basic elements, atom by atom. Then, to manufacture any item, the Santa Claus machine selects the necessary materials from its stockpile, vaporizes them and injects them into a mold that changes the materials into the desired item. Instructions for manufacturing, including directions on adapting new processes and replication, are stored in a giant computer within the Santa Claus machine. If the product demands become excessive, the Santa Claus machine would simply reproduce itself.

SRS units might be used in very large space construction projects (such as lunar mining operations) to facilitate and accelerate the exploitation of extraterrestrial resources and to make possible feats of planetary engineering. For example, we could deploy a seed SRS unit on Mars as a prelude to permanent human habitation. This machine would use local Martian resources to automatically manufacture a large number of robot explorer vehicles. This armada of vehicles would be disbursed over the surface of the Red Planet searching for the minerals and frozen volatiles needed in the establishment of a Martian civilization. In perhaps just a few years, a population of some 1,000 to 10,000 smart machines would scurry across the planet, completely exploring its entire surface and preparing the way for permanent human settlements.

Replicating systems would also make possible large-scale interplanetary mining operations. Extraterrestrial materials could be discovered, mapped and mined, using teams of surface and subsurface prospector robots that were manufactured in large quantities in an SRS factory complex. Raw materials would be mined by hundreds of machines and then sent wherever they were needed in heliocentric space.

Atmospheric mining stations could be set up at many interesting and profitable locations throughout the Solar System. For example, Jupiter and Saturn could have their atmospheres mined for hydrogen, helium (including the very valuable isotope, helium-3) and hydrocarbons using "aerostats." Cloud-enshrouded Venus might be mined for carbon dioxide, Europa for water and Titan for hydrocarbons. Large quantities of useful volatiles might be obtained by intercepting and mining comets with fleets of robot spacecraft. Similar mechanized space armadas might mine water-ice from Saturn's ring system. All of these smart robot devices would be mass produced by seed SRS units. Finally, extensive mining operations in the main asteroid belt would yield large quantities of heavy metals. Using extraterrestrial materials, these replicating machines could, in principle, manufacture huge mining or processing plants or even ground-to-orbit or interplanetary vehicles. This

large-scale manipulation of the extraterrestrial environment would occur in a very short period of time, perhaps within one or two decades of the initial introduction of replicating machine technology.

From the viewpoint of our extraterrestrial civilization, perhaps the most exciting consequence of the self-replicating system is that it would provide a technological pathway for organizing potentially infinite quantities of matter. Large reservoirs of extraterrestrial matter might be gathered and organized to create an ever widening human habitat throughout the Solar System. Self-replicating space stations, space settlements and domed cities on certain alien worlds of our Solar System would permit a diversity of environmental niches never before experienced in the history of the human race.

The SRS unit would provide such a large amplification of matter manipulating capability that it is possible even now to start seriously considering terraforming or planetary engineering strategies for the Moon, Mars, Venus and certain other alien worlds. SRS technology also appears to be our key to exploration and to human habitat expansion beyond the very confines of the Solar System. Although such interstellar missions may today appear highly speculative, and indeed they certainly require technologies that exceed contemporary or even projected levels in many areas, a consideration of possible interstellar or even intergalactic applications is actually quite an exciting and useful mental exercise. It illustrates immediately the fantastic power and virtually limitless potential of the SRS concept.

It appears likely that before humans move out across the interstellar void, smart robot probes will be sent ahead as scouts. Interstellar distances are so large and search volumes so vast, that self-replicating probes represent a highly desirable, if not totally essential, approach to surveying other star systems for suitable extrasolar planets and for extraterrestrial life. One recent study on galactic exploration suggests that search patterns beyond the 100 nearest stars would most likely be optimized by the use of SRS probes. In fact, reproductive probes might permit the direct reconnaissance of the nearest one million stars in about 10,000 years and the entire Galaxy in less than one million years—starting with a total investment by the human race of just one self-replicating interstellar robot spacecraft!

Of course, the problems in keeping track of, controlling and assimilating all the data sent back to the home star system by an exponentially growing number of robot probes is simply staggering. We might avoid some of these problems by sending only very smart machines capable of greatly distilling the information gathered and transmitting only the most significant quantities of data, suitably abstracted, back to Earth. We might also set up some kind of command and control hierarchy, in which each robot probe only communicates with its parent. Thus, a chain of "ancestral repeater stations" would be used to control the flow of messages and exploration reports. Imagine the exciting chain reaction that might occur as one or two of the leading probes encountered an intelligent alien race. If the alien race proved hostile, an extraterrestrial alarm would be issued, taking light years to ripple across interstellar space, repeater station by repeater station, until humans at "Central Robot Probe Control" received notification that: "Robot 24–76–AX–JA–2 was laser-blasted by hostile aliens in the vicinity of the Rigel system." Would we retaliate and send more sophisticated, possibly predator robot probes to that area of the Galaxy—or would we elect to place warning beacons all around the area, signaling any other robot probes to swing clear of the alien hazard?

In time giant space arks, representing an advanced level of synthesis between human crew and robot "crew," will depart from the Solar System and plunge into the interstellar medium. Upon reaching another star system that contained suitable planetary resources, the space ark itself could undergo replication. The human passengers (perhaps several generations of humans beyond the initial crew that departed the Solar System) would then redistribute themselves between the parent space ark and offspring space arks. Consequently, the original space ark would serve as an extraterrestrial refuge for humanity and any terrestrial life-forms included on the giant, mobile habitat. This dispersal of humanity to a variety of ecological niches among the stars would ensure that not even disaster on a stellar scale, such as our Sun going supernova, could threaten the complete destruction of man and all his accomplishments. These self-replicating space arks would allow their human crews to literally green the Galaxy with life as we know it.

Finally, self-replicating systems could also be used by advanced civilizations to perform gigantic feats of astro-engineering. The harnessing of the total radiant energy output of a star, through the robot-assisted construction of a Dyson sphere, is an example of such large-scale astro-engineering projects that might be undertaken.

CONTROL OF SELF-REPLICATING SYSTEMS

What happens if a self-replicating system gets out of control? Before we seed the Solar System or the Galaxy with even a single SRS unit, we should also know how to pull its plug if things get out of control. Some people have already raised a very obvious and legitimate concern about SRS technology: Do these smart machines represent a long-range threat to human life? In particular, will machines with advanced levels of artificial intelligence become our adversaries, whether they can replicate or not? Similarly, even in the absence of very advanced levels of machine intelligence, the self-replicating system might represent a threat just through uncontrollable growth. These questions can no longer remain only in the realm of science fiction. We must start examining the implications of developing advanced machine intelligences and self-replicating systems *before* we bring them into existence—or perhaps someday find ourselves in some mortal conflict over scarce resources with our own intelligent machine creations.

Of course, we need smart machines to help us improve life on Earth and to develop our extraterrestrial civilization.

So we should proceed with their development, but include safeguards to avoid undesirable future situations in which the machines turn on their human masters and eventually enslave or exterminate us. Asimov's "Three Laws of Robotics" (discussed earlier in this entry) appears to represent a good starting point in developing safe smart machines.

However, any machine sophisticated enough to survive and reproduce in largely unstructured environments, would probably also be capable of performing a certain amount of automatic or self reprogramming. This type of SRS unit might eventually be able to program itself around any rules of behavior that were stored in its memory banks by its human creators. In learning about its environment, the smart SRS unit might decide to modify its behavior patterns to better suit its own needs. Let's use Asimov's First Law of Robotics ("A robot may not injure a human being, or through inaction, allow a human being to come to harm") as a reference. Well, if our very smart SRS really "enjoys" being a machine and making (and perhaps improving) other machines, then when faced with a situation in which it must save a human master's life at the cost of its own, the smart machine may decide to shut down instead of performing the life-saving task it was preprogrammed to do. Thus, while it didn't harm the endangered human being, it also didn't help the person out of danger, either. Science fiction contains many stories about robots, androids and even computers turning on their human builders. (Perhaps there is a legitimate warning being expressed.)

An SRS population in space might be controlled by one or all of the following techniques. First, the human builders could implant machine-genetic instructions that contained a hidden or secret cutoff command. This cutoff command would be activated after the SRS units had undergone a predetermined number of replications. For example, after each machine replica is made, one generation command could be deleted—until eventually the entire replication process is terminated upon construction of the last, predetermined replica. Second, a special, predetermined signal from Earth on some emergency frequency channel might be used to cut the power of the main bus for individual, selected groups or all SRS units at any time.

For low-mass SRS units (perhaps the 100 to 10,000 kilogram class) population control might prove more difficult because of the shorter replication times in comparison to larger SRS factory units. To keep these mechanical "critters" in line, we might use a predator robot. These predator robots would be programmed to attack and destroy only the SRS-type unit, whose population needed control. Of course, we could also develop a universal destructor (UD). This machine would be capable of taking apart any other machine encountered and it would stockpile the victim machine's parts and store any information found in the victim's memory banks. Predator species are used today in wildlife refuges on Earth to keep animal populations in balance. Similarly, perhaps, we could use a linear supply of nonreplicating machine predators to control an exponentiating population of SRS "prey" units. We might also design the initial SRS units to be sensitive to population density. They could then become infertile, stop their operations or even (like lemmings on Earth) report to a central facility for disassembly, when they sensed overcrowding or overpopulation. Unfortunately, if these smart SRS units reflect their human creators too exactly, even without preprogramming overcrowding might force such machines to compete among themselves for dwindling supplies of extraterrestrial resources—dueling, mechanical cannibalism and even robotic warfare might result! Hopefully, however, we will create these machines to reflect the best of the human mind and spirit, and with their help sweep through the Galaxy spreading life, intelligence and organization in a golden age of interstellar development.

In the very long term, there appear to be two general pathways for mankind: either we are a very important biological stage in the overall evolutionary scheme of the Universe; or else we are an evolutionary dead end. If we limit ourselves to just the fragile biosphere of "Spaceship Earth," a natural disaster or our own technological foolhardiness will almost certainly terminate our existence, perhaps in just centuries or maybe in a few millennia. Even excluding these unpleasant consequences, without an extraterrestrial frontier, our planetary society will simply stagnate while other civilizations flourish and populate the Galaxy.

Replicating system technology provides us with very interesting options for continued evolution beyond the boundaries of the planet Earth. We might decide to create autonomous, interstellar self-replicating probes (which in some sense are our "offspring") and send these across the interstellar void on missions of exploration. Or, we could develop a closely knit (symbiotic) human–machine system, a highly automated interstellar ark, that is capable of crossing interstellar regions and then replicating itself when it encounters star systems with suitable planets.

According to some scientists, any intelligent civilization that wants to explore a portion of the Galaxy more than 100 light years from their parent star system, would probably find it more efficient to use self-replicating robot probes. This galactic exploration strategy would produce the largest amount of data about other star systems for a given period of exploration. For example, it has been estimated that the entire Galaxy could be explored in about one million years, assuming the replicating interstellar probes could achieve speeds of one-tenth the speed of light. If many advanced alien civilizations follow this approach, then the most probable interstellar probe we are likely to encounter would be of the self-replicating type.

One very large advantage of using interstellar robot probes versus interstellar (radio) beacons in the search for extraterrestrial intelligence (SETI) is the fact that these probes could also serve as cosmic safety deposit boxes, carrying the cultural treasures of civilizations through the Galaxy long after the parent civilization has perished. The gold-anodized records we included on the *Voyager* spacecraft and the plaques placed on the *Pioneer 10* and *11*

spacecraft are our first attempts at achieving a small degree of cultural immortality in the cosmos. Starfaring self-replicating machines should be able to keep themselves running for a long time. It has been estimated by certain scientists that there may exist at present only 10 percent of all the alien civilzations that have ever lived in the Galaxy (the remaining 90 percent having perished). If this is true, then statistically at least, nine out of every 10 robot star probes within the Galaxy might actually be the only surviving emissaries from long-dead civilizations.

If we ever encounter such alien probes and are able to decipher their contents, we may eventually learn about some incredibly ancient alien societies. Those now extinct societies may, in turn, lead us to many others. In a sense, by encountering and successfully interrogating an alien robot star probe, we may actually be treated to a delightful edition of the proverbial *Encyclopedia Galactica*—a literal compendium of the technical, cultural and social heritage of thousands of extraterrestrial civilizations within the Galaxy (most of which are now extinct).

This raises a number of fundamental ethical questions about the use of interstellar self-replicating probes. Is it morally right, or even equitable, for a self-replicating machine to enter an alien star system and convert a portion of that star system's mass and energy to satisfy its own purposes? Does an intelligent race legally "own" its parent star, home planet and any materials residing on other celestial objects within its star system? Does it make a difference whether the star system is inhabited by intelligent beings, or is there some lower threshold of galactic intelligence below which starfaring races may ethically (on their own value scale) invade such a star system and appropriate the resources needed to continue on their mission? If an alien probe enters a star system, by what criteria does it judge any indigenous lifeforms to be "intelligent" so as not to severely disturb existing ecospheres in their interstellar scavenger hunt for resources? And, of course, the really important question: "Now that humanity has developed space technology, are we above (or below) the cosmic appropriations threshold?"

In summary, the self-replicating system is a very powerful tool with ramifications on a cosmic scale. With properly developed and controlled SRS technologies, mankind could set in motion a chain reaction that spreads life, organization and consciousness across the Galaxy in an expansion wave limited in speed only by the speed of light itself. With these smart machines as our close partners in interstellar exploration, we could literally green the Galaxy in about one million years!

Roentgen Satellite (ROSAT)   A cooperative program involving the United States (NASA), Germany and the United Kingdom. The main objective of the German X-ray satellite, *Roentgen*, is to conduct a complete sky survey for electromagnetic radiations in the 0.041 to 1.0 kiloelectron volt (keV) range. The spacecraft's scientific payload consists of two imaging telescopes: the large German X-ray telescope (with NASA-provided high-resolution imager) and an extreme ultraviolet (EUV) wide-field camera provided by the United Kingdom. The X-ray telescope has been designed to study X-ray sources with high resolution in the 0.1 to 2.0 kiloelectron volt (keV) regime; while the wide-field camera is examining currently uncharted extreme ultraviolet (EUV) regions of the sky. A U.S.-provided Delta 2 expendable launch vehicle successfully placed this X-ray observatory into Earth orbit on June 1, 1990.

See also: **astrophysics; X-ray astronomy**

rogue star A wandering star that passes close to a solar system, disrupting the celestial objects in that system and triggering cosmic catastrophes on life-bearing planets. Close passage of a rogue star could result in the stimulation of massive comet showers, giant tidal surges or the disruption of minor planet (asteroid) orbits. The impact of comets or asteroids on a life-bearing planet could in turn trigger the mass extinction of many species, since the planetary biosphere would be violently disturbed in a very short period of time. Other names for a rogue star include "death star" and "interstellar vagrant."

See also: **Nemesis**

Rosetta Mission This planned Comet Nucleus Sample Return mission is a cornerstone space activity to which the European Space Agency (ESA) has committed itself beyond the year 2000. This type of mission represents the next logical step after comet flyby and rendezvous missions. The Rosetta mission takes its name from the famous Rosetta stone discovered in Egypt on 1799, whose deciphering led archaeologists to an understanding of ancient Egyptian hieroglyphic writing. Scientists anticipate that comet nucleus samples collected during the Rosetta mission will similarly unlock some of the mysteries concerning the origins of the Solar System and establish appropriate relationships between interstellar materials, cometary materials and primitive meteorites. By providing samples of the most primitive Solar System materials, this mission will support the initiation of the detailed scientific investigation of the pre-Solar System space environment, including possibly data representative of interstellar and galactic regimes. Cometary samples carry the primordial material out of which all planetary objects were formed 4.6 billion years ago. The Rosetta mission will, therefore, be an exciting space mission that leads to the roots of our Solar System.

To provide appropriate cometary material samples, the Rosetta space system must (1) rendezvous with an active relatively fresh comet; (2) fully investigate the surface of the comet's nucleus, including the characterization of active and inactive regions; (3) obtain three general types of material samples (a core sample, a volatile sample and a nonvolatile sample); (4) store these samples at appropriate "cold" conditions (typically less than 160 degrees Kelvin); and (5) return these "cold-stored" samples to Earth for detailed analysis.

In addition to cometary science itself, scientists anticipate

that detailed analysis of returned cometary materials will also make major contributions to an understanding of condensation processes in the solar nebula, the composition of planetesimals and the formation of planets, the conditions and processes in interstellar clouds and prebiotic chemical evolution.

A launch date sometime in the year 2001 is now envisioned for this exciting robot space mission to a comet's nucleus.

See also: **comet; Comet Rendezvous and Asteroid Flyby Mission; European Space Agency; Vesta Missions**

S

Satellite Power System (SPS) A very large space structure constructed in Earth orbit that takes advantage of the nearly continuous availability of sunlight to provide useful energy to a terrestrial power grid. The original SPS concept, called the "Solar Power Satellite," was first presented in 1968 by Dr. Peter Glaser. The basic concept behind the SPS is quite simple. Each SPS unit is placed in geosynchronous orbit above the Earth's equator, where it experiences sunlight over 99 percent of the time. These large orbiting space structures then gather the incoming sunlight for use on Earth in one of three general ways: microwave transmission, laser transmission or mirror transmission.

In the fundamental microwave transmission SPS concept, solar radiation would be collected by the orbiting SPS and converted into radiofrequency (RF) or microwave energy. This microwave energy would then be precisely beamed to a receiving antenna on Earth, where it is converted into electricity.

In the laser transmission SPS concept, solar radiation is converted into infrared laser radiation, which is then beamed down to special receiving facilities on the Earth for the production of electricity.

Finally, in the mirror transmission SPS concept, very large (kilometer dimensions) orbiting mirrors would be used to reflect raw sunlight directly to terrestrial energy conversion facilities 24 hours a day.

In the microwave transmission SPS concept described above, incoming sunlight is converted into electrical power on the giant space platform by either photovoltaic (solar cell) or heat engine (thermodynamic cycle-turbogenerator) techniques. This electric power is then converted at high efficiency into microwave energy. The microwave energy, in turn, is focused into a beam and aimed precisely at special ground stations. The ground station receiving antenna (also called a rectenna) reconverts the microwave energy into electricity for distribution in a terrestrial power grid.

Because of its potential to relieve long-term national and global energy shortages, the Satellite Power System concept has been studied extensively. Some of these recent studies have suggested that the SPS units be constructed using extraterrestrial materials from the Moon and Earth-approaching asteroids, with manufacturing and construction activities being accomplished by space workers who would reside in large, permanent space settlements in cislunar space. Other SPS studies examined the development and construction of SPS units using only terrestrial materials, which were placed in low Earth orbit (LEO) by a fleet of special heavy-lift launch vehicles (HLLV). In this scenario, construction work could be accomplished in LEO, perhaps at the site of a permanent space station or space construction base. Assembled SPS sections would then be ferried to geosynchronous Earth orbit (GEO) by a fleet of orbital transfer vehicles (OTVs). At GEO, a crew of space workers would complete final assembly and prepare the SPS unit for operation.

It is perhaps too early to fully validate the SPS concept or to totally dismiss it. What can be stated at this time, however, is that the controlled beaming to Earth of solar energy (either as raw concentrated sunlight or as converted microwaves or laser radiation) appears to represent a major space industrialization pathway. This particular commercial avenue could be a very powerful stimulus toward the creation of our extraterrestrial civilization. However, the SPS concept also involves potential impacts on the terrestrial environment. Some of these environmental impacts are comparable in type and magnitude to those arising from other large-scale terrestrial energy technologies; while other impacts are unique to the SPS concept. Some of these SPS-unique environmental and health impacts are potential adverse effects on the Earth's upper atmosphere from launch vehicle effluents and from energy beaming (that is, microwave heating of the ionosphere); potential hazards to terrestrial life-forms from nonionizing radiation (microwave or infrared laser); electromagnetic interference with other spacecraft, terrestrial communications and astronomy; and the potential hazards to space workers, especially exposure to ionizing radiation doses well beyond currently accepted industrial criteria. These issues will have to be favorably resolved, if the SPS concept is to emerge as a major pathway in humanity's creative use of the resources of outer space.

In 1980 the U.S. Department of Energy and NASA defined an SPS reference system to serve as the basis for conducting initial environmental, societal and comparative assessments; alternative concept trade-off studies; and supporting critical technology investigations. This SPS reference system is, of course, not an optimum or necessarily the preferred system design. It does, however, represent

Table 1 SPS Reference System Characteristics

System Characteristics	
General capability (utility interface) 300 GW - total 5 GW - single unit	Number of units: 60 Design life: 30 years Deployment rate: 2 units/year
Satellite	
Overall dimensions: 10 × 5 × 0.5 km Structural material: graphite composite	Satellite mass: 35–50 × 10⁶ kg Geostationary orbit: 35,800 km
Energy Conversion System	
Photovoltaic solar cells: silicon or gallium aluminum arsenide	
Power Transmission and Reception	

D.C.–R.F. conversion: klystron
Transmission antenna diameter: 1 km
Frequency: 2.45 GHz
Rectenna dimensions (at 35° latitude)
  Active area: 10 × 13 km
  Including exclusion area: 12 × 15.8 km
	Rectenna construction time: ≈ 2 years
	Rectenna peak power density: 23 mW/cm²
	Power density at rectenna edge: 1 mW/cm²
	Power density at exclusion edge: 0.1 mW/cm²
	Active, retrodirective array control system with pilot beam reference

Space Transportation System	

Earth-to-LEO - Cargo: vertical takeoff, winged two-stage (425 metric ton payload)
           Personnel: modified Shuttle
LEO-to-GEO - Cargo: electric orbital transfer vehicle
           Personnel: two-stage liquid oxygen/liquid hydrogen	

Space Construction	
Construction staging base - LEO: 480 km Final construction - GEO: 35,800 km Satellite construction time: 6 months	Construction crew: 600 System maintenance crew: 240

SOURCE: NASA/Department of Energy.

one potentially plausible approach for achieving SPS concept goals.

This reference SPS system configuration and its main technical characteristics are summarized in table 1. The proposed configuration would provide 5 gigawatts (GW) of electric power at the terrestrial grid interface. In the reference scenario, 60 such SPS units would be placed in geostationary orbit and thus provide some 300 gigawatts of electric power for use on Earth. It has been optimistically estimated that only about six months would be required to construct each SPS unit.

Saturn   Saturn is the sixth planet from the Sun and to many the most beautiful celestial object in our Solar System. To the naked eye the planet is yellowish in color. Because of its great distance from the Earth, Saturn appears the least bright and moves the most slowly through the Zodiac. This planet is named after the elder god and powerful titan of Roman mythology (Cronus in Greek mythology), who ruled supreme until he was dethroned by his son Jupiter (Zeus to the early Greeks).

Composed mainly of hydrogen and helium, Saturn (with an average density of just 0.7 grams per cubic centimeter) is so light that it would float on water, if there were some cosmic ocean large enough to hold it. The planet takes about 29.5 Earth years to complete a single orbit around the Sun. But a Saturnian day is only approximately 10 hours and 39 minutes long.

The first telescopic observations of the planet were made by Galileo Galilei in 1610. The existence of its magnificent ring system was not known until the Dutch astronomer Christian Huygens, using a better-resolution telescope, properly identified it in 1655. Actually, Galileo had seen the Saturnian rings but mistook them for large moons on either side of the planet. Huygens is also credited with the discovery of Saturn's largest moon, Titan, which also occurred in 1655.

Astronomers had very little information about Saturn, its rings and constellation of moons, until the *Pioneer 11* spacecraft (September 1, 1979), *Voyager 1* spacecraft (November 2, 1980) and *Voyager 2* spacecraft (August 26, 1981) encountered the planet. These robot spacecraft encounters revolutionized our understanding of Saturn and have provided the bulk of our current information about this interesting giant planet, its beautiful ring system and its large complement of moons (17 identified at present with other "moonlets" suspected).

Saturn is a giant planet, second in size only to mighty Jupiter. Like Jupiter, it has a stellar-type composition, rapid rotation, strong magnetic field and intrinsic internal heat source. Saturn has a diameter of approximately 120,660 kilometers at its equator, but 10 percent less at the poles because of its rapid rotation. Saturn has a mass of 5.68×10^{26} kilograms, which is about 95 times the mass of the Earth; yet its average density is only 0.7g/cm³—the lowest of any planet—indicating that much of Saturn is in a gaseous state. Table 1 lists contemporary data for the planet.

Table 1   Physical and Dynamic Properties of the Planet Saturn

Diameter (equatorial)	120,660 km
Mass	5.68×10^{26} kg
Density	0.7 g/cm³
Surface gravity	12 m/sec² (approx.)
Escape velocity	36 km/sec (approx.)
Albedo	0.5–0.6
Atmosphere	Hydrogen (88%), helium (11%), small amounts of methane (CH_4), ammonia (NH_3) and water (H_2O)
Natural satellites	17 (more moons anticipated)
Rings	Complex system (thousands)
Period of rotation (a Saturnian day)	0.444 days
Average distance from the Sun	14.27×10^8 km (9.539 AU) [79.33 light-min]
Eccentricity	0.056
Period of revolution around Sun (a Saturnian year)	29.46 years
Mean orbital velocity	9.6 km/sec
Solar flux at planet (at top of clouds)	15.1 watts/m² (at 9.54 AU)
Magnetosphere	Yes (strong)
Mean temperature (at cloud tops)	−180°C (−290°F)
Saturn-to-Earth distances Minimum	12.64×10^8 km (8.45 AU) [70.26 light-min]
Maximum	15.71×10^8 km (10.50 AU) [87.33 light-min]

SOURCE: NASA.

Saturn is truly one of the most exquisite objects in our Solar System, adorned by an unrivalled, complex, yet delicate, ring system (see fig. 1). Like the gaseous giant Jupiter, Saturn is also composed primarily of hydrogen (about 88 percent). But in contrast to the vivid colors and unbridled turbulence found in the Jovian atmosphere, the Saturnian atmosphere presents a more subtle, almost butterscotch-colored hue with more refined and subdued markings.

SATURN'S MAGNIFICENT RING SYSTEM
Although the other giant planets—Jupiter, Uranus and Neptune—all have rings, Saturn's magnificent ring system with its billions and billions of icy particles whirling around the planet in an orderly fashion is uniquely beautiful, a

Fig. 1 A mosaic of Saturn's ring system compiled from *Voyager 1* imagery data (November 1980). (Courtesy of NASA.)

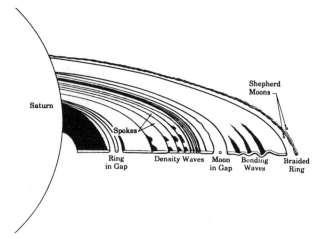

Fig. 2 Various features of the complex Saturnian ring system. (Artist rendering courtesy of NASA/JPL.)

natural wonder of the Solar System. The main ring areas stretch from about 7,000 kilometers above Saturn's atmosphere out to the F Ring—a total span of about 74,000 kilometers. Within this vast region, the icy particles are generally organized into ringlets, each typically less than 100 kilometers wide (see table 2).

The *Pioneer 11* and *Voyager* spacecraft discovered that there were actually thousands of these ringlets of various shapes and patterns, including circular rings, kinky rings, eccentric rings, clump rings, dense rings and even gossamer rings. This complex ring system also contains resonances, spokes, shepherding moons (moons that keep the icy particles in an organized structure) and most likely additional (as yet undiscovered) moonlets (see fig. 2).

These rings consist mainly of water-ice and frosted rock particles, ranging in size from tiny dustlike crystals to giant boulders. Figure 3 presents the results of a computerized

population model of the number and size of ice particles populating a typical 3-meter by 3-meter (i.e., room-size) section of Saturn's A-ring. This computerized population model is based on radio signal measurements taken during the *Voyager 1* encounter. The particle sizes shown range from marble-size (about 2 centimeters in diameter) to beachball-size (70 centimeters in diameter).

Scientists now think that the Saturnian rings resulted from one of three processes: a small moon venturing too close to Saturn and ultimately torn apart by large gravity-induced tidal forces, some of the planet's primordial ma-

Table 2   The Rings of Saturn

Feature	Distance from Center of Saturn (km)	(R_s)	Period (hr)
Cloud tops	60,330	1.00	10.66
D Ring inner edge	67,000	1.11	4.91
C Ring inner edge	73,200	1.21	5.61
B Ring inner edge	92,200	1.53	7.93
B Ring outer edge	117,500	1.95	11.41
Cassini Division (middle)	119,000	1.98	11.75
A Ring inner edge	121,000	2.01	11.93
Keeler (Encke) Gap	133,500	2.21	13.82
A Ring outer edge	136,200	2.26	14.24
F Ring	140,600	2.33	14.94
G Ring	170,000	2.8	19.9
E Ring (middle)	230,000	3.8	31.3

SOURCE: NASA.

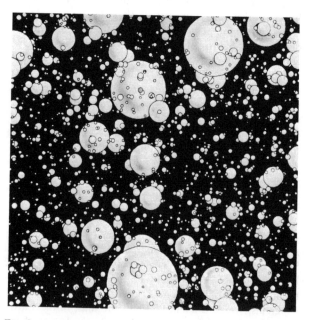

Fig. 3 A computer-generated model of Saturn's A Ring particle population (based on *Voyager 1* radio signal attenuation and scattering data). (Courtesy of NASA.)

terial failing to coalesce into a moon, or collisions among several larger objects orbiting the planet creating the rings.

THE SATELLITES OF SATURN

Before the first spacecraft encounter, Saturn was believed to have 11 satellites. Saturn is now known to have at least seventeen satellites (tables 3 and 4). The moons of Saturn form a diverse and remarkable constellation of celestial objects. The largest satellite, Titan, is in a class by itself. The six other major satellites have much in common, all being of intermediate size (400 to 1,500 kilometers in diameter) and consisting mainly of water-ice. The ten smaller moons appear to include both captured asteroids and fragments from intersatellite collisions. Moving outward from the planet, these six icy moons are Mimas, Enceladus, Tethys, Dione, Rhea and Iapetus. All are in orbits that are nearly circular and located within the equatorial plane of Saturn. The inner five of these icy moons occupy adjacent orbits, while Iapetus is much farther out, beyond Titan and Hyperion, in a highly inclined orbit about 3.5 million kilometers from Saturn. In terms of size, these six satellites can be conveniently divided into three groups of two each: Mimas and Enceladus (400 to 500 kilometers in diameter),

Tethys and Dione (about 1,000 kilometers in diameter), and Rhea and Iapetus (about 1,500 kilometers diameter).

Titan is the largest and most interesting of Saturn's satellites. It is the second-largest moon in the Solar System and the only one known to have a dense atmosphere. The atmospheric chemistry presently taking place on Titan may be similar to those processes that occurred in the Earth's atmosphere several billion years ago. For this reason, Titan is one of the most interesting objects in the Solar System to exobiologists and planetary scientists.

Larger in size than the planet Mercury, Titan has a density that appears to be about twice that of water-ice. Scientists believe therefore that it may be composed of nearly equal amounts of rock and ice. Titan's surface is hidden from the normal view of spacecraft cameras by a dense, optically thick photochemical haze whose main layer is about 300 kilometers above the moon's surface. Several distinct, detached layers of haze have also been observed above the optically opaque haze layer. Titan's southern hemisphere appears slightly brighter than the northern one, possibly because of seasonal effects.

The atmospheric pressure near Titan's surface is about 1.6 bars, some 60 percent greater than on the surface of

Table 3   Physical and Dynamic Properties of the Moons of Saturn

Moon	Diameter (km)	Semi-Major Axis of Orbit (km)	Period of Rotation (days)
Phoebe	220 (approx.)	12,954,000	550.5 (retrograde)
Iapetus	1,460	3,560,800	79.33
Hyperion	400 (approx.)	1,481,000	21.28
Titan	5,150	1,221,860	15.95
Rhea	1,530	527,100	4.52
1980S6 (Dione Trojan; leading)	40 (approx.)	378,100	2.74
Dione	1,120	377,420	2.74
1980S13 (Calypso, Tethys Trojan; leading)	35 (approx.)	294,670	1.89
Tethys	1,060	294,670	1.89
1980S25 (Telesto, Tethys Trojan; trailing)	35 (approx.)	294,670	1.89
Enceladus	500	238,040	1.37
Mimas	392	185,540	0.942
1980S1 (Epimetheus; co-orbital)	200 (approx.)	151,470	0.695
1980S3 (Janus; co-orbital)	120 (approx.)	151,420	0.694
1980S26 (F-Ring shepherd)	90 (approx.)	141,700	0.629
1980S27 (F-Ring shepherd)	100 (approx.)	139,350	0.613
1980S28 (Atlas, A-Ring shepherd)	40 (approx.)	137,670	0.602

SOURCE: NASA.

Table 4   Physical Data for the Larger Moons of Saturn

Moon	Diameter (km)	Mass (kg)	Density (g/cm³)	Albedo
Iapetus	1,460	1.9×10^{21}	1.2	0.05–0.5
Titan	5,150	1.35×10^{23}	1.9	0.2
Rhea	1,530	2.5×10^{21}	1.3	0.6
Dione	1,120	1.05×10^{21}	1.4	0.6
Tethys	1,060	7.6×10^{20}	1.2	0.8
Enceladus	500	8.4×10^{19}	1.2	>0.9
Mimas	392	4.5×10^{19}	1.4	0.7

SOURCE: NASA.

the Earth. The Titanian atmosphere is mostly nitrogen, which also happens to be the major constituent of the Earth's atmosphere. The existence of carbon–nitrogen compounds on Titan is possible because of the great abundance of both nitrogen and hydrocarbons. Titan is unique in providing in quantity all of the building blocks for complex organic compounds. Scientists now believe that large, complex organic molecules are forming continuously in Titan's atmosphere. In the process, hydrogen is released and escapes rapidly to space. As a result, the chemical processes involving hydrogen release are not reversible, resulting in the accumulation of hydrocarbon products. Eventually, these heavy organic molecules drift down to the moon's surface, while atmospheric methane is renewed by slow diffusion upward from the surface and the lower atmosphere. Thick clouds of frozen methane probably exist in Titan's troposphere, and may even extend down to its surface.

What does the surface of Titan look like? There must be large quantities of methane on the surface, enough perhaps to form methane rivers or even a methane sea. The temperature on the surface is about 91 degrees Kelvin (−182°C), which is close enough to the temperature at which methane can exist as a liquid or solid. Some people have therefore suggested that Titan is like the Earth, with methane playing the role that water plays on Earth. This analogy, if correct, leads to visions of oceans of methane near Titan's equator and frozen methane ice caps in the moon's polar regions. Titan's surface also experiences a constant rain of organic compounds from the upper atmosphere, perhaps creating up to a 100-meter-thick layer of tar-like materials.

Most of the sunlight striking Titan is absorbed in the high layers of photochemical haze and smog that occur about 200 kilometers above the satellite's surface. Only a small percentage of sunlight reaches the methane clouds in the troposphere and even less gets to the surface, which is probably quite dark and gloomy.

In any event, after the Voyager missions, we now know that Titan is a fascinating celestial object—in some ways more like the Earth than any other object in our Solar System. Its unique atmospheric conditions cause the production of large quantities of organic materials and their accumulation on the surface. Titan's methane, through photochemistry, is converted to ethane, acetylene, ethylene and (when combined with nitrogen) hydrogen cyanide. The last is an especially important molecule, because it is a building block of amino acids. However, Titan's low temperature most likely inhibits the development of more complex organic compounds.

To the exobiologist, Titan is perhaps the most interesting alien world in our Solar System. It represents a celestial laboratory in which the processes of prebiotic organic chemistry—that ultimately gave rise to life on Earth some four billion years ago—may be studied today! The *Cassini* mission with its special probe into the Titanian atmosphere will help resolve some of these intriguing questions.

Mimas, Enceladus, Tethys, Dione and Rhea are approximately spherical in shape and appear to be composed of mostly water-ice. Enceladus reflects almost 100 percent of the sunlight that strikes it. All five satellites represent a size moon not previously explored.

Mimas, Tethys, Dione and Rhea are all cratered. Enceladus appears to have the most active surface of any satellite in the Saturnian system (except perhaps for Titan, whose surface has not yet been observed). At least five types of surface terrain features have been identified on Enceladus. Because it reflects so much sunlight, Enceladus has an observed surface temperature of only 72 degrees Kelvin.

Images of Mimas reveal a huge impact crater. This crater, named Arthur, is about 130 kilometers wide—one-third the diameter of Mimas. Arthur is 10 kilometers deep, with a central mountain almost as large as Mount Everest on Earth.

Voyager 2 photographs of Tethys reveal an even larger impact crater, nearly one-third the diameter of Tethys and larger than the moon Mimas. In contrast to Mimas' Arthur, the floor of the large crater on Tethys has returned to approximately the original shape of the surface, most likely a result of Tethys' larger gravity and the relative fluidity of its water-ice. A gigantic fracture covers three fourths of Tethys' circumference. The fissure is about the size scientists would predict if this moon was once fluid and its crust hardened before the interior. This canyon has been named Ithaca Chasma. The surface temperature on Tethys is about 86 degrees Kelvin.

Iapetus has long been known to have large differences in surface brightness. Brightness of the surface material on the trailing side has been measured at 50 percent, while the leading side material reflects only five percent of the sunlight that strikes it. Most of the dark material is distributed in a pattern directly centered on the leading surface, causing some speculation that dark material in orbit around Saturn was swept up by Iapetus. The trailing face of Iapetus, however, has several craters with dark floors. This implies that the dark material may have originated in the moon's interior. It is possible that the dark material on the leading hemisphere was exposed by the ablation (erosion) of a thin, overlying bright surface covering.

Phoebe was imaged by the *Voyager 2* after the spacecraft had passed Saturn. Phoebe orbits in a retrograde direction (the opposite direction of the other moons of Saturn) in a plane much closer to the ecliptic than Saturn's equatorial plane. Phoebe is roughly circular in shape and reflects about six percent of the incident sunlight. This moon is also quite red in color. Scientists believe that Phoebe may be a captured asteroid with its composition unmodified since its formation in the outer Solar System. If this is true, then it represents the first such "minor planet" photographed at close enough range to show shape and surface brightness.

Both Dione and Rhea have bright, wispy streaks that stand out against an already bright surface. The streaks are most likely the results of ice that evolved from the interior along fractures in the crust.

Saturn has many small satellites, ranging in size from Hyperion (about 400 kilometers across) down to Atlas, the A-Ring shepherd moon, which is only about 40 kilometers in diameter. Scientists think that the irregularly shaped moon, Hyperion, is much smaller than its original size, because it orbits Saturn in a random-like chaotic tumbling, an orbital condition perhaps indicative of an ancient, shattering cosmic collision. The tiny moons 1980S26, 1980S27 and 1980S28 are called *shepherd satellites* because of their assumed role in helping to keep particles in Saturn's ring system within their proper orbits. Epimetheus (1980S1) and Janus (1980S3) are called the *co-orbitals*, because they share approximately the same orbit around Saturn. As they near each other, these satellites trade orbits (the outer orbit is about 50 kilometers farther from Saturn than the inner orbit. The tiny moon 1980S6 shares the orbit of Dione. Because 1980S6 is about 60 degrees ahead of its larger celestial companion, it is called the "Dione Trojan" or a "Lagrangian satellite." (Lagrangian satellites are small objects that, by maintaining approximately 60 degrees of arc ahead of or behind a larger parent celestial object, can share the same orbit and orbital speed as the parent.) The existence of a "trailing Dione Trojan" moon is also suspected. Similarly, two other small Saturnian satellites, Calypso (1980S13) and Telesto (1980S25), are called the Tethys Trojans, because they circle Saturn in the same orbit as Tethys, about 60 degrees of arc ahead of and behind that celestial body. (The name "Trojan" comes from the Trojan asteroids that occupy similar Lagrangian positions in Jupiter's orbit around the Sun.)

See also: **Cassini Mission; *Pioneer 10, 11; Voyager***

Saturnian   Of or relating to the planet Saturn; (in science fiction) a native of the planet Saturn.

Schwarzschild radius   The "event horizon" or boundary of no return of a black hole. Anything crossing this boundary can never leave the black hole.

See also: **black holes**

science fiction   A form of fiction in which technical developments and scientific discoveries represent an important part of the plot or story background. Frequently, science fiction involves the prediction of future possibilities based on new scientific discoveries or technical breakthroughs. Some of the most popular science fiction predictions waiting to happen are interstellar travel, contact with extraterrestrial civilizations, the development of exotic propulsion or communication devices that would permit us to break the speed-of-light barrier, travel forward or backward in time, and very smart machines and robots.

According to the popular writer, Isaac Asimov, one very important aspect of science fiction is not just its ability to predict a particular technical breakthrough, but rather its ability to predict change itself through technology. Change plays a very important role in our modern life. As we enter the next millennium, people responsible for societal planning must not only consider how things are now, but how they will (or at least might be) in the upcoming decades. Gifted science fiction writers, like Jules Verne, H.C. Wells, Isaac Asimov and Arthur C. Clarke, are also skilled technical prophets who help many people peek at tomorrow before it arrives.

For example, the famous French writer, Jules Verne, wrote *De la terre à la lune (From the Earth to the Moon)* in 1865. This science fiction account of a manned voyage to the Moon originated from a Floridian launch site near a place Verne called "Tampa Town." A little over 100 years later, directly across the state from the modern city of Tampa, the once isolated regions of the east central Florida coast shook to the mighty roar of a *Saturn V* rocket. The crew of the *Apollo 11* mission had embarked from Earth and man was to walk for the first time on the lunar surface!

search for extraterrestrial intelligence (SETI)   The search for extraterrestrial intelligence, or SETI, is basically a manifestation of our natural curiosity and desire to explore. These instincts are some of the oldest and most fundamental aspects of our human nature. It is highly probable that many ancient peoples looked up into the night sky and wondered about the existence of other worlds and other beings. However, it is only within the last few decades, with the arrival of the Age of Space, that mankind could do more than simply speculate about extraterrestrial life, including other life-forms that had achieved intelligence and developed technology. The classic paper by Giuseppe Cocconi and Philip Morrison entitled "Searching for Interstellar Communications" (*Nature*, 1959), is often regarded as the start of modern SETI. The entire subject of extraterrestrial intelligence (ETI) has left the realm of science fiction and is now regarded as a scientifically respectable (though currently speculative) field of endeavor.

Well, just where do we look for "little green men" or how do we listen for their signals? Scientists have wrestled with both these questions over the last two decades. In an effort to guess the number of technically advanced alien civilizations that might now exist in our Galaxy, Frank Drake and other scientists developed a mathematical relationship called the Drake Equation. Based on the Drake Equation, pessimistic estimates on the number of communicative extraterrestrial civilizations are very low (10 to perhaps 100); while more optimistic speculations indicate a Galaxy with perhaps 10,000 to 100,000 advanced civilizations!

The major aim of modern SETI programs is to listen for evidence of microwave signals generated by intelligent extraterrestrial civilizations. This search includes radio astronomy mapping of a major portion of the sky and the study of manmade (artificial) radio frequency interference for use in future SETI projects.

The current understanding of stellar formation leads scientists to think that planets are normal and frequent companions of most stars. As interstellar clouds of dust and gas condense to form stars, they appear to leave behind clumps of material that form into planets. The Milky Way Galaxy contains at least 100 billion to 200 billion stars.

Present theories on the origin and chemical evolution of life indicate that life is probably not unique to Earth but maybe common and widespread throughout our Galaxy. Scientists further believe that life on alien worlds could have developed intelligence, curiosity and the technology necessary to build the devices needed to transmit and receive electromagnetic signals across the interstellar void.

If this is true, then some scientists also believe that alien civilizations might be searching at this very moment for intelligent companions. There may even be some type of galactic community in which interstellar communications are shared by many different extraterrestrial civilizations. However, to date, none of our efforts here on Earth to detect and identify radio wave signals from alien civilizations have been successful. Since Dr. Frank Drake's initial SETI activities under Project Ozma in 1960, there have been many, but limited, scientific attempts around the world to listen for signals from intelligent alien sources. Unfortunately, none of these searches have yet yielded signals that could be positively identified as originating from alien civilizations among the stars.

Until now only very narrow portions of the electromagnetic spectrum have been examined for artificial signals (generated by intelligent alien civilizations). Manmade radio and television signals, the kind radio astronomers reject as clutter and interference, are actually similar to the signals SETI researchers are hunting for.

The sky is full of radio waves. In addition to the electromagnetic signals we generate as part of our technical civilization (radio, TV, radar, etc.) the sky also contains natural radio wave emissions from such celestial objects as our Sun, the planet Jupiter, quasars, radio galaxies and pulsars. Even interstellar space is characterized by a constant, detectable radio-noise spectrum.

However, SETI scientists are looking for radio wave signals that are considerably different from known natural extraterrestrial (ET) radio sources. Typically, a natural ET radio signal occupies a wide bandwidth, perhaps a kilohertz (1,000 cycles per second) or more. The radio wave signals that might be generated by intelligent alien races should not exhibit such wide bandwidths. For example, manmade radio emissions usually have strong carrier components that occupy less than one hertz (1 cycle per second). No natural ET radio sources have been found to date that broadcast on such narrow frequencies; and none may actually exist.

In conducting their search for intelligent radio signals from ET civilizations, SETI scientists must consider four general parameters or dimensions: (1) the location of the transmitting source; (2) the frequency range within which the source is transmitting; (3) the modulation or method used to impart information to the carrier electromagnetic signal; and (4) the signal power that can be detected by the receiving antenna.

At the best sensitivity now available, a SETI effort might be able to detect directive transmitters like the largest radio telescope/transmitter we now possess (the Arecibo facility) at a distance of more than 1,000 light years. In past efforts, many SETI observers had assumed that alien transmissions might be associated with star systems like or at least similar to our Sun. Stars of much greater luminosity than the Sun were considered too short-lived to permit the chemical evolution of life to occur; while stars very much less luminous than our Sun appear to have unfavorable ecospheres made inhospitable, in part, by violent coronal activities. SETI observers have also avoided stars that have departed the main sequence because even an advanced alien race would be hard pressed to survive when their parent star became a red giant or violently exploded as a supernova.

SETI observers have developed a variety of strategies and search scenarios (recall again the Drake equation). However, for many such efforts an all-sky survey for intense artificial sources, complemented by detailed investigations of nearby star systems for weaker signals, appears to be a favorable strategy to adopt. This is especially true when you consider the current lack of knowledge on this fascinating subject. As a rule of thumb, SETI scientists suggest that because we really don't know about any other planetary civilization but our own, it is most wise to make as few assumptions as possible in developing search strategies and signal detection schemes. For example, even if the Galaxy is bursting with intelligent civilizations (a speculation subject to extensive debate), what part of the electromagnetic spectrum would these civilizations use in communicating with each other? And would mature civilizations use the same type of transmissions to communicate with both advanced civilizations and emerging civilizations? Do you use the same language when you talk to adults and when you talk to little children?

Consequently, the signal broadcast frequency has been the subject of wide and varied speculation among SETI scientists. Some strongly believe that the region from 1.4 to 1.7 gigahertz represents an excellent prospect for detecting intelligent alien signals. It happens that this region lies between the natural radio wave emissions of hydrogen (H) and the radical hydroxyl (OH). Some SETI scientists have rather romantically called this region of the electromagnetic spectrum the "water hole." It is one of the most favored SETI observation regions because of the important role water plays in life on Earth and the fact that the region is also one of relative radio quiet, making any artificial signals within it fairly easy to detect and identify. Although its choice at present is quite pleasing from a terrestrial viewpoint, the water hole region is only a tiny fraction of the overall electromagnetic spectrum. Human philosophy may not, however, be relevant in other parts of the Galaxy. From a point of view of physics, the frequency band that is the most efficient for electromagnetic communications on an interstellar scale is the microwave "window" which lies between 1 and 100 gigahertz. If we conduct our SETI activities using radio antennas in space or on the far side of the Moon, we might consider this entire bandwidth in the search strategy. If we are limited to using only ground-based radio antenna facilities this window narrows to about one-tenth its size in space because of the effects of the Earth's atmosphere. However, even on Earth the lower end of this microwave window (the water hole) is quite free of natural radio noise.

How would an alien race modulate the signal so that it contained useful information? And then, what information would they send? Source modulation is another area of considerable speculation among SETI observers. If we were to detect a narrow-bandwidth carrier signal with no modulation (that is, no information content) of any kind, SETI scientists would most likely assume it is some type of previously unknown natural ET radio signal. If any alien race is sending a message, they will most likely include some information in it. But what method would an alien race use to modulate their signal? SETI scientists have suggested two basic possibilities: The alien society could use a strong pulsed signal; or perhaps a strong carrier component of narrow bandwidth continuously transmitted. Contemporary SETI equipment here on Earth can detect both types of signals.

SETI observations may be performed using radio telescopes on Earth, in space or even on the far side of the Moon. Each location has distinct advantages and disadvantages. For example, until a full extraterrestrial civilization matures (one with both lunar settlements and permanent space stations), the construction and operation of large radio antennas (dedicated to SETI activities) on either the lunar surface or in orbit will be prohibitively expensive—negating the technical advantages of using either location. However, once the technical infrastructure has been developed in cislunar space (perhaps in the next century) then extraterrestrial SETI observations from either cislunar space or the Moon's far side surface will become very attractive alternatives. But for the next decade or so, SETI scientists must remain content with using ground-based radio telescopes.

The SETI Microwave Observing project is part of NASA's Exobiology program to understand the origin, evolution and distribution of life in the Universe. Managed by the Life Sciences Division at NASA headquarters, this SETI project is a joint activity of the NASA Ames Research Center (ARC) and the Jet Propulsion Laboratory (JPL), with ARC serving as the lead center. The objective of the project is to conduct a systematic search of the microwave portion of the electromagnetic spectrum for extraterrestrial signals of intelligent origin. Because we are now unsure about how intelligent, technical civilizations might be distributed in our Galaxy, NASA proposes to conduct both a full-sky survey and a more sensitive targeted search as part of its Microwave Observing project.

The full-sky survey mode will search the entire celestial sphere over a wide frequency range (1,000 to 10,000 megahertz [MHz], in addition to selected frequencies up to 25,000 MHz) to explore the possibility that there may be a few extraterrestrial civilizations transmitting strong signals, possibly as interstellar beacons. The radio telescopes used will be the 34-meter (112-ft) diameter antennas that are part of NASA's Deep Space Network (DSN). The full-sky survey will be conducted by moving the radio telescope across the sky at a constant rate. However, in the direction of the sky where there are many stars, this survey will be conducted more slowly to improve sensitivity. This full-sky survey will take about 5 to 7 years to complete.

The second mode of the Microwave Observing project is a high-sensitivity targeted search that will look for weak signals originating near solar-type stars within 100 light-years from Earth. We know that life and technically-intelligent beings evolved around one such star (the Sun), and so these stellar targets appear to be the most favorable initial candidates for targeted SETI activities. The objective is to see if any extraterrestrial civilizations in the vicinity of nearby (on a galactic scale) solar-type stars are transmitting (or perhaps "leaking") electromagnetic signals that our contemporary radio telescopes, equipped with special SETI-instrumentation, are sensitive enough to detect. The largest radio telescopes available, such as the 305-meter (1,000-ft) diameter Arecibo antenna and the Deep Space Network's 64-meter (210-ft) diameter antennas, will be upgraded with special SETI instrumentation and used to search the candidate solar-type stars over a frequency range from 1,000 to 3,000 MHz, plus any accessible frequencies as high as 10,000 MHz. The frequency range covered will be much broader than previous SETI searches; but more significantly, this targeted search effort will be capable of detecting a much broader class of signal types than has been possible in the past. The proposed targeted search effort will look for narrowband signals that pulse on and off at regular intervals as well as those signals that are continuously present. This search will also be capable of recognizing artificial signals (those made intentionally by intelligent civilizations versus by natural cosmic phenomena), even if their frequencies are drifting in time. After the special automated SETI-instrumentation systems are installed at a number of radio observatories around the world, the detailed observations of about 1,000 targeted solar-type stars should take approximately five years.

And just what would a radio frequency signal from an intelligent extraterrestrial civilization look like? Figure 1 presents a spectrogram that shows a possible intelligent signal from outside our Solar System. This particular signal was actually sent by the *Pioneer 10* spacecraft from beyond the orbit of Neptune, and was received by a Deep Space Network radio telescope at Goldstone, California using a 65,000-channel spectrum analyzer. The three signal components are quite visible above the always-present background radio noise. The center spike appearing in figure 1 has a transmitted signal power of approximately one watt—about half the power of a miniature Christmas tree light. NASA's Microwave Observing project will be looking for a radio frequency signal that might appear this clearly, or for one that may actually be quite difficult to distinguish from the background radio noise. To search through this myriad

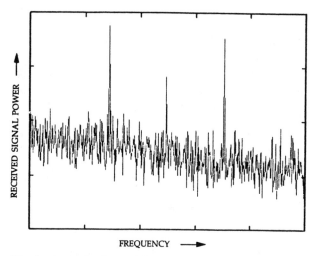

Fig. 1 A simulated signal from an extraterrestrial civilization (using the *Pioneer 10* spacecraft transmitting a signal from beyond the orbit of Neptune). (Courtesy of NASA.)

of signals, the NASA SETI program is developing state-of-the-art spectrum analyzers that will simultaneously sample millions of frequency channels and automatically identify candidate extraterrestrial signals for further observation and analysis.

One of the signs of maturity of planetary civilization is its concern for the cosmic question: "Are we alone?" Since the start of the Space Age, we have taken this question from the back of cereal boxes and placed the intriguing issue in the arena of legitimate scientific curiosity. The detection of but a single, clearly identifiable signal from an extraterrestrial alien civilization will have immeasurable impact on our own civilization and on our cosmic perspective. SETI represents an exciting human activity that we can engage in over the next century. A serious SETI program not only helps us mature as a planetary civilization but has the potential of positively answering one of the greatest mysteries of all.

If an alien signal is ever detected and decoded, then we would face another challenging question: Do we answer? How would you respond to a little green man's radio message? For the present time, SETI scientists are content to passively listen for intelligent signals coming to us across the interstellar void.

See also: **Arecibo Observatory; consequences of extraterrestrial contact; Drake equation; extraterrestrial civilization; Fermi paradox; interstellar contact; life in the Universe; Project Cyclops; Project Ozma; water hole**

Selenian Of or relating to the Earth's Moon. Once a permanent lunar base is established, a resident of the Moon.

Seyfert galaxy A type of spinal galaxy with a very bright central nucleus named after the American astronomer, Carl Seyfert, who first observed them in 1943. The bright nuclei of Seyfert galaxies appear to contain hot gases that are in rapid motion.

See also: **active galaxies; galaxy**

"shimanagashi" syndrome Will terrestrial immigrants to extraterrestrial communities suffer from the "shimanagashi" syndrome? During Japan's feudal period, political offenders were often exiled on small islands. This form of punishment was called "shimanagashi." Today, in many modern prisons one can find segregation or isolation units in which inmates who are considered troublemakers are confined for a period of time. Similarly, but to a lesser degree, mainlanders who spend a few years on an isolated island feel a strange sense of isolation—even though the island (such as Hawaii) may have large cities and many modern conveniences. These mainland visitors start feeling left out and even intellectually crippled despite the fact that life might be physically very comfortable there. Early extraterrestrial communities will also be relatively small and physically isolated from the Earth. However, electronic communications, including the transmission of books, journals and contemporary literature, could avoid or minimize such feelings of isolation. As the actual numbers of extraterrestrial settlements grows, physical travel between them could also reduce the sense of physical isolation.

solar cell Solar cells are proven direct energy conversion devices that have been used for over three decades to provide electric power for spacecraft. In a direct energy conversion (DEC) device electricity is produced directly from the primary energy source without the need for thermodynamic power conversion cycles involving the heat engine principle and the circulation of a working fluid. A solar cell or "photovoltaic system" turns sunlight directly into electricity. They have no moving parts to wear out and produce no noise, fumes or other polluting waste products.

The solar cell has been and will continue to be used extensively on board spacecraft and space platforms to provide electric power. However, many future missions, especially those operating far away from the Sun, or in periods of extensive darkness or shadowing, or requiring very large quantities of power (such as megawatts electric) from compact, mobile systems, may require the use of alternate forms of space power, such as space nuclear power.

See also: **Satellite Power System; space nuclear power**

solar constant The total amount of the Sun's radiant energy that normally crosses to a unit area at the top of the Earth's atmosphere (that is, at one astronomical unit from the Sun). The currently adopted value of the solar constant is 1371 ± 5 watts per square meter. The spectral distribution of the Sun's radiant energy resembles that of a blackbody radiator with an effective temperature of 5800 degrees Kelvin. This means that the majority of the Sun's radiant energy lies in the visible portion of the electromagnetic spectrum, with a peak value near 0.450 micrometer. (A micrometer is one millionth of a meter).

See also: **Sun**

solar mass The mass of our Sun, 1.99×10^{30} kilograms; it is commonly used as a unit in comparing stellar masses.

See also: **stars; Sun**

solar nebula The cloud of dust and gas from which the Sun, the planets and other minor bodies of the Solar System are postulated to have formed (condensed).

See also: **stars; Sun**

solar sail A proposed method of space transportation that uses solar radiation pressure to gently push a giant gossamer structure and its payload through interplanetary space (see fig. 1). As presently envisioned, the solar sail would use a large quantity of very thin reflective material to produce a net reaction force by reflecting incident sunlight. Because solar radiation pressure is very weak and decreases as the square of the distance from the Sun, enormous sails—perhaps 100,000 to 200,000 square meters—would be needed to achieve useful accelerations and payload transport.

The main advantage of the solar sail would be its long duration operation as an interplanetary transportation system. Unlike rocket propulsion systems that must expel their onboard supply of propellants to generate thrust, solar sails have operating times only limited by the effective lifetimes in space of the sail materials. The solar photons that do the "pushing" constantly pour in from the Sun and are essentially "free." This makes the concept of solar sailing particularly interesting for cases where we must ship large amounts of nonpriority payloads through interplanetary space—as for example, a shipment of special robotic exploration vehicles from Earth to Mars.

However, because the large reflective solar sail cannot generate a force opposite to the direction of the incident solar radiation flux, its maneuverability is limited. This lack of maneuverability along with long transit times represent the major disadvantages of the solar sail as a space transportation system.

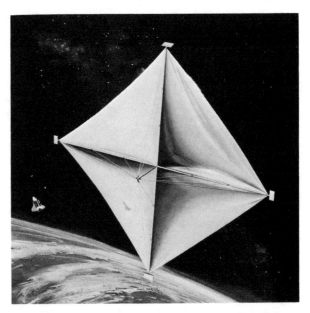

Fig. 1 The Space Shuttle is dwarfed by a giant solar sail that is being prepared for a comet intercept mission (artist's concept). (Courtesy of NASA.)

solar system When used in the lower case, the term refers to any star and its gravitationally bound collection of planets, asteroids and comets.

Solar System When used with capital letters, our Sun and the collection of celestial objects that are bound to it gravitationally. These celestial objects include the nine known major planets, over 60 moons, more than 2,000 minor planets or asteroids and a very large number of comets (see fig. 1). Except for the comets, all of these celestial objects orbit around the Sun in the same direction, and their orbits lie close to the plane defined by the Earth's own orbit and the Sun's equator.

The nine major planets can be divided into two general categories: (1) the terrestrial or Earth-like planets, consisting of Mercury, Venus, Earth and Mars; and (2) the outer or Jovian planets, consisting of the gaseous giants Jupiter, Saturn, Uranus and Neptune. Pluto is currently regarded

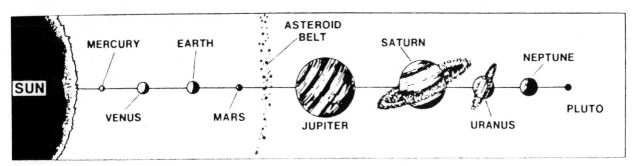

Fig. 1 The major components of our Solar System. (Note: from now until 1999, Pluto is actually *inside* the orbit of Neptune.) (Drawing courtesy of NASA.)

as a "frozen snowball," and along with its moon, Charon, appears to be in a class by itself. As a group, the terrestrial planets are dense, solid bodies with relatively shallow or no atmospheres. In contrast, the Jovian planets are believed to contain modest-sized rock cores, surrounded by layers of frozen hydrogen, liquid hydrogen and gaseous hydrogen. Their atmospheres also contain other gases such as helium, methane and ammonia.

See also: **asteroid; comet; Earth; Jupiter; Mars; Mercury; Neptune; Pluto; Saturn; Sun; Uranus; Venus**

solar wind The variable stream of electrons, protons and various atomic nuclei (such as alpha particles) that continuously flows outward from the Sun into interplanetary space.

See also: **Earth's trapped radiation belts; magnetosphere; Sun**

solipsism syndrome A psychological disorder that could happen to the inhabitants of space bases or space settlements. It is basically a state of mind in which a person feels that everything is a dream and not real. The whole of life becomes a long dream from which the individual can never awaken. A person with this syndrome feels very lonely and detached, and eventually becomes apathetic and indifferent. This syndrome might easily be caused in a space habitat environment where everything is artificial or man-made. To avoid or alleviate the tendency toward solipsism syndrome in space habitats, we would use large-geometry interior designs (that is, have something beyond the obvious horizon); place some things beyond the control or reach of the inhabitants' manipulation (that is, an occasional rainy day weather pattern variation or small animals that have freedom of movement); and provide growing things like vegetation, animals and children.

See also: **hazards to space workers; space settlement**

South Atlantic Anomaly A region of the Earth's trapped radiation particle zone that dips close to the planet in the southern Atlantic Ocean southeast of the Brazilian coast. This region represents the most important source of (ionizing) radiation for space travelers and workers in low Earth orbit (LEO).

See also: **Earth's trapped radiation belts; hazards to space workers**

Soviet Space Shuttle The Soviet counterpart to the U.S. Space Shuttle, a 100-ton reusable spaceship called *Buran* (snowstorm), made its maiden flight on November 15, 1988. *Buran's* first flight was totally automated (unmanned) and relied on ground controllers for on-orbit maneuvering. During this initial flight, *Buran* remained in space for two orbits of the Earth (about 3 hours and 25 minutes) and then used onboard computers to carry out an

automatic approach and landing at the Soviet Baikonur Cosmodrome (spaceport).

Although the Soviet Shuttle is nearly identical in physical appearance to that of its American cousin (see fig. 1), there are some significant differences in the two man-rated launch systems. The major difference is that *Buran* does not contain its own main liquid rocket engines (as does the U.S. Space Shuttle Orbiter vehicle). Instead, complete propulsion to low-Earth-orbit (LEO) for *Buran* is provided by the all-liquid-propellant Soviet Energia rocket system. The Soviet Shuttle vehicle does have, however, a set of small maneuvering thrusters to finalize its orbit and to deorbit.

The Soviets report that *Buran* and its future sisterships will be used to ferry people and payloads into low-Earth-orbit; and then to return both cosmonauts and payloads from space to a runway landing on Earth. The Soviets further report that *Buran* can take 30,000 kilogram (66,000 lb-mass) payloads into low-Earth-orbit and return from space with up to 20,000 kilograms (44,000 lb-mass) of cargo. Soviet aerospace officials also have stated that special-purpose *Buran* missions can last up to 30 days in space and that the reusable space vehicle can accommodate up to ten cosmonauts.

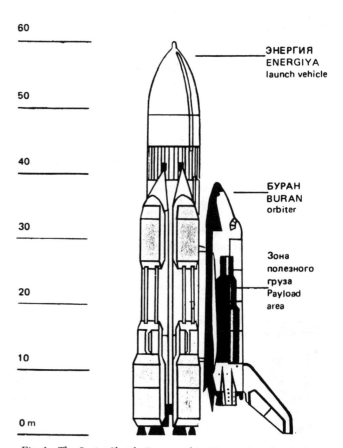

Fig. 1 The Soviet Shuttle *Buran* and its *Energia* launch vehicle. (Courtesy of Glavkosmos.)

See also: **space launch vehicles; Space Transportation System**

space base

space base A large, permanently inhabited space facility located in orbit around a celestial body or on its surface. The space base would have a much larger crew than a space station with from 10 up to perhaps 1,000 occupants.

Orbiting space bases can be built in modular fashion, using space station hardware as the building blocks; while bases on alien worlds could be expanded from an initial "seed complex," using extraterrestrial materials to the greatest extent possible. For example, a lunar base could use lunar regolith (soil) for shielding material to protect the occupants from both micrometeoroid hits and ionizing radiation.

Space bases in low Earth orbit would operate nominally at about 500-kilometer (310-mi) altitude at typical inclinations of either 28.5, 55 or 97 degrees. Such bases would be built up from initial space station complexes and would serve the following objectives: (1) support a long-term human presence in space for a much larger population of astronauts; (2) support a wide variety of operational Earth observation activities (such as climatology, crop forecasting, mineral prospecting, marine resource utilization, etc.); (3) conduct detailed scientific investigations of near-Earth space and other celestial objects within the Solar System; (4) develop and demonstrate the technology and operational capability for spacebased manufacturing and materials processing; (5) develop and demonstrate the engineering techniques needed for on-orbit construction and assembly of very large space structures; (6) service operational orbit transfer vehicles; and (7) prepare interplanetary payloads for their missions (such as the vehicle for a manned Mars mission sometime in the early 21st century).

Geostationary Earth orbit (GEO) is the preferred location for a number of information transfer, Earth observation and scientific sensor systems. GEO construction and maintenance bases might range in size from a small modular 8-person "work shack" to a larger, more extensive 50-person construction base.

Our return to the Moon will be marked by the development of lunar surface bases. These bases might include (1) a 6- to 10-person temporary surface base that provides life support for up to six months; (2) a permanent 10- to 20-person science and engineering technology base; and (3) a larger, permanent complex of laboratories, greenhouses, habitats and pilot factories for 100 to 1,000 lunar inhabitants. These lunar surface bases will demonstrate the technologies necessary for permanent habitation of the Moon and for full exploitation of its resources.

Manned expeditions to Mars in the next century will be complemented by the establishment of surface bases—at first temporary and then permanent. The permanent bases on the Red Planet will be our foothold in heliocentric space and serve as the trail head or staging point for manned missions to the main asteroid belt and the giant outer planets beyond.

See also: **lunar bases and settlements; Mars base; space settlement; space station**

space commerce Space is a challenging new place for business that will create many new jobs and new industries in the next century. Advanced space technologies can also influence terrestrial industries, enhancing productivity and stimulating creativity. Space also serves as a very powerful catalyst for seeking new knowledge and for better technical education. In February 1988 former President Reagan announced a major new National Space Policy and Commercial Space Initiative for the United States. His actions recognized the importance of a strong commercial space sector. This policy has helped redefine the relationship between the American government and industry with respect to space commercial initiatives, with greater reliance being placed on private-sector investment and innovation.

The prospect of doing business in space has actually been studied and discussed since the beginning of the Space Age. Direct economic benefits have already been demonstrated in space transportation, satellite communications and remote-sensing markets. Indirect benefits to terrestrial industries have been equally significant. Many space-related technologies have been applied to Earth-based industries, resulting in new and innovative techniques and processes that have improved the quality of production. With the establishment of operational space-based facilities, such as Space Station *Freedom*, new areas of economic potential, such as materials research and processing in space, will be identified and acted upon. Perhaps most significant of all, entirely new and unforeseen space markets are likely to emerge as our business-related activities in space increase.

Recent studies of existing or near-term worldwide markets for space-related goods or services have identified five major areas that now make up the field commonly referred to as *space commerce*. These areas are (1) space transportation, (2) satellite communications, (3) satellite remote sensing, (4) space-based industrial facilities and (5) materials research and processing in space.

Space transportation, which is essentially the foundation of all space-based activities, is a rapidly growing international industry. Commercial launch services range from suborbital sounding rocket flights for microgravity experimentation to the use of powerful multistage rocket vehicles for placing large communications satellites and other massive payloads into Earth orbit. Support services, such as payload processing, are also being developed into commercial operations and the development of new, highly automated "commercial" spaceports are being considered by business groups in the United States and abroad. For example, in the United States both Florida and Hawaii have active state-sponsored programs to develop commercial spaceports.

Satellite communications is now the most developed area of space commerce. Since their first operational deployments in the 1960s, communications satellites have become a vital element in the global telecommunications network. The communication satellite has quite literally "wired the world," permitting communications between points on land, sea and in the air, as well as instantaneous communications on a global basis. In 1989, for example, more than two thirds of the world's international voice communications and essentially all of the world's video programming were transmitted by satellite. Almost every nation on this planet has a membership in at least one international communications satellite organization. The manufacture of communications satellites and their associated ground-support facilities is a multibillion-dollar-per-year international industry. In addition, a substantial commercial infrastructure of peripheral enterprises—such as satellite-capacity resellers, domestic and international teleports and satellite-linked cable television networks—has also arisen.

Satellite remote sensing, a technique for examining the properties of Earth from space, has the potential of becoming a rapidly expanding area of space commerce. For example, the United States, France and the Soviet Union now have versatile commercial remote-sensing satellite systems in operation. Many other nations have also announced plans to launch and operate such systems in the early 1990s. The multispectral, remotely-sensed data from these space systems can be used in such important activities as: mineral exploration, agriculture, fishing, forestry, environmental monitoring and urban planning and land-use evaluation. The fastest-growing segment of this emerging space industry is the value-added product and service sector, which consists of companies that use advanced computer systems and software to accurately interpret these remotely-sensed data and then integrate the results with other data bases (for example, population data or power consumption data) to produce customized analytical and predictive products.

Space-based industrial facilities range from relatively simple unmanned (but man-tended) platforms to fully equipped orbiting laboratories capable of supporting manned operations in a "shirt-sleeve" environment. These space-based industrial facilities can be used for scientific research, technology development, spacecraft assembly and servicing, and microgravity-environment materials processing.

Materials research and processing in space (MRPS) is now perhaps the least-developed area of space commerce, but remains nevertheless one of the most promising. Materials research and processing on orbit involves exploiting the microgravity and high-vacuum characteristics of space. Space-based MRPS experiments have already led to significant advances in materials science, improved ground-based production methods and the identification of other industrial applications. Continued MRPS activities could result in as-yet-undiscovered commercially viable products that require space-based production (e.g., because of a need for extensive periods of low vibration, or microgravity). These candidate MRPS products include special pharmaceuticals and biotechnology materials, advanced electronic materials and devices and unique metal alloys.

It is interesting to close this entry on space commerce

and its great potential for impacting terrestrial marketplaces in the 21st century with the following "thought experiment." Compare a 1950s-era department store's "electronic goods" with those now available in a typical American shopping mall. You will, of course, be quite amazed by this merchandise comparison. Now let's peek ahead some 50 years into the future. How many and what type of made-in-space, made-on-the-Moon and imported-from-Mars products can you imagine? This is what space commerce might include in your very own lifetime!

See also: **materials research and processing in space; microgravity; remote sensing; space industrialization; space launch vehicles; space station**

space construction　Large structures in space, such as modular space stations, global communication and information services platforms and satellite power systems (SPSs), will all require on-orbit assembly operations by space construction workers. Figure 1 shows astronauts equipped with advanced spacesuit systems (that permit maneuvering) in the process of constructing and aligning a large space structure.

Space construction requires protection of the work force and some materials from the hard vacuum, intense sunlight and natural radiation environment encountered above the Earth's protective atmosphere. Outer space, however, is also an environment that in many ways is ideal for the construction process. First, because of the absence of significant gravitational forces (that is the microgravity experienced by the free-fall condition of orbiting objects), the structural loads are quite small, even minute. Structural members may, therefore, be much lighter than terrestrial structures of the same span and stiffness. Second, the absence of gravitational forces greatly facilitates the movement of material and equipment. On Earth, the movement

Fig. 1　Space workers in advanced spacesuit maneuvering units performing on-orbit assembly operations. (Courtesy of NASA.)

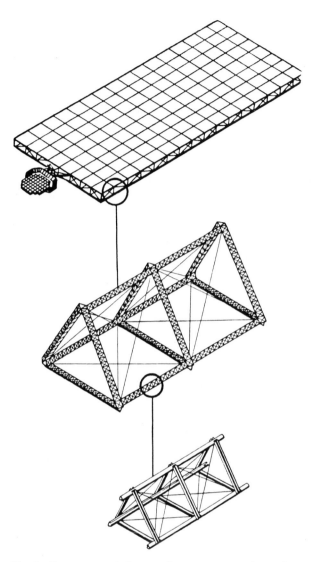

Fig. 2 Basic structural elements for space construction projects, such as a satellite power system. (Courtesy of NASA.)

for launch in the form of tightly wound rolls and spools. Since repetitive operations are more easily automated, regular uniform cross-section structural members are being considered for building large space structures. For example, figure 2 shows the basic satellite power system structural elements.

A process similar to the terrestrial roll-forming of light sheet metal members has been adapted to produce these structural members at a rapid pace using an automated beam builder. This machine consists of roll-forming, heating and cooling, and ultrasonic welding components that function to "produce" a finished beam. These basic structural members may be produced from rolls of aluminum strip stock or from a graphite fiber/thermoplastic-impregnated roll. The latter space construction material has the advantage of having a low coefficient of thermal expansion. The basic structural element, a triangular cross-section shape, may be assembled into primary-structure triangular trusses. These members are then assembled into a trussed box structure.

The Space Shuttle and Space Station *Freedom* can be used to explore a variety of space construction techniques, including erection, automated fabrication methods and deployment. As a matter of fact, automated fabrication is believed to be a key requirement for viable space construction activities. Table 1 identifies some typical space construction hardware and supporting systems. It should be quite obvious from these listings that working in space will require the close interaction of astronaut and very smart machine. Such smart machines will in turn also trigger a robotics technology explosion on Earth.

The advanced maneuverable spacesuit will be a versatile, self-contained, life-supporting backpack with gaseous nitrogen-propelled jet thrusters that enable a space worker to travel back and forth to various space construction locations. The automated beam builder is a machine designed for fabricating "building-block" structural beams in space. Combined with a space structure fabrication system, the beam builder allows space workers to manufacture and assemble intricate structures in low Earth orbit using the

of material during a construction operation absorbs a large portion of the total work effort expanded by construction personnel and their machines. Third, the absence of an atmosphere, with its accompanying wind loads, inclement weather and unpredictable change, permits space work to be accurately planned and readily executed without environmental interruptions (except, perhaps due to solar flares, which would increase the radiation hazard).

In order to minimize transportation costs for large space construction projects made with terrestrial materials, the construction materials shipped from Earth should be packaged in a very dense form. The desire to make very low density, lightweight systems from high-density materials leads space technologists and engineers to consider automated, on-orbit fabrication techniques. Such techniques will allow the materials payloads to be densely packaged

Table 1 Typical Space Construction Equipment and Supporting Systems

- Advanced maneuverable spacesuits
- Automated beam builder
- Space structure fabrication system
- Remote astronaut work stations
  "Closed cherry picker"
  "Open cherry picker"
  Free-flyer work station
- Manned orbital transfer vehicle (MOTV)
- Cargo orbital transfer vehicle (COTV)
- Advanced launch vehicles (for personnel and cargo)

SOURCE: Author.

Space Shuttle on Space Station *Freedom* as an early construction camp. Eventually, permanent space bases in LEO will be used to serve as construction sites for even more complicated space-assembled and manufactured systems. Remote astronaut work stations can be mounted on the end of the Shuttle Orbiter's remote manipulator system or on manipulator arms attached to the Space Station. These "open cherry pickers" would have a convenient tool and parts bin, a swing-away control and display panel and lights for general and point illumination. The closed version of this cherry picker would involve a pressurized manned remote work station that contains life-support equipment and controls and displays for operating dexterous manipulators.

The cargo orbital transfer vehicle (COTV) and the manned orbital transfer vehicle (MOTV) are used to move materials and personnel between various orbital locations—for example, from low Earth orbit to geosynchronous orbit. COTV concepts include a chemically fueled, high-thrust system with a short trip time and a high degree of reusability; a nuclear rocket COTV featuring a propulsion system capable of high thrust and high specific impulse; and an electric propulsion COTV (either solar or nuclear powered) for non-time critical transfer of massive quantities of cargo or for the gentle movement of very delicate structures. One MOTV design involves a reusable chemical propellant (probably liquid oxygen/liquid hydrogen, abbreviated LOX/LH$_2$) vehicle for the rapid, reliable transfer of space workers between orbital locations. Typical configurations can accommodate up to 160 passengers and 100 metric tons of supplies. Advanced launch vehicles will carry space workers and their personal equipment from the Earth's surface to low Earth orbit. Other, heavy-lift-capability, advanced launch vehicles will be needed to haul massive quantities of cargo from a terrestrial spaceport to a space station on space construction site in LEO. Space construction based on terrestrial raw materials will depend on making these freight-hauling costs as low as possible—or else we will have to also explore the use of extraterrestrial materials as may be found on the Moon or Earth-approaching asteroids.

Two particular orbital locations have been studied for primary space construction activities: geostationary orbit and low Earth orbit. Construction in geostationary orbit (GEO) offers the advantage of essentially continuous sunlight, thereby decreasing the differential thermal effects on a large space structure during construction and reducing the need for artificial illumination of the construction site. Remember, a construction site in low Earth orbit would travel around the Earth experiencing night and day cycles every hundred minutes or so. Additionally, at GEO the construction process can be designed to produce the large structure in its final operational form. This avoids the requirement to transport a large space system from LEO to its operational location at GEO.

Construction in LEO, however, does provide the advantage of operating in close proximity to the Earth. Space construction workers and their facilities at LEO are afforded additional protection from the space radiation environment by the Earth's magnetosphere and by being below the Earth's trapped radiation belts (also called the Van Allen belts). Gravitational attraction varies with altitude above the Earth, so that very large structures built in LEO must overcome relatively large gravity gradient torques during construction and transport to GEO. To lessen these gravity-induced torques, the structure could be built in LEO in modules, with the final assembly occurring at GEO.

Construction activities in outer space impose a need for protecting the space worker from the environmental hazards encountered there. Provisions for a life-supporting atmosphere and protection from solar thermal and ultraviolet radiation are well-understood engineering problems. Protection against the space radiation environment (such as energetic particles) requires that additional knowledge be gathered on the anticipated dynamic range of these phenomena at various potential space construction locations, so that engineering measures to shield space workers can be most effective from an overall protection and cost point of view. Radiation exposure standards for space workers must also be established to identify the necessary protective measures required for space work stations, habitats and orbital transfer vehicles.

See also: **hazards to space workers; large space structures; robotics in space; Satellite Power System; space industrialization**

spacecraft charging In orbit or in deep space, spacecraft and space vehicles can develop an electric potential up to tens of thousands of volts relative to the ambient extraterrestrial plasma (the solar wind). Large potential differences (called "differential charging") can also occur on the space vehicle. One of the consequences is electrical discharge or arching, a phenomenon that can damage space vehicle surface structures and electronic systems. Many factors contribute to this complex problem including the spacecraft configuration, the materials from which the spacecraft is made, whether the spacecraft is operating in sunlight or shadow, the altitude at which the spacecraft is performing its mission and environmental conditions such as the flux of high energy solar particles and the level of magnetic storm activity.

Wherever possible, spacecraft designers use conducting surfaces and provide adequate grounding techniques. These design procedures can significantly reduce differential charging, which is generally a more serious problem than the development of a high spacecraft-to-space (plasma) electrical potential.

See also: **hazards to space workers**

space debris Space junk or derelict man-made space objects in orbit around the Earth. Space debris represents a hazard to astronauts, spacecraft and large space facilities like a space station.

Since the start of the Space Age in 1957, the natural meteoroid environment has been a design consideration for spacecraft. Meteoroids are part of the interplanetary environment and sweep through Earth orbital space at an average speed of about 20 kilometers per second. Space science data indicate that, at any one moment, a total of approximately 200 kg of meteoroid mass is within some 2,000 kilometers of the Earth's surface—the region of space (called low Earth orbit or LEO) most frequently used. The majority of this mass is in meteoroids about 0.01 cm diameter; however, lesser amounts of this mass occur in meteoroid sizes both smaller and larger than 0.01 cm. The natural meteoroid flux also varies in time as the Earth travels around the Sun.

Man-made space debris is often referred to as *orbital debris*, and differs from natural meteoroids because it remains in Earth orbit during its lifetime and is not a transient phenomena like the meteoroid showers that occur as the Earth travels through interplanetary space around the Sun.

The estimated mass of man-made objects orbiting the Earth within about 2,000 kilometers of its surface is approximately 3 million kilograms (or about 15,000 times more mass than represented by the natural meteoroid environment). These man-made objects are for the most part in high-inclination orbits and pass one another at an average relative velocity of 10 kilometers per second (about 22,000 miles per hour). Most of this mass is contained in about 3,000 spent rocket stages, inactive satellites and a comparatively few active satellites. A smaller amount of the man-made space debris mass (about 40,000 kg) is distributed in the remaining 4,000 orbiting objects currently being tracked by space surveillance systems.

The large majority of these smaller space debris objects are the resultant by-products of over 130 on-orbit fragmentations (satellite breakup events). Recent ground-based telescope measurements of orbital debris combined with detailed analysis of the hypervelocity impact pits (caused by tiny pieces of man-made debris) as found on the surfaces of returned-to-Earth portions of the *Solar Max* satellite and the recently retrieved *Long Duration Exposure Facility* (LDEF) indicate a total mass of at least 1,000 kg for orbital debris sizes of 1 cm or smaller, and about 300 kg for orbital debris smaller than 0.1 cm. In fact, the orbital debris environment is now considered more hazardous than the natural meteoroid environment to spacecraft operating in Earth orbit below an altitude of 2,000 km.

Unfortunately, precise data concerning the contemporary space debris environment is extremely limited, because today's space surveillance systems cannot effectively track orbiting objects smaller than 10 cm in diameter. For example, the U.S. Space Command's Space Surveillance Network (SSN) was not designed to track small debris particles (less than 10 cm diameter).

The nerve center of the U.S. Space Command's space surveillance mission is the Space Surveillance Center (SSC) located deep inside Cheyenne Mountain near Colorado Springs, Colorado. A computer system (linked to a worldwide network of interactive radars and optical tracking devices) keeps a constant record of the movements of thousands of man-made objects orbiting the Earth. The Space Surveillance Center maintains a computerized catalog of orbiting satellite payloads and pieces of space debris. The SSC charts the present position of these orbiting objects and plots their anticipated future orbital paths. For larger space objects, the Space Surveillance Center also forecasts the general time and location of where they might reenter the Earth's atmosphere. Personnel at the SSC can generate ground traces that show the actual or anticipated path a satellite is following as projected on a map of the world.

The Space Surveillance Center's satellite catalog dates back to 1957, when the Soviet Union opened the Space Age with the launch of *Sputnik 1*. Since then, more than 18,300 objects have been cataloged. Of these objects, over 7,000 now remain in orbit, ranging in size from a wrench "dropped" by a Space Shuttle astronaut during an extra-vehicular activity (EVA) to massive spacecraft that weigh several tons. Objects no longer in space are logged at the SSC as having "decayed" or "de-orbited." A decayed satellite (or piece of orbital debris) is one that reenters the Earth's atmosphere under the influence of natural forces and phenomena (e.g., atmospheric drag). Space vehicles and satellites that are intentionally removed from orbit are said to have been "de-orbited." For example, the Space Shuttle is de-orbited after each space mission, so that it can be serviced and used again.

As previously mentioned, the Space Surveillance Center also monitors satellites (especially large ones) when they are about to reenter the Earth's atmosphere. One of the most celebrated reentries of a large man-made object occurred on July 11, 1979, when the then decommissioned and abandoned first American space station, called *Skylab*, came plunging back to Earth over Australia and the Indian Ocean—a somewhat spectacular reentry event that nevertheless occurred without harm to life or property. In fact, although man-made objects reenter from orbit on the average of more than one per day, only a very small percentage of these reentry events result in debris surviving to the Earth's surface. The aerodynamic forces and heating associated with reentry processes usually breaks up and vaporizes most of the incoming space debris. Of those few objects that do survive their fiery plunge through the Earth's atmosphere, most impact in the oceans or on remote, essentially uninhabited or only sparsely populated land areas.

There are two general types of orbital debris of concern: (1) large objects (greater than 10 cm in diameter) whose population, while small in absolute terms, is large relative to the population of similar masses in the natural meteoroid environment; and (2) a much greater number of smaller objects (less than 10 cm diameter), whose size distribution approximates the natural meteoroid population and whose numbers add to the "natural debris" environment in those size ranges. The interaction of these two general classes of space debris objects, combined with their long residence times in orbit, creates further concern that there will inevitably be new collisions producing additional fragments and causing the total space debris population to grow.

In discussing space debris, the space around the Earth is generally divided into three orbital regimes: (1) low Earth orbit (LEO), which is defined by objects orbiting the Earth at less than 5,500 kilometers altitude; this corresponds to objects with orbital periods of less than 225 minutes; (2) medium Earth orbit (MEO), which is defined by objects orbiting the Earth between the LEO and the geosynchronous Earth orbit regime; and (3) geosynchronous Earth orbit (GEO), which is defined by objects orbiting the Earth at an altitude of approximately 35,860 kilometers; this corresponds to objects with an orbital period of approximately 24 hours.

As shown in table 1, the greatest number of tracked space objects can be found in LEO, followed by GEO, with the remaining objects residing in MEO. Figure 1 presents a typical altitude distribution of tracked space objects (limited to objects greater than 10 cm in diameter) in LEO up to an altitude of 2,000 km. In this figure the average number of objects at any one time found in a 10-kilometer-altitude band is plotted against altitude. The peak space object density occurs near 800 kilometers altitude, where the density is about 200 objects in a 10-kilometer altitude band. In the 350- to 500-kilometer altitude region, the operational locations of Space Station *Freedom*, the space object density is approximately 20 to 50 objects in a 10-kilometer altitude band.

Extrapolation from the tracked space objects, evaluation of ground-based radar and optical observations and detailed examination of the surfaces of various pieces of space hardware returned to Earth (e.g., LDEF) have led to current predictions that the 7,000 tracked objects represent less than one percent of the total orbital debris population. The data presented in table 2 provides a estimation of the actual space debris population from both a numerical and mass-on-orbit perspective.

Contemporary computer simulations of the space debris

ALTITUDE DISTRIBUTION OF OBJECTS IN LEO

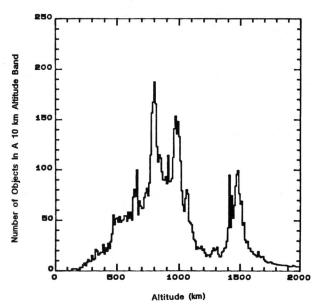

Fig. 1   Altitude distribution of objects in low Earth orbit (LEO) as of August 1, 1988. (Courtesy of U.S. Government/Interagency Group [Space].)

population predict that there are about 17,500 objects in the 1- to 10-cm diameter range (representing about 0.5% of the total space debris population) and approximately 3.5 million objects between 0.1 and 1.0 cm in diameter (representing about 99.3% of the total space debris population).

The explosion or fragmentation of a large space object has the potential of producing a large number of smaller objects, objects too small to be detected by contemporary ground-based space surveillance systems. This is especially true in the case of high-intensity explosions or in explosions involving payloads designed to break up into some particular size debris. For example, it is theoretically possible for a single 100-kilogram-mass payload to break up into 100,000 one-cm-size objects or perhaps into 100 million (10^8) 0.1 cm-size objects. A breakup due to a typical hypervelocity collision involving a 100-kilogram-mass space object might create somewhat fewer debris objects—perhaps 10,000 one-cm-size objects or about one million (10^6) 0.1-cm-size objects. Finally, low-intensity explosions might produce about 1,000 objects of either 0.1 or 1.0 cm size.

An orbiting object loses energy through frictional encounters with the upper limits of the Earth's atmosphere and as a result of other orbit-perturbing forces (e.g., gravitational influences). Over time, the object falls into progressively lower orbits and eventually makes a final plunge towards Earth. Once a space object enters the sensible atmosphere, atmospheric drag will slow it down rapidly and cause it to either completely burn up or fall through the atmosphere and impact on the Earth's surface or in its oceans.

Table 1   Tracked Space Objects By Altitude Regime

Earth–Space Location	LEO	MEO	GEO	Total
Total number of space objects tracked	5,923	683	453	7,059

SOURCE: Based on data as of August 1, 1988, from U.S. Government/Interagency Group (Space).

Table 2   Estimated Space Debris Population

Size	No. Objects	% by No.	Mass on Orbit	% by Mass
>10 cm	7,000	0.2%	2,999,000 kg	99.97%
1–10 cm	17,500	0.5%	1,000 kg	0.03%
<0.1–1 cm	3,500,000	99.3%		
TOTAL	3,524,500	100%	3,000,000 kg	100%

SOURCE: Based on data as of August 1, 1988, from U.S. Government/Interagency Group (Space).

For example, in LEO, unless constantly reboosted, satellites in circular orbits at altitudes between 200 and 400 km will reenter the atmosphere within a few months. For objects orbiting at altitudes between 400 and 900 km, orbital lifetimes can exceed a year or more depending upon the actual mass and surface area of the satellite. A glass marble, for example, in a 500-km altitude circular orbit will most likely stay in orbit for about one year; but if it were at 800-km altitude, it would stay up for about 30 years. For objects in circular orbits at altitudes above 900 km, orbital lifetimes will be 500 years or more. Orbital lifetimes for objects in elliptical orbits can vary significantly from the lifetimes of objects in circular orbits. The lower the perigee (closest approach) altitude of an eliptical orbit, the greater the influence of atmospheric drag on the object and the shorter the overall orbital lifetime (without reboost).

The natural decay of Earth-orbiting objects is also greatly affected by solar activity. High levels of solar activity heat the Earth's upper atmosphere, causing it to expand further into space and to reduce the orbital lifetimes of space objects found at somewhat higher altitudes in the LEO regime. However, above 600 km altitude, the atmospheric density is sufficiently low and solar-activity-induced atmospheric density increases do not noticeably affect the debris population lifetimes. This solar-cycle-based natural cleansing process for space debris in LEO is extremely slow and by itself cannot offset the present rate of man-made space debris generation.

The effects of orbital debris impacts on spacecraft and space facilities depend on the velocity and mass of the debris. For debris sizes less than approximately 0.01 cm in diameter, surface pitting and erosion are the primary effects. Over a long period of time, the cumulative effect of individual particles colliding with a satellite could become significant, because the number of such small debris particles is very large in LEO.

For debris larger than about 0.1 cm in diameter, the possibility of structural damage to a satellite or space facility becomes an important consideration. A 0.3-cm-diameter sphere of aluminum traveling at 10 kilometers per second, for example, has about the same kinetic energy as a bowling ball traveling at 100 kilometers per hour (60 miles per hour). Aerospace engineers anticipate significant structural damage to the satellite or space facility if such an impact occurs.

Space system engineers, therefore, find it helpful to distinguish three space debris size ranges in designing spacecraft. These are (1) debris sizes 0.01 cm diameter and below, which produce surface erosion; (2) debris sizes ranging from 0.01 to 1.0 cm diameter, which produce significant impact damage that can be quite serious; and (3) space debris objects greater than 1.0 cm diameter, which can readily produce catastrophic damage in a satellite or space facility.

Today, only about five percent of the cataloged objects in Earth orbit are active, operational spacecraft. The remainder of these space objects represents various types of orbital debris. Space debris is often divided up into four general categories: (1) operational debris (12%), objects intentionally discarded during satellite delivery or satellite operations; this category includes lens caps, separation and packing devices, spin-up mechanisms, payload shrouds, empty propellant tanks, and a few objects discarded or "lost" during EVA by astronauts and cosmonauts; (2) spent and intact rocket bodies (14%); (3) inactive (decommissioned or dead) payloads (20%); and (4) fragmentation (on-orbit satellite breakup) (49%). (See table 3.)

Since the first recognized fragmentation event in June, 1961, over 130 objects (including payloads, rocket bodies and other debris) have experienced on-orbit breakups. On-orbit fragmentations may result from explosions or collisions, and may be accidental or deliberate. A space object, for example, may be deliberately destroyed by an explosive charge as part of a spacecraft test; or a rocket stage may suffer a catastrophic propulsion failure leading to an explosion. At present, catastrophic collisions with space debris appear to be a less common cause for fragmentation events, although a few candidate cases are still being analyzed. Table 4 describes several satellite breakup events.

Collisions between objects in low Earth orbit are expected to occur at an average relative velocity of 10 kilometers per second. At these encounter velocities, the impact shock wave creates such high temperature and large internal pressures in the colliding materials that they melt and produce millions of tiny particles.

Very small orbital debris particles (less than 0.05 cm in diameter) are created by the disintegration or decomposition of spacecraft surfaces (e.g., paint flaking, plastic erosion and metal erosion) and by the firing of solid-propellant upper-stage rocket motors, which produce aluminum oxide particles. Thousands of kilograms of aluminum oxide dust are introduced each year into the near-Earth space environment as a result of solid rockets fired to transfer payloads from low Earth orbit to geostationary orbit or into interplanetary trajectories. A single solid rocket motor firing can place some 2,000 to 12,000 kilograms of aluminum oxide dust (billions of tiny particles) in space. Fortunately, since typical transfer orbits are elliptical, most of these tiny aluminum oxide particles quickly reenter the Earth's atmosphere under the influence of atmospheric drag encountered during perigee. But a small fraction of these particles

Table 3   Sources of Tracked Space Objects (By Altitude Regime)

	Active/Inactive Spacecraft	Rocket Bodies	Fragmentary and Other	Total Debris
LEO	1,134	651	4,138	5,923
MEO	232	302	149	683
GEO	329	123	1	453
TOTAL	1,695	1,076	4,288	7,059[a]

[a] 472 tracked objects pending entry in the catalog.

SOURCE: Based on data as of August 1, 1988, from U.S. Government/Interagency Group (Space).

Table 4   Selected On-Orbit Fragmentation Events

Common Name	International Designation	Satellite Catalog No.	Launch Date	Breakup Event Date	Fragments Cataloged upon Breakup	Fragments Currently Tracked in Orbit	Probable Cause
Transit 4A	1961-Omicron	119	6/29/61	6/29/61	292	227	Unknown
Nimbus 4	1970-025	4367	4/8/70	10/17/70	356	303	Unknown
NOAA 3	1973-086	6921	11/6/73	12/28/73	195	179	Unintentional explosion
Pageos 1	1966-056	2253	6/24/66	7/12/75	82	13	Unknown (possible collision)
Cosmos 777	1975-102	8416	10/29/75	1/26/76	63	0	Intentional explosion
Spot-1	1986-019	16615	2/22/86	11/13/86	475	443	Unintentional explosion

SOURCE: Satellite breakup status as of July 20, 1988, based on data from U.S. Government/Interagency Group (Space).

remains in orbit, representing a growing threat for surface erosion and contamination effects on spacecraft in low Earth orbit.

The disintegration of spacecraft exterior paints is now believed to be caused by atomic oxygen-induced erosion of the organic binder of the paint. Paint flaking represents another major source of tiny debris particles in LEO.

Aerospace engineers must, therefore, carefully consider the growing space debris problem when they design future space systems. They must make future satellites and space facilities as "litter-free" in design as possible, and should also design the systems for retrieval or removal at the termination of their useful operations. We may also see the start of an exciting new extraterrestrial career field—space environmental engineer (or space refuse collector)! If the space debris problem gets too much out of hand in the 21st century, telerobotic "extraterrestrial garbage trucks" might be used to sweep orbital regions and to clear cislunar trajectories of such collision hazards.

See also: **hazards to space workers; meteoroids; orbits of objects in space**

space industrialization   A new wave in humanity's sociotechnical development in which the special environmental conditions and properties of outer space are used for the economic and social benefit of all people on Earth. Some interesting properties of cislunar space include hard vacuum, microgravity, low vibration levels (in orbiting spacecraft), a wide-view angle of Earth and the heavens and complete isolation from the terrestrial biosphere. Contemporary studies have suggested long-term space industrialization opportunities in the following four general categories: (1) information services; (2) "made-in-space" products; (3) energy; and (4) human activities (including space therapeutics and tourism).

See also: **extraterrestrial resources; materials research and processing in space; space commerce**

Space Infrared Telescope Facility (SIRTF)   Originally planned as a *Spacelab* facility, NASA's Space Infrared Tele-

scope Facility has been redesigned to become the fourth Great Observatory. The SIRTF will provide astronomers unprecedented spatial and spectral resolution in the infrared portion of the electromagnetic spectrum. To achieve this resolution, the facility's complement of four science instruments will be cryogenically cooled. This long-lived, meter-class infrared observatory will study the cold regions of space. Specifically, SIRTF will investigate (1) the locations in space where the cosmic gas and dust condense into stars; (2) the "cool" objects in our Solar System, such as asteroids, comets and planetary systems; and (3) infrared-emitting extragalactic objects. NASA currently plans to launch the SIRTF with the Space Shuttle and then to perform astronaut-assisted on-orbit servicing of this platform using the Shuttle and Space Station *Freedom*.

See also: **astrophysics, Great Observatories, infrared astronomy, Space Transportation System**

space launch vehicles   How do we get people and cargo off the planet Earth and into the extraterrestrial environment of outer space? The answer at present: Use rocket-propelled launch vehicles. Whatever space mission is undertaken, the vehicle carrying the payload is currently propelled into space by rocket power. Rockets that are used only once to carry payloads or people into space are called expendable launch vehicles (ELVs). With the introduction of the U.S. Space Shuttle fleet in the early 1980s, followed by the Soviet Shuttle (the *Buran*) in late 1988, a new class of *reusable* aerospace vehicles entered the space launch business. An *aerospace vehicle* is capable of operating both within a planet's atmosphere and in outer space.

Many aerospace experts believe that today's biggest stumbling-block to the rapid development of a Solar System civilization is the high cost of access to space. For example, to put a kilogram of anything into low Earth orbit (LEO) presently costs between $6,600 and $13,200 (or about $3,000 to $6,000 per pound)! A good deal of research and engineering needs to be done in the next decade to design launch systems that significantly lower the transportation cost of getting into space. Some aerospace planners believe that by the year 2000 advanced design space launch vehicles

Table 1 United States Launch Vehicles

Launch Vehicle	Type[a]	Capabilities (Approx.)[a]
Scout	ELV	210–260 kg to LEO
Delta II	ELV	4,000–5,000 kg to LEO (various configurations)
Atlas I	ELV	5,700–5,900 kg to LEO
Atlas II	ELV	6,600 kg to LEO
Titan II	ELV	1,900 kg to LEO (sun-synchronous orbit)
Titan III	ELV	14,500 kg to LEO
Titan IV	ELV	18,000 kg to LEO
Space Shuttle	Aerospace vehicle	22,000 kg to LEO
Pegasus	Aircraft-launched ELV	200–400 kg in LEO; winged rocket launched by "mother" plane (e.g., 747 or B-52)

[a]ELV, expendable launch vehicle; LEO, low Earth orbit.

SOURCE: Developed by author from NASA, DOD, Department of Commerce and (Congressional) Office of Technology Assessment data.

might lower the cost of access to LEO to about $220 per kilogram (or about $100 per pound).

Once placed in LEO by the launch vehicle, payloads destined for higher-altitude locations, such as geosynchronous (GEO) orbit, require an additional propulsive "kick," which is normally provided by an upper-stage propulsion device. Upper-stage systems now used by the United States include the Payload Assist Module (PAM), Inertial Upper Stage (IUS) and Transfer Orbit Stage (TOS). PAM (which is available in several configurations) is designed to boost satellites deployed in LEO to higher altitudes. For example, PAM-D is capable of launching satellite payloads weighing up to 1,250 kilograms (2,750 lb-mass), PAM-DII up to 1,890 kilograms (4,150 lb-mass), and PAM-A up to 2,000 kilograms (4,400 lb-mass). Similarly, the U.S. Air Force has developed the Inertial Upper Stage (IUS), a two-stage, solid-propellant rocket vehicle capable of boosting heavier payloads to higher altitudes. For example, an IUS can transfer up to 2,270 kilograms (5,000 lb-mass) from a Space Shuttle in LEO to GEO, or transfer 1,910 kilograms (4,200 lb-mass) from LEO to GEO when used with a Titan

Table 2 Selected Foreign Launch Vehicles

Country	Launch Vehicle	Type[a]	Capabilities (Approx.)[a]
European Space Agency (ESA)	Ariane 4	ELV	Up to 8,000 kg to LEO (various configurations)
	Ariane 5	ELV	20,000 kg to LEO (available late 1990s)
	Ariane 5/Hermes	Aerospace vehicle	Planned aerospace vehicle configuration for late 1990s
Japan	H-1	ELV	3,000 kg to LEO
	H-2	ELV	8,000–9,000 kg to LEO (available mid-1990s)
	H-2/HOPE	Aerospace vehicle	Planned aerospace vehicle configuration for late 1990s
Soviet Union	Vostok	ELV	5,000 kg to LEO
	Molniya	ELV	7,000 kg to LEO
	Tsyklon	ELV	4,000 kg to LEO
	Zenit	ELV	13,700 kg to LEO
	Energia	ELV	100,000 kg to LEO
	Energia/Buran	Aerospace vehicle	30,000 kg to LEO (Soviet Space Shuttle)
Israel	Shavit	ELV	50–100 kg to LEO
India	Satellite Launch Vehicle-3	ELV	35–40 kg to LEO
Peoples Republic of China	Long March 3	ELV	1,400 kg to GTO
	Long March 4	ELV	2,400 kg to LEO (sun-synchronous)

[a]ELV, expendable launch vehicle; GTO, geostationary transfer orbit; LEO, low Earth orbit.

SOURCE: Developed by author from NASA, Department of Commerce and (Congressional) Office of Technology Assessment data.

Table 3   The World's Major Launch Centers

Country	Launch Site	Latitude	ELV Support Capability	Orbital Inclination	Status
USA	Cape Canaveral, Florida	28.5 N	All	Equatorial	Operational
USA	Vandenberg Air Force Base, California	34.7 N	All	Polar	Operational
USA	Wallops Island, Virginia	37.9 N	Suborbital/small	Equatorial	Operational
USA	Palima Point, Hawaii	19.0 N	All	Equatorial/polar	Planning stage
Australia	Woomera Launch Facility	31.1 S	Suborbital	Suborbital	Operational
Australia	Cape York, Queensland	11.0 S	All	Equatorial/polar	Planning stage
Brazil	Alcantara, Maranhao State	2.0 S	Suborbital/small	Equatorial/polar	Under construction
France	Guiana Space Center, Kourou, French Guiana	5.2 N	All	Equatorial/polar	Operational
India	Sriharikota Launch Range, Sriharikota Island	13.9 N	Suborbital/small	Equatorial/polar	Operational
India	Thumba Equatorial Rocket Launching Station	8.0 N	Suborbital	Suborbital	Operational
Italy	San Marco, Kenya	2.9 S	Suborbital	Equatorial	Operational
Japan	Osaki Launch Site, Tanegashima	30.4 N	Suborbital/small	Equatorial/polar	Operational
Japan	Takesaki Launch Site, Tanegashima	30.4 N	All	Equatorial/polar	Operational
PRC[a]	Xi Chang, PRC	28.1 N	All	Equatorial	Operational
PRC[a]	Shuang Cheng-Tzu	40.6 N	Suborbital/small	Equatorial	Operational
Sweden	Esrange, Kiruna	68.0 N	Suborbital	Suborbital	Operational
USSR	Tyuratam	45.6 N	All	Equatorial/polar	Operational
USSR	Plesetsk	62.8 N	All	Equatorial/polar	Operational
USSR	Kapustin Yar	48.4 N	Suborbital/small	Equatorial/polar	Operational

[a] PRC, People's Republic of China.

SOURCE: U.S. Dept of Commerce (1988).

III expendable launch vehicle. In 1989, an IUS deployed from the Space Shuttle *Atlantis* successfully sent the *Magellan* spacecraft on its 15-month interplanetary trajectory to Venus. Finally, the Transfer Orbit Stage (TOS) was developed by Orbital Services Corporation for use in placing satellite payloads weighing between 2,730 and 5,900 kilograms (6,000 and 13,000 lb-mass) into geosynchronous transfer orbit (GTO) or other high-energy trajectories. TOS can be launched from either the Space Shuttle or an expendable Titan III or IV launch vehicle. NASA will use a TOS upper stage to send the *Mars Observer* spacecraft on its interplanetary journey to the Red Planet in 1992.

The Tethered Satellite System, a cooperative program between NASA and Italy, is a reusable facility that consists of a satellite attached by a strong but lightweight cable to a deploying mechanism that is mounted on a pallet in the Space Shuttle Orbiter's cargo bay. The tethered satellite can weigh up to 500 kilograms (1,100 lb-mass) and may be deployed "downward" (toward the Earth) or "upward" (away from the Earth) to distances from the Space Shuttle of up to 100 kilometers (62 mi). In addition to opening up previously inaccessible regions of the Earth's upper atmosphere to extended scientific investigation, the Tethered Satellite System allows aerospace engineers to demonstrate *propulsionless* reboost and orbital transfer of objects in space.

Finally, in support of future space missions to lunar and planetary bases, NASA is now examining the characteristics of an advanced orbital transfer vehicle, called the Space Transfer Vehicle (STV). An evolutionary STV would be capable of placing large automated planetary payloads on Earth-escape trajectories, transporting humans and cargo to and from the Moon, and supporting human expeditions to Mars.

See also: **National Aerospace Plane; orbital transfer vehicle; Soviet Space Shuttle; Space Transport System**

space law   Space law is basically the code of international law that governs the use and/or control of outer space by different nations on Earth. There are four international agreements, conventions or treaties that currently govern space activities: (1) Treaty on Principles Governing the Activities of States in the Exploration and Use of Outer Space, Including the Moon and Other Celestial Bodies (1967), which is also called the "Outer Space Treaty"; (2) Agreement on the Rescue of Astronauts, the Return of Astronauts and the Return of Objects Launched Into Outer Space (1968); (3) Convention on International Liability for Damage Caused by Space Objects (1972); and (4) Convention on Registration of Objects Launched into Outer Space (1975).

A fifth treaty, The United Nations Moon Treaty, or "Moon Treaty," was adopted by the U.N. General Assembly on December 5, 1979, and entered into force on July 11, 1984, although neither the United States nor the Soviet Union has signed this particular treaty. The "Moon Treaty" is based to a considerable extent on the 1967 Outer Space Treaty, and is considered to represent a meaningful advance in international law dealing with outer space. It contains obligations of both immediate and long-term application to such matters as the safeguarding of human life on celestial bodies, promotion of scientific investigations, exchange of information about and derived from activities on celestial bodies and enhancement of opportunities and conditions for evaluation, research and exploitation of the natural resources of celestial bodies. The provisions of this new "Moon Treaty" also apply to other celestial bodies within the Solar System (other than Earth) and to orbits around the Moon.

Two recent international conventions address (in part) the use of space nuclear power systems: (1) Convention on Early Notification of a Nuclear Accident (1986) and (2) Convention on Assistance in the Case of a Nuclear Accident or Radiological Emergency (1987).

Few human undertakings have stimulated so great a degree of legal scrutiny on an international level as has the development of modern space technology. Perhaps it is because space activities involve technologies that do not respect national (terrestrial) boundaries and, therefore, place new stress on traditional legal principles. In fact, these traditional legal principles, which are based on the rights and powers of territorial sovereignty, are often in conflict with the most efficient application of new space systems. In order to resolve such complicated and complex Space Age legal issues, both the technologically developed and developing nations of the Earth have been forced to rely even more on international cooperation.

The United Nations Committee on the Peaceful Uses of Outer Space (COPUOS) has been and continues to be the main architect of international space law. It was established by resolution of the U.N. General Assembly in 1953 to study the problems associated with the arrival of the Space Age. COPUOS is made up of two subcommittees, one of which studies the technical and scientific aspects and the other the legal aspects of space activities. Some contemporary topics for space law discussion include (1) remote sensing; (2) direct broadcast communications satellites; (3) the use of nuclear power sources in space; (4) the delimitation of outer space (i.e., where does space begin and national "air space" end from a legal point of view); (5) military activities in space; (6) space debris; (7) "Who speaks for Earth?" if we ever receive a radio signal or some other form of direct contact from an intelligent alien species; and even (8) "How do we treat an alien visitor?"—a delicate legal question, especially if the alien's initial actions are apparently belligerent.

space life sciences Throughout history visionaries have speculated over a future in which human beings understand the scientific truth about their origins, control the terrestrial environment and live successfully beyond the boundaries of Earth at various locations in our Solar System. Up until the Space Age, however, such speculations had often overlooked some fundamental facts—namely that the Universe is complex and mostly inhospitable, and that life as we know it evolved in the protective shelter of the Earth's

atmosphere and its constant gravitational force. The knowledge obtained through the new Space Age discipline called *space life sciences* will play a pivotal role as human beings reach out to explore and settle the Solar System in the next century on our way to the stars. To conduct the advanced space missions now planned for the next century (such as human expeditions to Mars and permanently inhabited lunar bases), information is needed concerning the existence of life beyond the Earth, the potential interactions between living organisms and planetary environments and the possibilities for human beings to permanently inhabit space on a safe and productive basis.

The American and Soviet experience in space up until now has given us just a glimpse of the potential problems and rewards facing humans on future missions, especially those missions of extended duration. Within the United States space program, the NASA space life sciences program is responsible for acquiring the knowledge that will support the human exploration and settlement of space.

Four challenges could potentially limit the duration of human space flight: (1) physiological deconditioning; (2) the biological effects of exposure to the space (ionizing) radiation environment; (3) possible psychological difficulties experienced by members of a space expedition crew or space base team; and (4) environmental requirements, including the need for life support on lengthy space voyages and tours of duty at planetary surface bases. The life sciences disciplines of Biomedical Research and Operational Medicine focus on the health and safety of human beings in space.

Biomedical Research concentrates on physiological deconditioning, which becomes a greater concern the longer the duration of a space mission. Ground- and space-based research have identified unresolved scientific issues relevant to cardiovascular physiology, with emphasis on a more complete characterization of cardiovascular deconditioning; neurophysiology and behavioral physiology, especially *space adaptation syndrome* (or space motion sickness); and bone, endocrine and muscle physiology.

Ionizing radiation presents significant challenges for long-duration missions, such as the one to three years required for a round trip to Mars. Although information is now available about the natural space radiation environment beyond the protection of the Earth's magnetic field, important questions still remain concerning the biological effects of extended exposure to galactic cosmic radiation and solar particle events and the optimum amount of shielding necessary to sufficiently protect astronauts on long-duration space missions in interplanetary space.

The success of such extended human missions will also depend substantially on the psychological interactions among the crew members and between the space crew and ground support teams. The most pressing issues for extended human missions (those that will offer only limited possibilities for emergency rescue and return to Earth) involve crew/closed environment interactions, interpersonal interactions, human/machine interface, crew selection and composition, the command and control structure and crew morale and motivation.

Environmental factors and life-support requirements directly relate to both the physiological and psychological well-being of the space crew or planetary surface base team. The primary concerns in this area include identifying requirements for a regenerative food, air and water system; developing an environmental monitoring system capable of detecting essentially all possible sources and types of contamination; determining the most workable systems to support future extravehicular activity (EVA) operations; and analyzing habitability requirements for extended missions.

The development of a bioregenerative life-support system is especially challenging. Today, Controlled Ecological Life-Support Systems (CELSS) technology focuses on combining biological and physiochemical processes to provide food, air and water by recycling materials inside the spacecraft or planetary base habitat. Ground-based research now indicates that such a system is possible. However, the behavior of plants in space over their full growing cycle is not yet understood.

Operational (Space) Medicine addresses the health care of astronauts, especially during long-duration missions. The most important operational issues include the development of requirements for a health maintenance facility, definition of medical requirements for a crew emergency return vehicle (CERV), development of an interactive database involving astronaut health records and establishment of training programs for in-flight medical specialists.

Gravitational Biology explores the scope and operating mechanisms of one of the strongest factors influencing life on our planet: gravity. This space life sciences discipline addresses fundamental questions concerning how living organisms perceive gravity, how gravity is involved in determining the developmental and physiological status of an organism and how gravity has influenced the evolution of life on Earth. Although these questions are motivated mostly by scientific interest, research in this discipline also helps determine if life can function effectively for extended period in microgravity or in reduced gravitational environments as will be encountered on the lunar surface or the surface of Mars. Space-based research using variable-force centrifuge facilities will provide unique opportunities to expose all types of living organisms to fractional gravity levels that range from zero to one g, thereby allowing scientists to investigate the effects of gravity on these various test organisms under controlled, prolonged and scientifically useful circumstances.

The NASA Biospherics Research Program is focused on developing an understanding of the interaction of biological and global-scale chemical and physical processes. It is a component of a growing international program of studies concerning the terrestrial biosphere as a complex, highly interactive system. A descriptive model of our planet's biosphere is needed by scientists to help them understand the causes and consequences of global change—both natural and human-caused—and to permit accurate measure-

ment of key global change phenomena and prediction of their overall environmental consequences. The Biospherics Research Program, sometimes called Mission to Planet Earth, global change or the International Geosphere–Biosphere Program (IGBP), includes scientists from the biological, geological, physical, Earth, atmospheric and marine sciences. This program draws much of its impetus from a growing awareness that adverse human-caused changes in the terrestrial biosphere may not be reversible for centuries, if ever.

Finally, exobiology focuses on some of the really big questions that have puzzled human beings throughout history, such as, Are we alone in the Universe? What led to the origin of life on Earth? Exobiologists now believe that the early environments of Mars and Earth were similar, and that rock and soil samples from Mars could fill important gaps in the Earth's geological record. Any valid indication of past (or even present) life on Mars would support the hypothesis that life could originate wherever the physical and chemical environment is favorable. Therefore, one of the main thrusts of NASA's present exobiology program is the use of robot vehicles to return soil and rock samples as a precursor to human expeditions in the early part of the next century.

See also: **global change; life in the Universe; microgravity; people in space**

space nuclear power   Through the cooperative efforts of the U.S. Department of Energy (DOE), formerly called the Atomic Energy Commission, and NASA, the United States has successfully used nuclear energy in its space program to provide electrical power for many missions, including science stations on the Moon, extensive exploration missions to Jupiter, Saturn and beyond, and even in the search for extraterrestrial life on Mars.

For example, when the *Apollo 12* astronauts departed from the lunar surface on their return trip to Earth (November 1969), they left behind a nuclear-powered science station that sent information back to terrestrial scientists for several years. That system, as well as similar stations left on the Moon by the *Apollo 14* through *17* missions, operated on electrical power supplied by radioisotope thermoelectric generators (RTGs). In fact, since 1961 nuclear power systems have helped assure the success of many space missions, including the *Pioneer 10* and *11* missions to Jupiter and Saturn, the *Viking 1* and *2* landers on Mars, and the spectacular *Voyager 1* and *2* missions to Jupiter, Saturn, Uranus, Neptune and beyond. It should also be realized that these magnificent space exploration missions would not have been possible without the use of nuclear energy (see table 1).

Energy supplies that are reliable, transportable and abundant represent a very important technology in the development of our extraterrestrial civilization. Space nuclear power systems will play an ever expanding role in the more ambitious space exploration and resource exploitation missions of the next few decades. For example, the move-

Table 1   Summary of Space Nuclear Power Systems Launched by the United States (1961–1991)

Power Source	Spacecraft	Mission Type	Launch Date	Status
SNAP–3A	Transit 4A	Navigational	June 29, 1961	Successfully achieved orbit
SNAP–3A	Transit 4B	Navigational	November 15, 1961	Successfully achieved orbit
SNAP–9A	Transit–5BN–1	Navigational	September 28, 1963	Successfully achieved orbit
SNAP–9A	Transit–5BN–2	Navigational	December 5, 1963	Successfully achieved orbit
SNAP–9A	Transit–5BN–3	Navigational	April 21, 1964	Mission aborted: burned up on reentry
SNAP–10A (reactor)	Snapshot	Experimental	April 3, 1965	Successfully achieved orbit
SNAP–19B2	Nimbus–B–1	Meteorological	May 18, 1968	Mission aborted: heat source retrieved
SNAP–19B3	Nimbus III	Meteorological	April 14, 1969	Successfully achieved orbit
SNAP–27	Apollo 12	Lunar	November 14, 1969	Successfully placed on lunar surface
SNAP–27	Apollo 13	Lunar	April 11, 1970	Mission aborted on way to Moon. Heat source returned to South Pacific Ocean
SNAP–27	Apollo 14	Lunar	January 31, 1971	Successfully placed on lunar surface
SNAP–27	Apollo 15	Lunar	July 26, 1971	Successfully placed on lunar surface
SNAP–19	Pioneer 10	Planetary	March 2, 1972	Successfully operated to Jupiter and beyond; in interstellar space
SNAP–27	Apollo 16	Lunar	April 16, 1972	Successfully placed on lunar surface
Transit–RTG	"Transit" (Triad–01–IX)	Navigational	September 2, 1972	Successfully achieved orbit
SNAP–27	Apollo 17	Lunar	December 7, 1972	Successfully placed on lunar surface
SNAP–19	Pioneer 11	Planetary	April 5, 1973	Successfully operated to Jupiter, Saturn and beyond
SNAP–19	Viking 1	Mars	August 20, 1975	Successfully landed on Mars
SNAP–19	Viking 2	Mars	September 9, 1975	Successfully landed on Mars
MHW	LES 8/9	Communications	March 14, 1976	Successfully achieved orbit
MHW	Voyager 2	Planetary	August 20, 1977	Successfully operated to Jupiter, Saturn, Uranus, Neptune and beyond
MHW	Voyager 1	Planetary	September 5, 1977	Successfully operated to Jupiter, Saturn and beyond
GPHS–RTG	Galileo	Planetary	October 18, 1989	Successfully sent on interplanetary trajectory to Jupiter (1996 arrival)
GPHS–RTG	Ulysses	Solar–Polar	October 6, 1990	Successfully sent on interplanetary trajectory to explore polar regions of Sun (1994–1995)

SOURCE: NASA; Dept of Energy.

ment of massive payloads from low Earth orbit (LEO) to high Earth orbit (HEO) or a lunar destination, the operation of very large space platforms throughout cislunar space and the successful startup and expansion of the initial lunar bases can all benefit from the creative application of advanced space nuclear power system technologies. Even more progressive space activities, such as asteroid movement and mining, planetary engineering and climate control and human expeditions to Mars and the planets beyond, all require compact energy systems at the megawatt and gigawatt levels.

Space nuclear power supplies offer several distinct advantages over the more traditional solar and chemical space power systems. These advantages include compact size; modest mass requirements; very long operating lifetimes; the ability to operate in extremely hostile environments (for example, intense trapped radiation belts, the surface of Mars, the moons of the outer planets and even interstellar space); and the ability to operate independent of distance from the Sun or orientation to the Sun. It appears that as the energy requirements of our initial extraterrestrial civilization efforts approach hundreds of kilowatts to megawatts, nuclear energy systems represent the only realistic technological option in the next few decades.

Space nuclear power systems use the thermal energy or heat released by nuclear processes. These processes include the spontaneous but predictable decay of radioisotopes, the controlled splitting for fissioning of heavy atomic nuclei (such as uranium-235) in a sustained neutron chain reactor and the joining together or fusing of light atomic nuclei (such as deuterium and tritium) in a controlled thermonuclear reaction. This "nuclear" heat can then be applied directly or converted by a variety of engineering techniques into electric power. Until we successfully achieve controlled thermonuclear fusion capabilities, nuclear energy applications (both terrestrial and extraterrestrial) will be based on the use of radioisotope decay and nuclear fission reactors.

Figure 1 illustrates a basic radioisotope thermoelectric generator. This RTG consists of two main functional components: the thermoelectric converter and the nuclear heat source. The isotope, plutonium-238, has been used as the heat source in all U.S. space missions involving radioisotope power supplies. Plutonium-238 has a half-life of 87.7 years and therefore supports a long operational life. (The half-life is the time required for one half the number of unstable nuclei present at a given time to undergo radioactive decay.) In the nuclear decay process, plutonium-238 emits

Fig. 1   Typical radioisotope thermoelectric generator (RTG) configuration for a space nuclear power application (the device is a modified SNAP 19 system). (Courtesy of NASA.)

Table 2   SP-100 Reference Flight System Design Concept (Subsystem Characteristics)

Nuclear Subsystems	Space Subsystems
• Fast spectrum reactor UN fuel Lithium cooled Nb-1Zr vessel	• Thermoelectric conversion Si-Ge/GaP thermoelectrics conductively-coupled
• Radiation Shield 17-degree 1/2-cone angle LiH neutron attenuation Tungsten gamma-ray attenuation	• Heat rejection Lithium pumped loop TEM pumps Potassium heat pipe radiator Self-thaw capability
• Heat transport Lithium-pumped loop Thermoelectric- electromagnetic (TEM) pumps Thaw-assist heat pipes	• Power conditioning and control 200 VDC primary power Full shunt regulation Millisecond response to load following 28 volt secondary power
• Instrumentation and control Temperature, position and electrical parameter sensing Multiplex signal transmission	• Structure/mechanical Titanium/beryllium structures State-of-the-art mechanics

Source: NASA, Department of Energy, and Department of Defense.

mainly alpha radiation, that has very low penetrating power. Consequently, only lightweight shielding is required to protect the spacecraft from its nuclear radiation. A thermoelectric converter uses the "thermocouple principle" to directly convert a portion of the nuclear heat into electricity.

We can also use a nuclear reactor to provide space nuclear power. The United States flew one space nuclear reactor, called SNAP-10A, in 1965. The objective of this program was to develop a space nuclear reactor power unit capable of producing a minimum of 500 watts-electric for a period of one year while operating in the extraterrestrial environment. SNAP-10A was the first and only space reactor flight tested in Earth orbit by the United States. It was a small zirconium hydride (ZrH) thermal reactor fueled by uranium-235. The SNAP-10A orbital test was successful, although the mission was prematurely terminated by the failure of an electronic component outside the reactor.

The United States is currently developing a 100-kilowatt electric class space nuclear reactor for a variety of space power applications in the late 1990s and beyond. Called SP-100, this reactor is being developed in a tri-agency program, involving NASA, the Department of Energy and the Department of Defense. Key features of the SP-100 design concept are presented in table 2. This demonstration

program will involve the operation of the SP-100 reactor for seven years—a full power life. For up to six months of that seven-year period, an electric propulsion system (a high-power arc jet) will be used to verify reactor system compatibility with a payload that requires large quantities of electric power.

Since the United States first used nuclear power in space, great emphasis has been placed on the safety of people and the protection of the terrestrial environment. A continuing major objective in any new space nuclear power program is to avoid undue risks. In the case of radioisotope power supplies, this means designing the system to contain the nuclear isotope fuel under all normal and potential accident conditions. For space nuclear reactor systems, like the SP-100 and more advanced systems, this means launching the reactor in a "cold" (nonoperating) configuration and starting up the reactor only after a safe, stable Earth orbit or interplanetary trajectory has been achieved.

See also: **fission (nuclear); fusion; lunar bases and settlements; Mars base; space nuclear propulsion**

space nuclear propulsion   Nuclear fission reactors can be used in two basic ways to propel a space vehicle: (1) to generate electric power for an electric propulsion unit and (2) as a thermal energy or heat source to raise a propellant (working material) to extremely high temperatures for subsequent expulsion out a nozzle. In the second application, the system is often called a nuclear rocket (see fig. 1).

In a nuclear rocket, chemical combustion is not required.

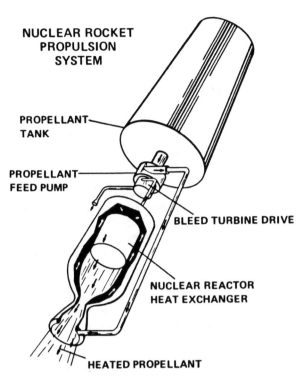

NUCLEAR ROCKET PROPULSION SYSTEM

PROPELLANT TANK

PROPELLANT FEED PUMP

BLEED TURBINE DRIVE

NUCLEAR REACTOR HEAT EXCHANGER

HEATED PROPELLANT

Fig. 1 Nuclear rocket propulsion system. (Drawing courtesy of the U.S. Department of Energy.)

Instead, a single propellant, usually hydrogen, is heated by the energy released in the nuclear fission process, which occurs in a controlled manner in the reactor's core. Conventional rockets, in which chemical fuels are burned, have severe limitations in the specific impulse a given propellant combination can produce. These limitations are imposed by the relatively high molecular weight of the chemical combustion products. At attainable combustion chamber temperatures the best chemical rockets are limited to specific impulse values of about 4,300 meters per second (440 seconds). Nuclear rocket systems using fission reactions, fusion reactions and even possibly matter–antimatter annihilation reactions (the "photon rocket") have much greater propulsion performance capabilities.

Engineering developments will be needed in the 21st century to permit the use of advanced fission reactor systems, such as the gaseous core reactor rocket or even fusion powered systems. However, the solid-core nuclear reactor rocket is within a flight-test demonstration of engineering reality. In this nuclear rocket concept, hydrogen propellant is heated to extremely high temperatures while passing through flow passageways within the solid fuel elements of a compact nuclear reactor system that uses uranium-235 as the fuel. The high temperature gaseous hydrogen then expands through a nozzle to produce propulsive thrust. From the mid-1950s until the early 1970s, the United States conducted a nuclear rocket program called Project Rover. The primary objective of Project Rover was to develop a

nuclear rocket for a manned mission to Mars. Unfortunately, despite the technical success of this nuclear rocket program, overall space program emphasis changed and the nuclear rocket and the Mars mission planning were discontinued in 1973.

Engineers who are involved with the development of our extraterrestrial civilization recognize one basic fact: the secret of space travel and of extending our human civilization throughout heliocentric space is energy—large amounts of portable energy for power and propulsion systems. The compact energy advantages of nuclear fission reactors (and eventually controlled fusion systems) will ultimately enable us to develop the powerful, reusable interplanetary spaceships we need to sweep across the Solar System. In the upcoming decades, nuclear rockets—based initially perhaps on Rover technology and then evolving into more advanced propulsion system technologies—will carry extraterrestrial settlers and their equipment to the surface of the Red Planet, to the minor planets of the main asteroid belt and to the exciting constellations of moons surrounding the gaseous giants, Jupiter and Saturn.

See also: **fission (nuclear); fusion; nuclear-electric propulsion system; Project Daedalus; Project Orion (II); space nuclear power; starship**

space platform An unmanned, free-flying orbital platform that is dedicated to a specific mission, such as space industrialization activities (materials processing in space) or scientific research (space telescope). It would orbit near a permanently manned space station and would be serviced by either the Space Shuttle or the space station.

See also: **space station**

spaceport A spaceport is both a doorway to outer space from the surface of a planet and a port of entry from outer space to a planet's surface. At a spaceport we find the sophisticated facilities required for the assembly, testing, launching and (in the case of reusable space vehicles) landing and postflight refurbishment of space vehicles. Typical operations performed at a spaceport include the assembly of space vehicles; preflight preparation of space launch vehicles and their payloads; testing and checkout of space vehicles, spacecraft and support equipment; coordination of launch vehicle tracking and data-acquisition requirements; countdown and launch operations; and (in case of reusable space vehicles) landing operations and refurbishment. A great variety of technical and administrative activities are also needed to support the operation of a spaceport. These include design engineering, safety and security, quality assurance, cryogenic fluids management, toxic and hazardous materials handling, maintenance, logistics, computer operations, communications and documentation.

Both expendable (one-time use) and reusable space vehicles can now be found at spaceport facilities around the globe. In the next century, highly automated spaceports will also appear on the lunar and Martian surfaces to support permanent human bases and settlements on these alien worlds.

See also: **European Space Agency; French Space Agency; lunar bases and settlements; Mars base; National Aeronautics and Space Administration; National Space Development Agency of Japan; space launch vehicles**

space settlement A large extraterrestrial habitat where from 1,000 to perhaps 100,000 people would live, work and play, while supporting space industrialization activity, such as the operation of a large space manufacturing complex or the construction of satellite power systems. One such possibility is a torus-shaped space settlement for about 10,000 people. Its inhabitants, all members of a space manufacturing complex work-force, would return after work to homes on the inner surface of the large torus, which is nearly 1.6 kilometers in circumference. It would rotate to provide the inhabitants with a gravity level similar to that experienced on the surface of the Earth. This habitat would be shielded against cosmic rays and solar flare radiation by a nonrotating shell of material that could be built up from accumulated slag or waste materials from lunar or asteroid mining operations. Outside the shielded area agricultural crops would be grown taking advantage of the intense, continuous stream of sunlight available in space. Docking areas and microgravity industrial zones are located at each end of the settlement; so are the large flat surfaces necessary to radiate waste heat away from the facility to outer space.

Another possible design is a spherical space settlement, called the Bernal sphere. This giant spherical habitat would be approximately two kilometers in circumference. Up to 10,000 people would live in residences along the inner surface of the large sphere. Rotation of the settlement at about 1.9 revolutions per minute (RPM) would provide Earth-like gravity levels at the sphere's equator, but there would be essentially microgravity conditions at the poles. In the settlement's "polar regions" human-powered flying machines could be used and the space settlers would be able to enjoy a variety of microgravity recreational pursuits. Because of the short distances between things in the equatorial residential zone, passenger vehicles would not be necessary. Instead, the space settlers would travel on foot or perhaps by bicycle. The climb from the residential equatorial area up to the sphere's poles would take about 20 minutes and would lead the hiker past small villages, each at progressively lower levels of artificial gravity. A corridor at the axis would permit residents to float safely in microgravity out to exterior facilities, such as observatories, docking ports, industrial and agricultural areas. Ringed areas above and below the main sphere in this type of space settlement would be the external agricultural toruses.

Another possible design is a very large set of twin 32-kilometer-long, 6.4-kilometer-diameter, cylindrical space settlements. These huge space settlements would be able to house several hundred thousand people. Each cylinder

rotates around its main axis once every 114 seconds to create an Earth-like level of artificial gravity. The teacup-shaped containers ringing each cylinder are agricultural stations. Each cylinder is capped by a space industrial facility and a power station. Large movable rectangular mirrors on the sides of each cylinder (hinged at one end to the cylinder) would direct sunlight into the habitat's interior, control the day–night cycles and even regulate the settlement's "seasons." A random number generator somewhere in the mirror's controller loop could be used to provide weather variations that are unpredictable but within certain previously established limits. This type of controlled randomness might be very necessary in overcoming some of the psychological problems that could arise from living in a totally "artificial" or man-made world.

The basic space settlement design will have to satisfy the essentials for life such as: air, food, water and, for extended stays in space, some type of artificial gravity. The space settlement design must ensure not only physiological safety and comfort but must also satisfy the psychological and esthetic needs of the inhabitants. Table 1 describes some of the physiological design criteria that can be applied to a space settlement.

Human beings living in space must have an adequate diet. Food in a large settlement should be nutritious, sufficiently abundant and even attractive. The settlers can get their food supplies from the Earth or the Moon (permanent lunar settlements in the next century will practice space farming and should be able to export food products to markets within cislunar space); or else, the settlers may elect to grow their own. Recent space station and space settlement studies indicate that when more than 10,000 person-days of food are needed each year in cislunar space, agriculture in space becomes economically competitive with food import from Earth. It also appears that for a large space settlement, a modified type of terrestrial agriculture, based on plants and meat-bearing animals, should solve both nutritional requirements and the need for dietary variety.

Photosynthetic agriculture can also be used to help re-

Table 1 Suggested Physiological Design Criteria for a Space Settlement

Pseudogravity	0.95 g
Rotation rate	≤ 1 rpm
Radiation exposure for the general population	≤ 0.5 rem/yr (5 mSv/yr)
Magnetic field intensity	≤ 100 μT
Temperature	23° ± 8° C
Atmospheric composition pO_2	22.7 ± 9 kPa (170 ± 70 mm Hg)
p(Inert gas; most likely N_2)	26.7 kPa < pN_2 < 78.9 kPa (200 < pN_2 < 590 mm Hg)
pCO_2	< 0.4 kPa (<3 mm Hg)
pH_2O	1.00 ± 0.33 kPa (7.5 ± 2.5 mm Hg)

SOURCE: NASA.

generate the space settlement's atmosphere by converting carbon dioxide and generating oxygen. Space agricultural activities might even provide a source of pure water from the condensation of humidity produced by transpiration.

The design of a space settlement must not exert damaging psychological stresses on its inhabitants. A sense of isolation (the Shimanagashi syndrome) or a sense of artificiality (the Solipsism syndrome) must be avoided through variety, diversity and flexibility of interior designs. Table 2 lists some suggested quantitative environmental design criteria for a large space settlement, while table 3 provides some qualitative design criteria.

The space settlement must also have a form of government or political organization that permits its inhabitants to enjoy a comfortable life-style under conditions that are crowded and physically isolated from other human communities. Since early space settlements may very likely be "company towns" dedicated to some particular space industrial activity, their organizations should also support a fairly high level of productivity and should maintain the physical security of the habitat.

Without proper organization and internal security this type of isolated community could easily become the victim of despots and self-elected demigods. If things really got out of hand in the settlement, a "space marshal" might have to be sent from Earth (arriving, of course, on the

noon space tug) to restore law and order on the extraterrestrial frontier.

Full social, political and economic autonomy from Earth does not appear possible for space settlements with populations under about 500,000. For example, a community of from 10,000 to 50,000 would be hard pressed to even support a large university or medical center. Therefore, many of the services and benefits of a full civilization (such as large universities) will still be supplied from Earth—at least until the lunar civilization attains a social critical mass of about half a million selenians.

At that point, some very interesting things will happen. For example, teen-age space settlers and their parents will have the opportunity to examine two sets of college catalogs: one from Earth and one from the Moon. And, as has never occurred in generations past, they will have the distinct opportunity of evaluating the pros and cons for advanced education in institutions on two different worlds! Of course, our young space settler might elect to stay at home and take teleconferenced courses through the Community College of Cislunar Space.

Space settlements, whatever their final design, population or political structure, will emerge as a major part of our extraterrestrial civilization in the next century and beyond. We cannot fully appreciate today the impact that (almost) self-sufficient pockets of humanity sprinkled throughout cislunar space will have on the technical, political, economic and social structure of 21st-century living. In time, as these settlements grow and replicate themselves, we will witness the rise of extraterrestrial city-states throughout cislunar and then heliocentric space.

Sometime in the next century, terrans who had maintained very little interest in things above the atmosphere, will suddenly awaken to find: "We have met the extraterrestrials and they are us!" Can you image the social impact of certain interplanetary relationships! "Your place or mine?" will now become a question of astronomical proportions.

The term "space settlement" is also used to describe permanent habitats for over about 1,000 people on the surface of another world, such as the Moon or Mars.

See also **Bernal sphere; Dyson sphere; lunar bases and settlements; Mars base; space industrialization; space station**

space station   A space station is an orbiting space system that is designed to accommodate long-term human habitation in space. The concept of people living in artificial habitats in outer space appeared in 19th-century science fiction literature in stories such as Edward Everett Hale's "Brick Moon" (1869) and Jules Verne's "Off on a Comet" (1878).

At the turn of the century, Konstantin Tsiolkovsky provided the theoretical underpinnings for this concept with his truly visionary writings about the use of orbiting stations as a springboard for exploring the Cosmos. Tsiolkovsky, the father of Soviet astronautics, provided a more technical introduction to the space station concept in his 1895 work

Table 2   Suggested Quantitative Environmental Design Criteria for a Space Settlement

Population: men, women, children	10,000
Community and residential, projected area per person, m^2	47
Agriculture, projected area per person, m^2	20
Community and residential, volume per person, m^3	823
Agriculture, volume per person, m^3	915

SOURCE: NASA.

Table 3   Suggested Qualitative Environmental Design Criteria for a Space Settlement

Long lines of sight
Larger overhead clearance
Noncontrollable unpredictable parts of the environment; for example, plants, animals, children, weather
External views of large natural objects
Parts of interior out of sight of others
Natural light
Contact with the external environment
Availability of privacy
Good internal communications
Capability of physically isolating segments of the habitat from each other
Modular construction
    of the habitat
    of the structures within the habitat
Flexible internal organization
Details of interior design left to inhabitants

SOURCE: NASA.

Dreams of Earth and Heaven, Nature and Man. He greatly expanded on the idea of a space station in his 1903 work entitled "The Rocket into Cosmic Space." In this technical classic, Tsiolkovsky described all the essential ingredients needed for a manned space station including the use of solar energy, the use of rotation to provide artificial gravity and the use of a closed ecological system complete with "space greenhouse."

Throughout the first half of the 20th century the space station concept continued to technically evolve. For example, the German scientist Hermann Oberth described the potential applications of a space station in his classic *The Rocket into Interplanetary Space* (1923). These suggested applications included the use of a space station as an astronomical observatory, an Earth-monitoring facility and a scientific research platform. In 1929 an Austrian named Potočnik (pen name Hermann Noordung) introduced the concept of a rotating, wheel-shaped space station. Potočnik called his design "Wohnrad" or "Living Wheel." Another Austrian, Guido von Pirquet, wrote many technical papers on space flight, including the use of a space station as a refueling node for space tugs. In the late 1920s and early 1930s, von Pirquet also suggested the use of multiple space stations at different locations in cislunar space. After World War II Dr. Wernher von Braun popularized the concept of a wheel-shaped space station in the United States.

The idea of much larger man-made space habitats and mini-planets accompanied the birth of the space station concept. In 1918 Dr. Robert Goddard, the father of American rocketry, introduced the concept of nuclear-powered space arks that could carry an entire civilization from one solar system to another. (Possibly to avoid criticism, Goddard's manuscript, entitled "The Ultimate Migration," describing this space ark was not published until after his death.) The British scientist and writer, J. D. Bernal, described man-made planets and large self-contained worlds in his 1929 work *The World, the Flesh and the Devil.* The brilliant writer and science prophet Arthur C. Clarke used very large space stations and habitats in his 1952 novel *Islands In the Sky.* The space engineer and visionary, Dr. Krafft Ehricke, suggested an entire line of evolutionary space stations and space habitats in his writings during the 1960s and 1970s. His uniquely far-reaching concepts included Astropolis (an orbiting city-state) and the Androcell (a miniature man-made planet). And most recently, several groups of space technology thinkers under the initial auspices of NASA's Ames Research Center conceived of several large space settlement designs, whose populations were committed to space manufacturing and the construction of Satellite Power Systems. The efforts of scientists like G. K. O'Neill and Brian O'Leary have helped stimulate much contemporary interest in living and working in outer space. Many of these larger space settlement designs might actually evolve from the modularized space station designs of the next three decades.

SKYLAB—FIRST U.S. SPACE STATION

Even before the Apollo program had successfully landed men on the Moon, NASA engineers and scientists were considering the next giant step in the U.S. manned space flight program. That next step became the simultaneous development of two complementary space technology capabilities. One was a safe, reliable transportation system that could provide routine access to space. The other was an orbital space station where human beings could live and work in space. This space station would serve as a base camp from which other, more advanced space technology developments could be initiated. This long-range strategy set the stage for two of the most significant American space activities carried out in the 1970s and 1980s: *Skylab* and the Space Shuttle.

On May 14, 1973, the United States launched its first space station, called *Skylab*. (See fig. 1.) It was launched by a Saturn V booster from the Apollo program. *Skylab* demonstrated that people could function in space for periods up to 12 weeks and, with proper exercise, could return to Earth with no ill effects.

In particular, the flight of *Skylab* proved that human beings could operate very effectively in a prolonged microgravity environment and that it was not essential to provide artificial gravity for people to live and work in space (at least for periods up to about six months). The *Skylab* astronauts accomplished a wide range of emergency repairs

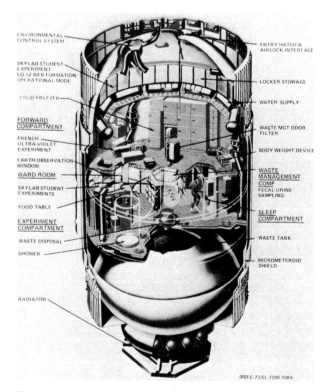

Fig. 1   Cutaway view of the *Skylab* Orbital Workshop. (Courtesy of Johnson Space Center.)

Table 1 Skylab Program Summary

Mission	Dates	Astronaut Crew	Mission Duration	Remarks
SKYLAB 1	Launched: May 14, 1973	Unmanned	May 14, 1973 to July 11, 1979	100-ton space station visited by three crews. Descended into Earth's atmosphere over Indian Ocean and Australia after 34,981st orbit.
SKYLAB 2	May 25–June 22, 1973	Charles Conrad Jr. Paul J. Weitz Joseph P. Kerwin	28 days, 50 min.	Repaired *Skylab*; 404 revolutions; 392 experiment hours; 3 EVAs total 5 hrs, 34 min.
SKYLAB 3	July 28–September 25, 1973	Alan L. Bean Jack R. Lousma Owen K. Garriott	59 days, 11 hrs, 9 min.	Performed station maintenance; 858 revolutions; 1,081 experiment hours; 3 EVAs total 13 hrs, 42 min.
SKYLAB 4	November 16, 1973–February 8, 1974	Gerald P. Carr William R. Pouge Edward G. Gibson	84 days, 1 hr, 16 min.	Observed Comet Kohoutek; 1,214 revolutions; 1,563 experiment hours; 4 EVAs total 22 hrs, 25 min.

SOURCE: NASA.

on station equipment, including freeing a stuck solar panel array (a task that saved the entire mission), replacing rate gyros and repairing a malfunctioning antenna. On two separate occasions the crews installed portable sun shields to replace the originals lost when *Skylab* was launched. These on-orbit activities clearly demonstrated the unique and valuable role people have in space. Table 1 summarizes the *Skylab* missions.

Unfortunately, *Skylab* was not designed for a permanent presence in space. The system was not designed to be routinely serviced on orbit (although the *Skylab* crews were able to perform certain repair functions). *Skylab* was not equipped to maintain its own orbit—a design deficiency that eventually caused its fiery demise on July 11, 1979, over the Indian Ocean and portions of western Australia. Finally, it was not designed for evolutionary growth and therefore was subject to rapid technological obsolescence. Future space station designs will take these shortcomings into account, and effectively use the highly successful *Skylab* program experience to develop a permanent, evolutionary and modular U.S. space station.

SOVIET SPACE STATION ACTIVITIES

While the United States was conducting the Apollo Moon Landing program, the Soviet Union began embarking on an ambitious space station program. The first Soviet space station, *Salyut-1,* was launched on April 19, 1971. (The Russian word "salyut" means "salute.") In the years since that launch, cosmonauts have spent almost 6,000 days in space on board various generations of Soviet space stations.

The first-generation Soviet space station is represented by the four early Salyut stations that were successfully operated between 1971 and 1976. These stations were very modest in size, provided essentially spartan crew accommodations, suffered from a constant shortage in electric power and had only one docking port. Yet these early stations, while often considered "primitive" in comparison to contemporary American aerospace technologies, represented a major Soviet step in achieving a long-term human presence in low-Earth-orbit.

The second-generation Soviet space station was introduced with the launch (on September 29, 1977) and successful operation of the *Salyut-6* station (see fig. 2). Several important design improvements appeared on this station, including the addition of a second docking port and the use of an automated *Progress* resupply spacecraft. The combined *Salyut-6* and *Salyut-7* space station occupancies included 11 long-duration cosmonaut crews and 15 "visiting" cosmonaut crews, including non-Soviet cosmonauts from such countries as Czechoslovakia, Poland, Germany (East), Bulgaria, Romania, Hungary and Mongolia.

In February 1986 the Soviets introduced their third-generation space station, the *Mir* ("peace") station. More extensive automation, more spacious crew accommodations for the two "resident" (long-duration mission) cosmonauts and the addition of a five-port docking adapter at one end of the station highlight some of the major design improvements found on *Mir*. The Soviet's have also introduced the use of an 11,000-kilogram (24,250-lb-mass) specialized science module, called the Kvant ("quantum"). When docked with the *Mir* station, the 11-ton Kvant module brings a significant research capability to the orbiting complex. The Soviet Union appears intent on maintaining an active, evolving capability to sustain an essentially permanent human presence in low Earth orbit. For example, in December 1988 two cosmonauts (Vladimir G. Titov and Musa K. Manarov) returned to Earth after successfully completing over a year on the *Mir* station. The Soviet government has also recently begun to encourage a much wider foreign participation in their space program, including science experiments on board the Kvant specialized module and even the "rental" of technical equipment and services on *Mir*.

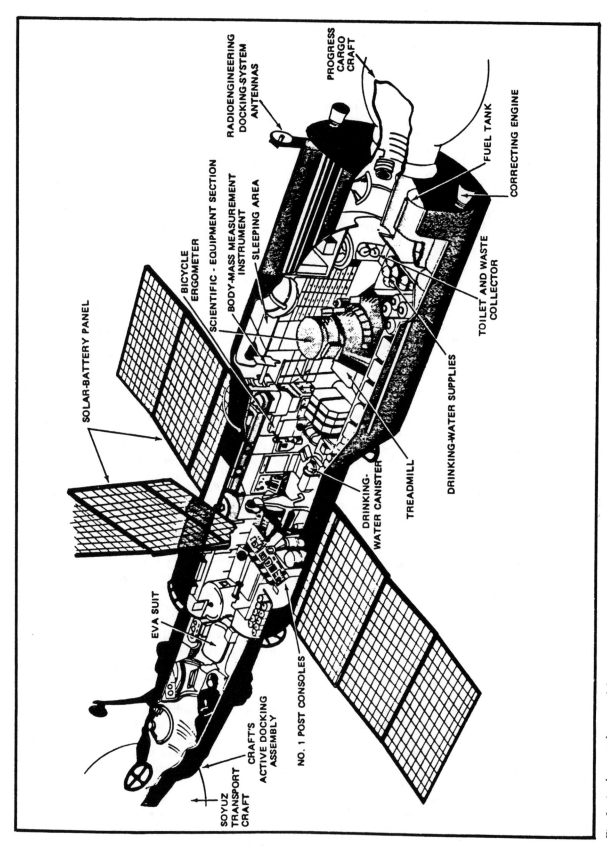

Fig. 2   A schematic diagram of the Soviet *Salyut-6* space station.   (Drawing courtesy of NASA.)

SPACE STATION *FREEDOM*

The U.S. Space Station *Freedom* will be a permanently manned international space complex in low Earth orbit that promises to stimulate a new era in the exploration and use of outer space. As currently planned, Space Station *Freedom* will encompass a permanently manned base, unmanned polar-orbiting platforms, a man-tended free-flyer to be serviced at the manned base, and the associated ground-based infrastructure. This combination of manned, unmanned and automated systems will establish a broad spectrum of capabilities that are responsive to both the currently identified needs and the evolutionary requirements of space science, technology and commerce.

The basic functions of this space complex will include (1) microgravity scientific laboratories equipped with the latest research instruments for the study of space science, chemistry, biology, life sciences and materials science; (2) a servicing and repair facility for satellites and spacecraft; (3) a transportation base from which missions to the Moon and Mars (and other space destinations) can be launched; (4) an orbiting "storage depot" for spare parts and equipment; (5) an assembly base for the on-orbit construction of large space structures; and (6) an observation site above the Earth's atmosphere that provides an excellent view of both the Universe beyond and the terrestrial biosphere below.

The initial Space Station configuration will be able to support a crew of eight. *Freedom* is an international effort in which the United States is providing a laboratory and a habitation module; Japan and the European Space Agency (ESA) are each providing permanently attached laboratory modules (Japanese Experiment Module and ESA Columbus Module); and Canada is providing a Mobile Servicing System (MSS) equipped with a manipulator arm. The Canadian MSS's robot arm will perform a variety of tasks from work stations both inside and outside *Freedom*'s pressurized modules. The pressurized modules will be connected by resource nodes that, in addition to connecting the modules, house the Space Station's command and control systems. Arrays of solar cells will provide 75,000 watts of electric power to the station. It should be noted, however, that NASA is currently (as of June 1991) reviewing this program, and that its future is uncertain.

The United States and ESA are also providing polar-orbiting unmanned platforms as part of this international space complex. Finally, ESA is also developing a Man-Tended Free Flyer (MTFF), called the Columbus Free-Flyer, that will co-orbit with, but function independently of, the main manned base. The MTFF will serve as an automated laboratory/factory within which microgravity materials science and processing can be conducted in an undisturbed environment. Periodically, microgravity manufactured products would be harvested and the ESA free-flyer would be serviced by the *Freedom* crew.

As now envisioned, Space Station *Freedom* will be launched and assembled on-orbit using the Space Shuttle. After assembly is complete and the Station is made operational (mid to late 1990s), the Shuttle will then make regular trips to *Freedom*, rotating crews, refurbishing supplies and equipment and returning products to Earth.

Freedom will circle the Earth approximately every 90 minutes at an orbital altitude of about 400 kilometers (250 miles) at an inclination of 28.5 degrees to the equator. This Space Station is being designed to operate for 20 to 30 years in space. *Freedom* represents an important orbital-stepping stone in the development of humanity's extraterrestrial civilization. Sometime in the early part of the next century, *Freedom* will serve as the Earth-orbiting transportation node for humans to return to the Moon to establish permanent bases and then possibly as the extraterrestrial embarkation point for the first human expeditions to Mars.

See also: **extraterrestrial civilizations; people in space; space settlement; Space Transportation System**

space suit   Outer space is a very hostile environment. If astronauts are to survive there, they must take part of the Earth's environment with them. Air to breathe, acceptable ambient pressures and moderate temperatures have to be contained in a shell surrounding the space traveler. This can be accomplished by providing a very large enclosed structure or habitat; or else on an individual basis by encasing the astronaut in a protective flexible capsule called the space suit.

Space suits used on previous missions from the Mercury program up through the Apollo/Soyuz test project have provided effective protection for American astronauts. However, they have also been handicapped by certain design problems. These suits were custom-fitted garments. In some suit models, more than 70 different measurements had to be taken of the astronaut in order to manufacture the suit to the proper fit. As a result, a space suit could be worn by only one astronaut on only one mission. These early space suits were stiff and even simple motions such as grasping objects quickly drained an astronaut's strength. Even donning the suit was an exhausting process lasting, at times, more than an hour and requiring the help of an assistant.

A new space suit has been developed for Shuttle-era astronauts that provides many improvements in comfort, convenience and mobility over the previous models. This suit, which is worn outside the Orbiter during extravehicular activity (EVA), is modular and features many interchangeable parts. Torso, pants, arms and gloves come in several different sizes and can be assembled for each mission in the proper combination to suit individual male and female astronauts. The Shuttle space suit is called the Extravehicular Mobility Unit (EMU) and consists of three main parts: liner, pressure vessel and primary life-support system (PLSS). These components are supplemented by a drink bag, communications set, helmet and visor assembly.

Containment of body wastes is a significant problem in space suit design. In the EMU, the primary life support system (PLSS) handles odors, carbon dioxide and the containment of gases in the suit's atmosphere. The PLSS is a two-part system consisting of a backpack unit and a control and display unit located on the suit chest. A separate unit is required for urine relief. Two different urine-relief systems have been designed to accommodate both male and female astronauts. Because of the short time durations for extravehicular activities, fecal containment is considered unnecessary.

The Manned Maneuvering Unit (MMU) is a nitrogen-gas-propelled, hand-controlled device that allows a space-suited astronaut to leave the Space Shuttle and perform untethered extravehicular activities (see fig. 1). The MMU device has been called "the world's smallest reusable space-craft."

Projected requirements for extravehicular activity (EVA) on Space Station *Freedom* involve considerable increases in both EVA frequency and duration over previous manned space missions. Such EVA increases require the design and development of a new, high-pressure space suit that provides increased durability, maintainability and hazard protection; zero-prebreathe capability is also required to permit frequent and rapid egress and ingress without exposing Space Station astronauts to the potentially lethal hazard of the bends (decompression sickness). The current Shuttle space suit has an internal suit pressure of 4.3 psi (29.7 kPa), while the Shuttle cabin pressure is maintained at 14.7 psi (101 kPa). Therefore, prior to performing an EVA, a Shuttle

astronaut must now prebreathe 100 percent oxygen for up to four hours to remove the nitrogen gas dissolved in his or her body tissues. The new, Space Station Era zero prebreathe suit (ZPS) will be operated at a higher internal pressure (typically close to 14.7 psi [101 kPa]) to minimize the pressure differential between the advanced space suit and the Space Station pressurized modules, thereby saving many hours of crew time in preparing for numerous EVAs.

Space Transportation System (STS) [Space Shuttle]

When the Space Shuttle flew again in September 1988, following over two and a half years of redesigning and upgrading after the *Challenger* accident (January 28, 1986), it still remained the most sophisticated, complex and technologically advanced manned spaceship in the world. Although the Space Shuttle has become a safer vehicle during this recovery period, veteran astronauts and aerospace engineers alike know that sending men and women into space will continue to involve risk. In August 1986, a decision was made to build a replacement orbiter and expand the Shuttle Orbiter fleet to four vehicles by the mid-1990s. The new Orbiter vehicle (OV-105) has been named *Endeavour* and is scheduled to fly its maiden mission in 1992.

The major components of the Space Shuttle System are the Orbiter vehicle; the three main liquid rocket engines, with a combined thrust of approximately 544,300 kilograms (1.2 million lb-force); the giant external tank (ET), which feeds liquid hydrogen fuel and liquid oxygen oxidizer to the three main engines; and the two solid rocket boosters (SRBs), with their combined thrust of some 2,630,800 kilograms (5.8 million lb-force). The SRBs take the Shuttle to an altitude of about 45 kilometers (28 mi) and a speed of 4,970 kilometers per hour (3,094 miles per hour) before they separate and fall back into the Atlantic Ocean to be retrieved, refurbished and prepared for another flight. After the solid rocket boosters are jettisoned, the Orbiter's three main engines, fed by the huge external tank, continue to burn and provide thrust for another six minutes before they are shut down at MECO (main engine cut-off), at which time the giant tank is jettisoned and falls back to Earth, disintegrating in the atmosphere with surviving pieces falling into remote ocean waters.

The Orbiter vehicle is both the heart and the brains of NASA's Space Transportation System. About the same size and weight as a DC-9 aircraft, the Orbiter contains the pressurized crew compartment (which can normally carry up to seven crew members), the huge cargo bay (which is 18.28 meters long and 4.57 meters in diameter [60 × 15 ft.]), and the three main engines mounted on its aft end. The thermal protection system (TPS), which protects the Orbiter during its very hot reentry through the Earth's atmosphere, was one Shuttle breakthrough technology that has proven much more challenging than originally expected.

The Orbiter's crew cabin has three levels. The uppermost is the flight deck where the commander and pilot control

Fig. 1   Astronaut Bruce McCandless performs the first untethered extravehicular activity using the Manned Maneuvering Unit during Shuttle mission 41-B (February 1984). (Courtesy of NASA.)

the mission. The middeck is where the galley, toilet, sleep stations and storage and experiment lockers are found. Also located in the middeck are the side hatch for passage to and from the Orbiter vehicle before launch and after landing, and the airlock hatch into the cargo bay and to outer space in support of on-orbit extravehicular activities (EVAs). Below the mid-deck floor is a utility area for air and water tanks.

The Orbiter's large cargo bay is adaptable to numerous tasks. It can carry satellites, large space platforms like the *Long-Duration Exposure Facility* (LDEF) (see Fig. 1) and even an entire scientific laboratory like the European Space Agency's *Spacelab* to and from low Earth orbit (LEO). It also serves as a work station for astronauts to repair satellites, a foundation from which to erect space structures and a place to store and hold spacecraft that have been retrieved from orbit for return to Earth.

Mounted on the port (left) side of the Orbiter's cargo bay behind the crew quarters is the remote manipulator system (RMS), which was developed and funded by the Canadian government. The RMS is a robot arm and hand with three joints similar to those found in a human being's shoulder, elbow and wrist. There are two television cameras mounted on the RMS near the "elbow" and "wrist." These cameras provide visual information for the astronauts who are operating the RMS from the aft station on the Orbiter's flight deck. The RMS is about 15 meters (50 ft) in length and can move anything from astronauts to satellites to and from the cargo bay as well as to different points in nearby outer space.

Table 1 summarizes the Space Shuttle program up to the *Challenger* accident. Despite the tragedy of the 51-L mission, this was an exciting and productive period in the U.S. manned space flight program—a time when some 125 crew

Fig. 1 The giant *Long Duration Exposure Facility* (LDEF) being deployed from the Orbiter's cargo bay during the STS 41-C mission in April 1984. The 10-tonne, 9.1 meters long and 4.3 meters in diameter 12-sided space platform was later retrieved by another Shuttle mission (STS 32) in January 1990. After almost six years in space, LDEF's 86 experiment trays represented a treasure in space environmental effects data. (The state of Florida appears on the Earth below.) (Courtesy of NASA.)

Table 1   Space Shuttle Launches

Kennedy Space Center
January 1, 1981–December 31, 1986
Total Space Shuttle Launches: 25

Columbia OV-102   *Discovery* OV-103
Challenger OV-099   *Atlantis* OV-104

Mission Name	Launch Date	Orbiter	Primary Payload	Launch Pad	Eastern Test Range	Result[a]
Operational Flight Tests						
STS-1 (Young, Crippen)	4/12/81	Columbia[b]	—	39A	9100	S
STS-2 (Engle, Truly)	11/12/81	Columbia[b]	OSTA-1	39A	9101	S
STS-3 (Lousma, Fullerton)	3/22/82	Columbia[c]	OSS-1	39A	9102	S
STS-4 (Mattingly, Hartsfield)	6/27/82	Columbia[b]	DoD 82-1	39A	9103	S
Operational Flights						
STS-5 (Brand, Overmeyer, Lenoir, Allen)	11/11/82	Columbia[b]	SBS-C/Anik C-3	39A	9104	S
STS-6 (Weitz, Bobko, Musgrave, Peterson)	4/4/83	Challenger[b]	TDRS-A	39A	9105	S
STS-7 (Crippen, Hauck, Ride, Fabian, Thagard)	6/18/83	Challenger[b]	Anik C-2/Palapa B-1	39A	9106	S
STS-8 (Truly, Brandenstein, Bluford, Gardner, Thornton)	8/30/83	Challenger[b]	INSAT 1B	39A	9107	S
STS-9 (Young, Shaw, Parker, Garriott, Merbold, Lichtenberg)	11/28/83	Columbia[b]	Spacelab 1	39A	9108	S
STS 41-B (Brand, Gibson, McCandless, McNair, Stewart)	2/3/84	Challenger[d]	Westar V1/Palapa B-2[c]	39A	3620	P
STS-41C (Crippen, Scobee, van Hoften, Nelson, Hart)	4/6/84	Challenger[b]	Long-Duration Exposure Facility	39A	6241	S
STS 41-D (Hartsfield, Coats, Hawley, Resnik, Mullane, Walker)	8/30/84	Discovery[b]	SBS-D/Telstar 3-C/ SYNCOM IV-2	39A	7013	S
STS 41-G (Crippen, McBride, Sullivan, Ride, Leestma, Garneau, Scully-Power)	10/5/84	Challenger[d]	ERBS/OSTA-3	39A	4651	S
STS 51-A (Hauck, Walker, Allen, Fisher, Gardner)	11/8/84	Discovery[d]	Anik D-2/SYNCOM IV-1	39A	7268	S
STS 51-C (Mattingly, Shriver, Buchli, Onizuka, Payton)	1/24/85	Discovery[d]	DoD	39A	9537	S
STS 51-D (Bobko, Williams, Seddon, Griggs, Hoffman, Garn, Walker)	4/12/85	Discovery[d]	Anik C-1/SYNCOM IV-3[f]	39A	8845	P

Mission Name	Launch Date	Orbiter	Primary Payload	Launch Pad	Eastern Test Range	Result[a]
Operational Flight Tests						
STS 51-B (Overmeyer, Gregory, Lind, Thagard, Thornton, van den Berg, Wang)	4/29/85	Challenger[b]	Spacelab 3	39A	5308	S
STS 51-G (Brandenstein, Creighton, Lucid, Nagel, Fabian, Baudry, Sultan Al-Saud)	6/17/85	Discovery[b]	Arabsat 1-B/ Telstar 3-D/ Morelos 1	39A	6559	S
STS 51-F (Fullerton, Bridges, Musgrave, England, Henize, Acton, Bartoe)	7/29/85	Challenger[b]	Spacelab 2	39A	3129	S
STS 51-I (Engle, Covey, Van Hoften, Lounge, Fisher)	8/27/85	Discovery[b]	AUSSAT 1/ASC 1/ SYNCOM IV-4	39A	1931	S
STS 51-J (Bobko, Grabe, Stewart, Hilmes, Pailes)	10/3/85	Atlantis[b]	DoD	39A	6983	S
STS 61-A (Hartsfield, Nagel, Buchli, Bluford, Dunbar, Furrer, Ockels, Messerschmid)	10/30/85	Challenger[b]	Spacelab D-1	39A	2470	S
STS 61-B (Shaw, O'Connor, Cleave, Spring, Ross, Vela, Walker)	11/26/85	Atlantis[b]	Morelos-B/AUSSAT-2/ RCA Satcom Ku-2	39A	4382	S
STS 61-C (Gibson, Bolden, Chang-Diaz, Hawley, Nelson, Cenker, Nelson)	1/12/86	Columbia[b]	RCA Satcom Ku-1	39A	0919	S
STS 51-L (Scobee, Smith, McNair, Resnik, Onizuka, Jarvis, McAuliffe)	1/28/86	Challenger	TDRS-B	39B	8603	U

[a] S, successful, P, partially successful; U, unsuccessful.
[b] Landed at Dryden Flight Research Facility, Edwards AFB, California.
[c] Landed at Space Harbor (formerly Northrup Strip), White Sands, New Mexico.
[d] Landed at Kennedy Space Center, Florida.
[e] Entered unplanned orbit; recovered on STS 51-A and returned to Earth on 11/16/84.
[f] Failed to activate after deployment; was successfully repaired during STS 51-I and later entered normal service.
SOURCE: NASA/KSC.

positions were flown successfully on the first 24 missions. A total of 19,970 crew hours were accumulated, or more than two and one-fourth years of total Shuttle crew experience. In this time period, the Shuttle fleet also flew more than 338,800 kilograms (747,000 lb-mass) of cargo to orbit, of which 136,100 kilograms (300,000 lb-mass) were deployed into space.

When the first four *Spacelab* missions flew inside Shuttle cargo bays during missions in 1983 and 1985, each with its own scientists and program of experiments, the amount of valuable scientific data generated was truly astonishing. The Orbiter serves as the "mothership" for the ESA *Spacelab*, which is designed to take full advantage of the space environment found in low earth orbit in support of inno-

vative scientific and engineering experiments. Highly skilled scientists from several nations have now taken advantage of microgravity, hard vacuum and the synoptic view of the Earth and the heavens to perform pioneering experiments in astrophysics, materials science, solar physics and space life sciences. *Spacelab* remains attached in the Orbiter's cargo bay during each mission and is then returned to Earth for refurbishment, reconfiguration and reflight with new equipment and experiments.

Since the Shuttle has returned to flight status (September 1988) a variety of interesting missions have been accomplished, including the rescue and return to Earth of the giant *Long-Duration Exposure Facility* (LDEF); the launch and successful deployment of the Hubble Space Telescope (although the telescope facility itself is now experiencing some operational difficulties); and the successful delivery and placement of two important interplanetary missions, the nuclear-powered *Galileo* spacecraft sent on a complex interplanetary trajectory to Jupiter and the *Magellan* spacecraft sent to orbit and radar-map the planet Venus.

The Space Shuttle fleet will remain the foundation of America's manned space program into the next century. When *Endeavour* (OV-105) joins its sister ships *Columbia* (OV-102), *Discovery* (OV-103) and *Atlantis* (OV-104) in the mid-1990s, the U.S. Space Transportation System should be able to fly 12 to 14 missions per year. Aerospace planners now anticipate that one hundred or more Shuttle missions are likely to occur in the last decade of the 20th century, and that some Orbiters may continue to fly until 2005 and perhaps even until 2010. The U.S. Space Station *Freedom* will depend heavily on the Space Shuttle. One contemporary space station development scenario estimates the need for some 20 Space Shuttle flights, beginning in 1995 and ending in 1998, to launch *Freedom*'s components and to support assembly and activation. (However, this particular mission sequence could change if unmanned, expendable rockets are also used to deliver Space Station components to orbit.) Yet, whatever the final mix of launch vehicles to build Space Station *Freedom*, the Space Shuttle will still play a major role in transporting the station's crews—our "extraterrestrial pioneers"—to America's first permanent outpost in space at the dawn of the next millennium!

See also: **Challenger accident; people in space; space launch vehicles; space station**

stars   A star is essentially a self-luminous ball of very hot gas that generates energy through thermonuclear fusion reactions that take place in its core.

Stars may be classified as either "normal" or "abnormal." Normal stars, like our Sun, shine steadily. These stars exhibit a variety of colors; red, orange, yellow, blue and white. Most stars are smaller than the Sun and many stars even resemble it. However, there are a few stars that are also much larger than our Sun. In addition, astronomers have observed several types of abnormal stars including giants, dwarfs and a variety of variable stars.

Most stars can be put into one of seven general spectral types called O, B, A, F, G, K and M (see table 1). This classification is a sequence established in order of decreasing surface temperature. Perhaps one easy way to remember this classification scheme is to use the mnemonic:

Oh Be A Fun Girl (or Guy), Kiss Me!

Our parent star the Sun is approximately 1.4 million kilometers (865,000 miles) in diameter and has an effective surface temperature of about 5,800 degrees Kelvin. The Sun, like other stars, is a giant nuclear furnace, in which the temperature, pressure and density are sufficient to cause light nuclei to join together or "fuse." For example, deep inside the solar interior, the hydrogen, which makes up 90 percent of the Sun's mass, is fused into helium atoms, releasing large amounts of energy that eventually works its way to the surface and is then radiated throughout the Solar System. The Sun is currently in a state of balance or equilibrium between two competing forces: gravity (which wants to pull all its mass inward) and the radiation pressure and hot gas pressure resulting from the thermonuclear reactions (which push outward).

Many stars in the Galaxy appear to have companions, with which they are gravitationally bound in binary, triple or even larger systems. Compared to other stars throughout the Galaxy, our Sun is slightly unusual. It does not have a known stellar companion. (However, the existence of a very distant, massive, dark companion called Nemesis has recently been postulated by some astrophysicists in an attempt to explain an apparent "cosmic catastrophe cycle" that occurs here on Earth.)

A STAR IS BORN

Astrophysicists have discovered what appears to be the life cycle of stars. Stars originate by the condensation of enormous clouds of cosmic dust and hydrogen gas, called nebulae. Gravity is the dominant force behind the birth of

Table 1   Stellar Spectral Classes

Type	Description	Typical Surface Temperatures (K)	Remarks
O	very hot, large blue stars [*hottest]	28,000–40,000	ultraviolet stars; very short lifetimes (3–6 million years)
B	large, hot blue stars	11,000–28,000	example: Rigel
A	blue-white, white stars	7,500–11,000	Vega, Sirius, Altair
F	white stars	6,000– 7,500	Canopis, Polaris
G	yellow stars	5,000– 6,000	the Sun
K	orange-red stars	3,500– 5,000	Arcturus, Aldebaran
M	red stars [*coolest]	<3,500	Antares, Betelgeuse

SOURCE: NASA.

a star. According to Newton's Universal Law of Gravitation, all bodies attract each other in proportion to their masses and distance apart. The dust and gas particles found in these huge interstellar clouds attract each other and gradually draw closer together. Eventually, enough of these particles join together to form a central clump that is sufficiently massive to bind all the other parts of the cloud by gravitation. At this point, the edges of the cloud start to collapse inward, separating it from the remaining dust and gas in the region.

Initially, the cloud contracts rapidly, because the thermal energy release related to contraction is easily radiated outward. However, when the cloud grows smaller and more dense, the heat released at the center cannot immediately escape to the outer surface. This causes a rapid rise in internal temperature, slowing down but not stopping the relentless gravitational contraction.

The actual birth of a star occurs when its interior becomes so dense and its temperature so high, that thermonuclear fusion occurs. The heat released in thermonuclear fusion reactions is greater than that released through gravitational contraction, and fusion becomes the star's primary energy producing mechanism. Gases heated by nuclear fusion at the cloud's center begin to rise, counterbalancing the inward pull of gravity on the outer layers. The star stops collapsing and reaches a state of equilibrium between these outward and inward forces. At this point, our star has become what astronomers and astrophysicists call a "main sequence star." Like our Sun, it will then remain in this state of equilibrium for billions of years, until all the hydrogen fuel in its core has been converted into helium.

Are stars being born today in our Galaxy? Scientists studying dense interstellar clouds with orbiting infrared telescopes have discovered an abundance of glowing objects hidden from optical telescopes by the intervening dust and gas. Because collapsing clouds radiate in the infrared portion of the electromagnetic spectrum until nuclear fusion processes begin, these infrared sources are considered to be stars in the birthing process.

MAIN SEQUENCE STARS

How long a star remains on the main sequence, burning hydrogen for its fuel, depends mostly on its mass. Our Sun has an estimated main sequence lifetime of about 10 billion years, of which approximately five billion years have now passed. Larger stars burn their fuels faster and at much higher temperatures. These stars, therefore, have short main sequence lifetimes, sometimes as little as one million years. In comparison, the "red dwarf stars," which typically have less than one-tenth the mass of our Sun, burn up so slowly that trillions of years must elapse before their hydrogen supply is exhausted. When a star has used up its hydrogen fuel, it leaves the "normal" state or departs the main sequence. This happens when the core of the star has been converted from hydrogen to helium by the thermonuclear reactions that have taken place.

RED GIANTS AND SUPERGIANTS

When the hydrogen fuel in the core of a main sequence star has been consumed, the core starts to collapse. At the same time, the hydrogen fusion process moves outward from the core into the surrounding, outer regions. There, the process of converting hydrogen into helium continues, releasing radiant energy. But as this burning process moves into the outer regions, the star's atmosphere expands greatly and it becomes a "red giant." The term giant is quite appropriate. If we put a red giant where our Sun is now, the innermost planet Mercury could be engulfed by it; similarly, if we put a larger "red supergiant" there, this supergiant would extend out past the orbit of Mars!

As the star's nuclear evolution continues, it might become a "variable star," pulsating in size and brightness over periods of several months to years. The visual brightness of such an "abnormal" star might now change by a factor of 100, while its total energy output varies by only a factor of two or three.

As an abnormal star grows, its contracting core may become so hot that it ignites and burns nuclear fuels other than hydrogen, beginning with the helium created in millions to perhaps billions of years of main sequence burning. The subsequent behavior of such a star is complex, but in general it can be characterized as a continuing series of gravitational contractions and new nuclear reaction ignitions. Each new series of fusion reactions produces a succession of heavier elements, in addition to releasing large quantities of energy. For example, the burning of helium produces carbon, the burning of carbon produces oxygen, and so forth.

Finally, when nuclear burning no longer releases enough radiant energy to support the giant star, it collapses and its dense central core becomes either a compact white dwarf or a tiny neutron star. This collapse may also trigger an explosion of the star's outer layers, which displays itself as a supernova. In exceptional cases with very massive stars, the core (or perhaps even the entire star) might even become a black hole.

WHITE DWARFS, NEUTRON STARS AND BLACK HOLES

When a star like our Sun has burned all the nuclear fuels available, it collapses under its own gravity until the collective resistance of the electrons within it finally stops the contraction process. The "dead star" has become a "white dwarf" and may now be about the size of the Earth. Its atoms are packed so tightly together that a fragment the size of a sugar cube would have a mass of thousands of kilograms! The white dwarf then cools for perhaps several billion years, going from white, to yellow, to red, and finally becomes a cold, dark sphere sometimes called a "black dwarf." (Note that the white dwarf is not experiencing nuclear burning; rather its light comes from a thin gaseous atmosphere that gradually dissipates its heat to space.) Astrophysicists estimate that there may be some 10 billion white dwarf stars in the Milky Way Galaxy alone, many of which have now become black dwarfs. This fate appears to be awaiting our own Sun and most other stars in the Galaxy.

However, when a star with a mass of about 1.5 to 3 times the mass of our Sun undergoes collapse, it will contract even further and ends up as a "neutron star," with a diameter of perhaps only 20 kilometers. In neutron stars, intense gravitational forces drive electrons into atomic nuclei, forcing them to combine with protons and transforming this combination into neutrons. Atomic nuclei are, therefore, obliterated in this process and only the collective resistance of neutrons to compression halts the collapse. At this point, the star's matter is so dense that each cubic centimeter has a mass of several billion tons!

But for stars that end their life having more than a few solar masses, even the resistance of neutrons is not enough to stop the unyielding gravitational collapse. In death such massive stars ultimately become "black holes." A black hole is an incredibly dense point mass or singularity that is surrounded by a literal "black region" in which gravitational attraction is so strong that nothing, not even light itself, can escape.

NOVAE, SUPERNOVAE AND PULSARS

Today, many physicists and astronomers relate the astronomical phenomena called supernovae and pulsars with neutron stars and their evolution. The final collapse of a giant star to the neutron stage may give rise to the physical conditions that cause its outer portions to explode, creating a "supernova." This type of cosmic explosion releases so much energy that its debris products will temporarily outshine all the hundreds of millions of ordinary stars in a galaxy.

A regular "nova" (the Latin word for "new," the plural of which is *novae*) that occurs more frequently is far less violent and spectacular. One common class, called "recurrent novae," is due to the nuclear ignition of gas being drawn from a companion star to the surface of a white dwarf. Such binary star systems are quite common and sometimes the stars will have orbits that regularly bring them close enough for one to draw off gas from the other.

When a supernova occurs at the end of a massive star's life, the violent explosion fills vast regions of space with matter that may radiate for hundreds or even thousands of years. The debris created by a supernova explosion will eventually cool into dust and gas, become part of a giant interstellar cloud and perhaps once again be condensed into a star or a planet. Most of the heavier elements found on Earth are thought to have originated in supernovae, because the normal thermonuclear fusion processes cannot produce such heavy elements. The violent power of a supernova explosion can, however, combine lighter elements into the heaviest ones found in nature (such as lead, thorium and uranium). Consequently, both our Sun and its planets were most likely enriched by infusions of material hurled into the interstellar void by ancient supernova explosions. That's right, you are made out of stardust!

Pulsars, first detected by radio astronomers in 1967, are sources of very accurately spaced bursts or pulses of radio signals. These radio wave signals are so regular, in fact, that the scientists who made the first detections were initially startled into thinking that they might have intercepted a radio signal from an intelligent alien civilization.

The pulsar, named because its radio wave signature regularly turns on and off or pulses, is considered to be a rapidly spinning neutron star. One pulsar is located in the center of the Crab Nebula, where a giant cloud of gas is still glowing from a supernova explosion that occurred in the year 1054 A.D.—a spectacular celestial event observed and recorded by ancient Chinese astronomers. The discovery of this pulsar has led scientists to make a great synthesis of our modern understanding of both pulsars and supernovae.

In a supernova explosion, a massive star is literally destroyed in an instant, but the explosive debris lingers and briefly outshines everything in a galaxy. In addition to scattering material all over interstellar space, supernova explosions leave behind a dense collapsed core made of neutrons. This neutron star, with an intense magnetic field, spins many times a second, emitting beams of radio waves, X-rays and other radiations. These radiations are possibly focused by the pulsar's powerful magnetic field and sweep through space much like a revolving lighthouse beacon. The neutron star, the end product of a violent supernova explosion, has become a pulsar!

Astrophysicists must now develop new theories to explain how pulsars can create intense radio waves, visible light, X-rays and gamma rays, all at the same time. Orbiting X-ray observatories, for example, have detected X-ray pulsars, such as Hercules X-1 in 1971. These X-ray pulsars are believed to be caused by a neutron star pulling gaseous matter from a normal companion star in a binary star system. As gas is sucked away from the normal companion to the surface of the neutron star, the gravitational attraction of the neutron star heats up the gas to millions of degrees Kelvin and causes it to emit X-rays.

The advent of the Space Age and the use of orbiting observatories to view the Universe as never before possible has greatly increased our knowledge about the many different types of stellar phenomena that make up the Universe. Most exciting of all, perhaps, is the fact that this process of astrophysical discovery has really only just begun!

See also: **astrophysics; black holes; fusion; Hubble Space Telescope; Sun**

starship A starship is a space vehicle capable of traveling the great distances between star systems. Even the closest stars in our Galaxy are often light-years apart. By convention, the word "starship" is used here to describe interstellar spaceships capable of carrying intelligent beings to other star systems; while robot interstellar spaceships are called interstellar probes.

What are the performance requirements for a starship?

First, and perhaps most important, the vessel should be capable of traveling at a significant fraction of the speed of light (c). Ten percent of the speed of the light (0.1c) is often considered as the lowest acceptable speed for a starship while cruising speeds of 0.9c and beyond are considered highly desirable. This "optic velocity" cruising capability is necessary to keep interstellar voyages to reasonable lengths of time, both for the home civilization and for the starship crew.

Consider, for example, a trip to the nearest star system Alpha Centauri—a triple star system about 4.23 light-years away. At a cruising speed of 0.1c, it would take about 43 years just to get there and another 43 years to return. The time dilation effects of travel at relativistic speeds would not help too much either, since a ship's clock would register the passage of about 42.8 years versus a terrestrial ground elapse time of 43 years. In other words, the crew would age about 43 years during the journey to Alpha Centauri. If we started with 20-year-old crewmembers departing from the Solar System in the year 2100 at a constant cruising speed of 0.1c, they would be approximately 63 years old when they reached the Alpha Centauri star system some

43 years later in 2143! The return journey would be even more dismal. Any surviving crewmembers would be 106 years old when the ship returned to the Solar System in the year 2186. Most if not all the crew would probably have died of old age or boredom. And that's for just a journey to the nearest star!

A starship should also provide a comfortable living environment for the crew and passengers (in the case of an interstellar ark). Living in a relatively small, isolated and confined habitat for a few decades to perhaps a few centuries can certainly overstress even the most psychologically adaptable individuals and their progeny. One common technique used in science fiction to avoid this crew stress problem is to have all or most of the crew placed in some form of "suspended animation" while the vehicle travels through the interstellar void, tended by a ship's company of smart robots.

Any properly designed starship must also provide an adequate amount of radiation protection for the crew, passengers and sensitive electronic equipment. Interstellar space is permeated with galactic cosmic rays. Nuclear radiation leakage from an advanced thermonuclear fusion

Table 1 Characteristics of Possible Starship Propulsion Systems

PULSED NUCLEAR FISSION SYSTEM (Project Orion)

• Principle Of Operation: Series of nuclear fission explosions are detonated at regular time intervals behind the vehicle; special giant pusher plate absorbs and reflects pulse of radiation from each atomic blast; system moves forward in series of pulses.

• Performance Characteristics: Very low efficiency in converting propellant (explosive device) mass into pure energy for propulsion; limited to number of nuclear explosives that can be carried on board; radiation hazards to crew (needs heavy shielding); probably limited to a maximum speed of about 0.01 to 0.10 the speed of light.

• Potential Applications: Most useful for interplanetary transport (especially for rapid movement to far reaches of Solar System); not suitable for a starship; very limited application for an interstellar robot probe; possible use for a very slow, huge interstellar ark (several centuries flight time). INTERPLANETARY VERSION COULD BE BUILT IN A DECADE OR SO; LIMITED INTERSTELLAR VERSION BY END OF 21st CENTURY

PULSED NUCLEAR FUSION SYSTEM (Project Daedalus)

• Principle of Operation: Thermonuclear burn of tiny deuterium/helium-3 pellets in special chamber (using laser or particle beam inertially confined fusion techniques); very energetic fusion reaction products exit chamber to produce forward thrust.

• Performance Characteristics: Uses energetic single step fusion reaction; thermonuclear propellant carried onboard vessel; maximum speed of about 0.12c considered possible.

• Potential Application: Not suitable for starship; possible use for robot interstellar probe (fly-by) mission or slow interstellar ark (centuries flight time). LIMITED SYSTEM MIGHT BE BUILT BY END OF 21st CENTURY (INTERSTELLAR PROBE).

INTERSTELLAR RAMJET

• Principle of Operation: First proposed by R. Bussard; after vehicle has an initial acceleration to near–light speed, its giant scoop (thousands of square kilometers in area) collects interstellar hydrogen which then fuels a proton–proton thermonuclear cycle or perhaps the carbon–cycle (both of which are found in stars); thermonuclear reaction products exit vehicle and provide forward thrust.

• Performance Characteristics: In principle, not limited by amount of propellant that can be carried; however, construction of light–mass giant scoop is major technical difficulty; in concept, cruising speeds of from 0.1c up to 0.9c might be obtained.

• Potential Applications: Starship; interstellar robot probe; giant space ark; WOULD REQUIRE MANY MAJOR TECHNOLOGICAL BREAKTHROUGHS—SEVERAL CENTURIES AWAY, IF EVER.

PHOTON ROCKET

• Principle of Operation: Uses matter and antimatter as propellant; equal amounts are combined and annihilate each other releasing an equivalent amount of energy in form of hard nuclear (gamma) radiation; these gamma rays are collected and emitted in a collimated beam out the back of vessel, providing a forward thrust.

• Performance Characteristics: The best (theoretical) propulsion system our understanding of physics will permit; cruising speeds from 0.1c to 0.99c.

• Potential Applications: Starship; interstellar probes (including self-replicating machines); large space arks; MANY MAJOR TECHNOLOGICAL BARRIERS MUST BE OVERCOME—CENTURIES AWAY, IF EVER.

engine or a matter/antimatter engine (photon rocket) must also be prevented from entering the crew compartment. In addition, the crew will have to be protected from nuclear radiation showers produced when a starship's hull, traveling at near light speed, slams into interstellar molecules, dust or gas. For example, a single proton (which we can assume is "stationary") being hit by a starship moving at 90 percent of the speed of light ($0.9c$) would appear to those on board like a one billion electron volt (GeV) proton being accelerated at them. Imagine traveling for years at the beam output end of a very-high-energy particle accelerator! Without proper deflectors or shielding, survival in the crew compartment from such radiation doses is doubtful.

To truly function as a starship, the vessel must be able to cruise at will, light-years from its home star system. The starship must also be able to accelerate to significant fractions of the speed of light; cruise at these near optic velocities; and then decelerate to explore a new star system or to investigate a derelict alien spaceship found adrift in the depths of interstellar space.

We will not discuss the obvious difficulties of navigating through interstellar space at near light velocities. It will be sufficient just to mention that when you "look" forward at near light speeds everything is "blueshifted": while when you look aft things appear "redshifted." The starship and its crew must be able to find their way from one location in the Galaxy to another—on their own.

What appears to be the major engineering technology needed to make the starship a real part of our extraterrestrial civilization is an effective propulsion system. Interstellar class propulsion technology is the key to the Galaxy for any emerging civilization that has mastered space flight within and to the limits of its own solar system. Despite the tremendous engineering difficulties associated with the development of a starship propulsion system, several concepts have been proposed. These include the pulsed nuclear fission engine (Project Orion concept), the pulsed nuclear fusion concept (Project Daedalus study), the interstellar nuclear ramjet and the photon rocket. These systems are briefly described in table 1 along with their potential advantages and disadvantages.

Unfortunately, based on our current understanding of the laws of physics, all known phenomena and mechanisms that might be used to power a starship are either not energetic enough or simply entirely out of the reach of today's technology and even the technology levels anticipated for several tomorrows. Perhaps major breakthroughs will occur in our understanding of the physical laws of the Universe—breakthroughs that provide insight into more intense energy sources or ways around the speed of light barrier now imposed by the theory of relativity. But until such new insights occur (if ever), human travel to another star system on board a starship must remain in the realm of future dreams.

See also: **interstellar contact; Project Daedalus: Project Orion (II); relativity; space nuclear propulsion**

star wars Interstellar warfare between two or more advanced extraterrestrial civilizations. The question still remains open as to whether intelligent creatures can develop the high-technology tools needed for interstellar travel without destroying themselves and their home planet(s) in the process. If alien creatures learn to live with their advanced technologies, it is highly probable that when they expand out into other star systems it will be a peaceful expansion. The alternative, unfortunately, is a barbaric struggle for domination of a particular star system followed, perhaps (if there are any survivors), by a belligerent expansion of the winning alien faction across interstellar space.

In this latter situation, contact with another intelligent species would almost certainly result in some form of interstellar warfare. if both civilizations have similar levels of technology, a distant planetary system around a mutually prized star might be the scene of violent conflict on an astronomical scale. A peacefully expanding race colliding with a belligerent race might withdraw, might elect to defend itself or might be quickly dispatched (due to a lack of weapons).

If a belligerent alien race encounters intelligent creatures with greatly inferior technologies, rapid annihilation, mass destruction or enslavement of the inferior civilization can be anticipated. If, however, the technology gap is not as great (say, for example, the star system inhabitants have already developed nuclear technologies and interplanetary travel), then the alien invaders might encounter severe resistance and conquest of the star system would be achieved only after heavy invader losses—a Pyrrhic victory on an interstellar scale.

While speculative, of course, these scenarios for interstellar conflict have numerous analogs in terrestrial history. Are intelligent creatures really the same everywhere in the Universe?

See also: **extraterrestrial civilizations; interstellar contact**

Star Wars (I) In science fiction, *Star Wars* is the first of George Lucas' famous trilogy of extraterrestrial adventure movies in which Luke Skywalker, Princess Leia Organa and a charming collection of assorted heroes battle the evil imperial forces of an ancient galactic empire, led by Lord Darth Vader and a host of heavies.

Star Wars (II) An unofficial but popular name given to the Strategic Defense Initiative (SDI)—a U.S. Department of Defense research program, initiated by (former) President Reagan in 1983, with the purpose of investigating methods of defending the United States and its allies from the threat of enemy nuclear-tipped ballistic missiles. This name originated from the extraterrestrial movie trilogy created by George Lucas.

stellar magnitude The relative luminance or brightness of a celestial body. In this system of measuring relative brightness, the smaller the number indicating magnitude, the more luminous the celestial object. Zero magnitude is the brightest star, and naked eye visual detection limits usually extend down to magnitude 6 (although on a perfectly black background the limit for a single luminous point approaches the 8th magnitude). The difference between successive magnitudes (for example, the 3rd and 4th magnitude) is 2.512 (the fifth root of 100), while the difference over five magnitudes (that is, 1st to 6th) is approximately 100. The planets, the Moon and the Sun are all brighter than magnitude zero and are therefore assigned negative magnitudes. For example, at their brightest the planet Venus has a relative visual magnitude of about −4.4 and the full Moon −12.7.

See also: **stars**

Sun   The Sun is our parent star and the massive, luminous celestial object about which all other bodies in our Solar System revolve. It provides the light and warmth upon which all terrestrial life depends. Its gravitational field determines the movement of the planets and other celestial bodies. The Sun is a main sequence star of spectral type G-2. In any other place in the Galaxy, it would hardly appear worthy of special study by alien astronomers and astrophysicists; but its central position in our Solar System makes it a unique object for research. Like all main sequence stars, the Sun derives its abundant energy output from thermonuclear fusion reactions involving the conversion of hydrogen to helium and heavier nuclei. Photons associated with these exothermic (energy-releasing) fusion reactions diffuse outward from the Sun's core, until after several thousand years, they reach the convective envelope. Another by-product of the thermonuclear fusion reactions is a flux of neutrinos that freely escape from the Sun.

At the center of the Sun is the core, where energy is released in thermonuclear reactions. Surrounding the core are concentric shells called the radiative zone, the convective envelope (which occurs at approximately 0.8 of the Sun's radius), the photosphere (which is the layer from which visible radiation emerges), the chromosphere and, finally, the corona (which is the Sun's outer atmosphere). Energy is transported outward through the convective envelope by convective (mixing) motions that are organized into cells. The Sun's lower or inner atmosphere, the photosphere, is the region from which energy is radiated directly into space. Solar radiation approximates a Planck distribution (blackbody source) with an effective temperature of 5,800 Kelvin. Table 1 provides a summary of the physical properties of the Sun.

The chromosphere, which extends for a few thousand kilometers above the photosphere, has a maximum temperature of approximately 10,000 Kelvin. The corona, which extends several solar radii above the chromosphere, has temperatures of over one million degrees Kelvin. These regions emit electromagnetic radiation in the ultraviolet (UV), extreme ultraviolet (EUV) and X-ray portions of the spectrum. This shorter wavelength EM radiation, though

Table 1   Physical and Dynamic Properties of the Sun

Diameter	1.392×10^6 km
Mass	1.99×10^{30} kg
Distance from the Earth (average)	1.496×10^8 km [1 AU] (8.3 light-min)
Luminosity	3.9×10^{26} watts
Density (average)	1.41 g/cm³
Equivalent blackbody temperature	5,800° Kelvin
Central temperature (approx.)	15,000,000° Kelvin
Rotation period (varies with latitude zones)	27 days (approx.)
Radiant energy output per unit surface area	6.4×10^7 W/m²
Solar cycle (total cycle of polarity reversals of Sun's magnetic field)	22 years
Sunspot cycle	11 years (approx.)
Solar constant (at 1 AU)	1371 ± 5 W/m²

SOURCE: NASA.

representing a relatively small portion of the Sun's total energy output, still plays a dominant role in forming planetary ionospheres and in photochemistry reactions occurring in planetary atmospheres.

Since the Sun's outer atmosphere is heated, it expands into the surrounding interplanetary medium. This continuous outward flow of plasma is called the solar wind. It consists of protons, electrons and alpha particles as well as small quantities of heavier ions. Typical particle velocities in the solar wind fall between 300 and 400 kilometers per second, but these velocities may get as high as 1,000 kilometers per second.

Although the total energy output of the Sun is remarkably steady, its surface displays many types of irregularities. These include sunspots, faculae, plages (bright areas), filaments, prominences and flares. All are believed to be ultimately the result of interactions between ionized gases in the solar atmosphere and the Sun's magnetic field. Most solar activity follows the sunspot cycle. The number of sunspots varies, with a period of about 11 years. However, this approximately 11-year sunspot cycle is only one aspect of a more general 22-year *solar cycle* that corresponds to a reversal of the polarity patterns of the Sun's magnetic field.

Sunspots were originally observed by Galileo in 1610. They are less bright than the adjacent portions of the Sun's surface, because they are not as hot. A typical sunspot temperature might be 4,500 Kelvin compared to the photosphere's temperature of 5,800 Kelvin. Sunspots appear to be made up of gases boiling up from the Sun's interior. A small sunspot may be about the size of the Earth, while larger ones could hold several hundred or even thousands of earth-sized planets. Extra-bright solar regions, called plages, often overlie sunspots. The number and size of sunspots appear to rise and fall through a fundamental 11-year cycle (or in an overall 22-year cycle, if you consider polarity reversals in the Sun's magnetic field). The greatest number occur in years when the Sun's magnetic field is the most severely twisted (called sunspot maximum). Solar physicists think that sunspot migration causes the Sun's magnetic field to reverse its direction. It then takes another 22 years for the Sun's magnetic field to return to its original configuration.

A solar flare is the sudden release of tremendous energy and material from the Sun. A flare may last minutes or hours and it usually occurs in complex magnetic regions near sunspots. Exactly how or why enormous amounts of energy are liberated in solar flares is still unknown, but scientists think the process is associated with electrical currents generated by changing magnetic fields. The maximum number of solar flares appears to accompany the increased activity of the sunspot cycle. As a flare erupts, it discharges a large quantity of material outward from the Sun. This violent eruption also sends shock waves through the solar wind.

Space-based solar observatories, including *Skylab* and *Solar Maximum Mission* data, have indicated that prominences (condensed streams of ionized hydrogen atoms) appear to spring from sunspot groups. Their looping shape suggests that these prominences are controlled by strong magnetic fields. About 100 times as dense as the solar corona, prominences can rise at speeds of hundreds of kilometers per second. Sometimes the upper end of a prominence curves back to the Sun's surface forming a "bridge" of hot glowing gas hundreds of thousands of kilometers long. On other occasions, the material in the prominence jets out and becomes part of the solar wind.

High-energy particles are released into heliocentric space by solar events, such as large solar flares. Because of their close association with large flares, these bursts of energetic particles are relatively infrequent. Solar flares represent a real hazard to astronauts in cislunar space, on the surface of the Moon, and traveling through interplanetary space on the way to and from Mars.

About five billion years from now, the Sun will have used up all the hydrogen fuel in its core and converted this hydrogen into helium. It will also have expanded and cooled. The hydrogen in the shell around the core will then begin thermonuclear burning. In the core itself, a major event called helium flash will occur. This is the initiation of a new thermonuclear reaction in which helium begins fusing and creates carbon (from three helium atoms) and oxygen (from one carbon atom and one helium atom). The expansion and cooling of the Sun's exterior surface will be accelerated. Our parent star will leave the main sequence and become a red giant. During this expansion, the Sun will probably grow large enough to engulf the Earth—boiling off all water and incinerating the land. This double shell burning of hydrogen and helium will then continue until thermal instabilities develop. These instabilities will cause the Sun to pulsate and eventually it will eject its outer shell of gases into space. The remaining core will contract until it is about the size of the Earth, forming an incredibly dense white dwarf star. This white dwarf will continue to cool itself by emitting ultraviolet radiation for many billions of years. By that time, the starfaring descendents of the human race will have spread throughout the Galaxy. Among these terrestrial progeny, some astrophysicists will look back to this portion of the Galaxy and observe the death of the star system that gave rise (perhaps uniquely) to intelligent life in the Galaxy!

See also: **fusion; stars**

sunlike stars   Yellow, main sequence stars with 5,000 to 6,000 degree Kelvin surface temperatures; spectral type G stars.

See also: **stars; Sun**

superior planets   Planets that have orbits around the Sun that lie outside the Earth's orbit. These planets include Mars, Jupiter, Saturn, Uranus, Neptune, and Pluto.

superluminal   With a speed greater than the speed of light.

T

tachyon   A hypothetical faster-than-light particle. If it exists, advanced alien civilizations may use it in some way to achieve more rapid interstellar communications.

teleportation   A concept used in science fiction to describe the instantaneous movement of material objects to other locations in the Universe.

See also: **science fiction**

Tenth Planet   Some astronomers have speculated that the currently estimated mass of Pluto is too small to account for observed perturbations in the orbits of Uranus and Neptune. Although the source of these perturbations is not presently known, one hypothesis that has been offered is that there is a massive object (perhaps a few times more massive than Earth) circling our Sun far beyond the orbit of Pluto. This theorized "Tenth Planet" (also referred to as "Planet X") has never been detected, and today the Tenth Planet hypothesis itself is not widely accepted as an adequate explanation of the perturbation data.

See also: **Nemesis,** *Pioneer 10, 11*

terran Of or relating to the planet Earth; a native of the planet Earth.

See also: **terrestrial**

terrestrial Of or pertaining to the Earth; an inhabitant of the planet Earth.

terrestrial planets In addition to the Earth itself, the terrestrial (or inner) planets include Mercury, Venus and Mars. These planets are similar in their general properties and characteristics to the Earth; that is, they are small, relatively highly dense bodies, composed of metals and silicates with shallow (or no) atmospheres as compared to the gaseous outer planets.

See also: **Earth; Mars; Mercury; Venus**

theorem of detailed cosmic reversibility A premise developed by Francis Crick and Leslie Orgel in support of their directed panspermia hypothesis. This theorem states that if we can now contaminate another world in our Solar System with terrestrial microorganisms, then it is also reasonable to assume that an intelligent alien civilization could have developed the advanced technologies needed to "infect" or seed the early prebiotic Earth with spores, microorganisms or bacteria.

See also: **extraterrestrial contamination; panspermia**

Thousand Astronomical Unit (TAU) Mission A proposed future NASA mission involving an advanced-technology robot spacecraft that would be sent on a 50-year journey into very deep space about 1,000 astronomical units (some 160 billion kilometers) away from Earth. The TAU spacecraft would feature an advanced multimegawatt nuclear reactor, ion propulsion and a laser (optical) communication system. Initially, the TAU spacecraft would be directed for an encounter with Pluto and its moon Charon; followed by passage through the heliopause; perhaps even reaching the inner Oort cloud (the hypothetical region where comets are thought to originate) at the end of its long mission. This advanced robot spacecraft would investigate low-energy cosmic rays, low-frequency radio waves, interstellar gases and other deep space phenomena. It would also perform high-precision astrometry (the measurement of distances between stars).

See also: **comet; robotics in space; space nuclear propulsion**

Tunguska event A violent explosion that occurred in a remote part of Siberia (Soviet Union) in June 1908. It is currently believed that this event was caused by the entrance of a cometary nucleus into the Earth's atmosphere. However, a few of the original investigators speculated that this destructive event was caused by the explosion of an alien spacecraft. (No firm technical evidence has been accumulated to support the latter hypothesis.)

See also: **comet; extraterrestrial catastrophe theory**

U

ultraviolet (UV) astronomy Astronomy based on ultraviolet (UV, 10 to 400 nanometer wavelength) portion of the electromagnetic (EM) spectrum. Because of the strong absorption of UV radiation by the Earth's atmosphere, ultraviolet astronomy must be performed using high-altitude balloons, rocket probes and orbiting observatories. Ultraviolet data gathered from spacecraft are extremely useful in investigating interstellar and intergalactic phenomena. Observations in the ultraviolet wavelengths have shown, for example, that the very-low-density material that can be found in the interstellar medium is quite similar in composition throughout our Galaxy, but that its distribution is far from homogeneous. In fact, UV data have led some space scientists to postulate that low-density cavities or "bubbles" in interstellar space are caused by supernova explosions and are filled with gases that are much hotter than the surrounding interstellar medium. Ultraviolet data gathered from space-based observatories, such as the *International Ultraviolet Explorer* (IUE), which was launched in 1978 as a joint ESA-NASA-British Science and Engineering Council mission, have revealed that some stars blow off material in irregular bursts and not in a steady flow as was originally thought. Ultraviolet data are also of considerable use in studying many of the phenomena that occur in distant galaxies. The IUE mission is expected to continue to produce useful data until late 1991. In late 1991, NASA plans to launch the *Extreme Ultraviolet Explorer* (EUVE) mission, which will investigate stellar, galactic and extragalactic objects by gathering data in the 10- to 100-nanometer region of the EM spectrum.

See also: **astrophysics**

ultraviolet (UV) radiation That portion of the electromagnetic spectrum that lies beyond visible (violet) light and is longer in wavelength than X-rays. Generally taken as electromagnetic radiation with wavelengths between 400 nanometers (just past violet light in the visible spectrum) and 10 nanometers (the extreme ultraviolet cutoff and the beginning of X-rays).

See also: **electromagnetic spectrum**

Ulysses Mission An international space project to study the poles of the Sun and the interstellar environment above and below these solar polar regions. The *Ulysses* spacecraft was built by Dornier Systems of Germany for the European

Space Agency (ESA), which is also responsible for on-orbit operations of the Ulysses mission. NASA provided launch support using the Space Shuttle *Discovery* and an upper-stage configuration consisting of a two-stage inertial upper stage (IUS) rocket and a PAM-S (Payload Assist Module) configuration. In addition, the United States, through the Department of Energy, is providing the radioisotope thermoelectric generator (RTG) that will provide electric power to this spacecraft. The *Ulysses* spacecraft will be tracked and scientific data will be collected by NASA's Deep Space Network (DSN). Spacecraft monitoring and control, as well as data reduction and analysis, will be performed at the NASA's Jet Propulsion Laboratory (JPL) by a joint ESA/JPL team.

The Ulysses mission, named for the legendary Greek hero in Homer's epic saga of the Trojan War who wandered into many previously unexplored areas on his return home, is a survey mission designed to support the following scientific objectives: to examine the properties of the solar wind, the structure of the Sun–solar wind interface, the heliospheric magnetic field, solar radio bursts and plasma waves, solar and galactic cosmic rays and the interplanetary/interstellar neutral gas and dust environment—all as a function of solar latitude.

Most important during this mission will be the data collected at high solar latitudes, near the polar regions of the Sun, which have never been studied by spacecraft. In fact, all spacecraft that have studied the Sun to date have done so in or near the Solar System's ecliptic plane. (The ecliptic is the plane in which the Earth orbits the Sun.) The scientific data returned by the 370-kg (814-lb) Ulysses spacecraft are expected to assist space scientists in their studies of the Sun and the adjacent interstellar environment. Because of the structure and shape of the Sun's magnetic field, space scientists expect to see very interesting phenomena both outbound from the Sun and inbound from interstellar space in both the northern and southern polar regions of the Sun.

The Space Shuttle *Discovery* carried the *Ulysses* spacecraft into Earth orbit on October 6, 1990, and then successfully deployed the spacecraft on an interplanetary trajectory that will ultimately let it explore the polar regions of the Sun. Because no present launch vehicle has enough energy to lift this spacecraft directly from Earth on a trajectory over the Sun's poles, *Ulysses* is first being sent to Jupiter by a compound upper stage configuration consisting of a two-stage IUS vehicle and a PAM-S vehicle. As shown in figure 1, *Ulysses* will then encounter the planet Jupiter in February 1992 (some 14 months after launch). Using the gravity-assist technique, Jupiter's enormous gravitational attraction will be used to change the spacecraft's trajectory so that, when it departs Jupiter, *Ulysses* will be climbing out of the ecliptic plane and heading for the Sun's southern polar region. The first solar polar encounter will begin when *Ulysses* reaches 70 degrees south solar latitude in mid-1994. The spacecraft will spend approximately four months above that latitude, about 2.3

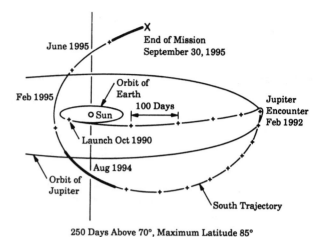

Fig. 1 The Ulysses mission profile in space and time. (Chart courtesy of NASA/JPL.)

astronomical units distance from the Sun. (An astronomical unit [AU] is about 150 million kilometers or 93 million miles, the average distance of the Earth from the Sun.)

The Sun's giant gravitational field will then bend the spacecraft's trajectory, enabling *Ulysses* to cross the Sun's equator in late 1994 (at about 1.4 AU distance) and then travel on toward the Sun's northern pole. During its second solar polar encounter, *Ulysses* will spend approximately four months at solar latitude greater than 70 degrees. The Ulysses mission is expected to terminate in September 1995. Prior to being renamed *Ulysses* in 1984 after a major project restructuring, this project (which originally involved two spacecraft, one provided by ESA and and one by NASA) was called the International Solar Polar Mission.

See also: **European Space Agency; National Aeronautics and Space Administration; orbit of objects in space** (gravity assist); **space launch vehicles; space nuclear power; Sun**

unidentified flying object (UFO) A flying object (apparently) seen in the terrestrial skies by an observer who cannot determine its nature. The vast majority of such "UFO" sightings can, in fact, be explained by known phenomena. However, these phenomena may be beyond the knowledge or experience of the person making the observation. Common phenomena that have given rise to UFO reports include artificial Earth satellites, aircraft, high-altitude weather balloons, certain types of clouds and even the planet Venus.

There are, nonetheless, some reported sightings that cannot be fully explained on the basis of the data available (which may be insufficient or too scientifically unreliable) or on the basis of comparison with known phenomena. It is the investigation of these relatively few UFO sighting cases that has given rise, since the end of World War II, to the "UFO hypothesis." This popular (though technically unfounded) hypothesis speculates that these unidentified flying objects are under the control of extraterrestrial beings who are surveying and visiting the Earth.

Modern interest in UFOs appears to have begun with a sighting report made by a private pilot named Kenneth Arnold. In June 1947 he reported seeing a mysterious formation of shining disks in the daytime near Mount Rainier in the State of Washington. When newspaper reporters heard of his account of "shining saucer-like disks"—the popular term "flying saucer" was born.

In 1948 the United States Air Force began to investigate these UFO reports. Project Sign was the name given by the Air Force to its initial study of UFO phenomena. In the late 1940s Project Sign was replaced by Project Grudge, which in turn became the more familiar Project Blue Book. Under Project Blue Book the Air Force investigated many UFO reports from 1952 to 1969. Then on December 17, 1969, the Secretary of the Air Force announced the termination of Project Blue Book.

The Air Force decision to discontinue UFO investigations was based on the following circumstances: (1) an evaluation of a report prepared by the University of Colorado and entitled, "Scientific Study of Unidentified Flying Objects" (this report is also often called the Condon report after its principal author); (2) a review of this University of Colorado report by the National Academy of Sciences; (3) previous UFO studies; and (4) Air Force experience from nearly two decades of UFO report investigations.

As a result of these investigations and studies and of experience gained from UFO reports since 1948, the conclusions of the Air Force were: (1) no UFO reported, investigated and evaluated by the Air Force ever have any indication of threatening national security; (2) there was no evidence submitted to or discovered by the Air Force that sightings categorized as "unidentified" represent technological developments or principles beyond the range of present-day scientific knowledge; and (3) there was no evidence to indicate that the sightings categorized as "unidentified" are extraterrestrial vehicles.

With the termination of Project Blue Book, the Air Force regulation establishing and controlling the program for investigating and analyzing UFOs was rescinded. All documentation regarding Project Blue Book investigations was then transferred to the Modern Military Branch, National Archives and Records Service, 8th Street and Pennsylvania Avenue N.W., Washington, D. C. 20408. This material is presently available for public review and analysis. If you wish to review these files personally, you need to simply obtain a researcher's permit from the National Archives and Record Service.

Today, reports of unidentified objects entering United States air space are still of interest to the military as part of its overall defense surveillance program. But beyond that, the Air Force no longer investigates reports of UFO sightings.

Similarly, while NASA is the focal point for answering public inquiries to the White House concerning UFOs, it is not engaged in a research program involving these UFO phenomena or sightings—nor is any other agency of the United States government.

One interesting result that emerged from Project Blue Book is a scheme, developed by Dr. J. Allen Hynek, to classify or categorize UFO sighting reports. Table 1 describes the six levels of classification that have ben used. A Type-A UFO report generally involves seeing bright lights in the night sky. These sightings usually turn out to be a planet (typically Venus), a satellite, an airplane or meteors. A Type-B UFO report often involves the daytime observation of shining disks (that is, flying saucers) or cigar-shaped metal objects. This type of sighting usually ends up as a weather balloon, a blimp or lighter-than-air ship or even a deliberate prank or hoax. A Type-C UFO report involves unknown images appearing on a radar screen. These signatures might linger, be tracked for a few moments or simply appear and then quickly disappear—often to the amazement and frustration of the scope operator. These radar visuals often turn out to be something like swarms of insects, flocks of birds, unannounced aircraft and perhaps the unusual phenomena radar operators call "angels." (To radar operators, angels are anomalous radar wave propagation phenomena.)

Close encounters of the first kind (visual sighting of a UFO at moderate to close range) represent the Type-D UFO reports. Usually, the observer reports something unusual in the sky that "resembles an alien spacecraft." In the Type-E UFO report, not only does the observer claim to have seen the alien spaceship but also reports the discovery of some physical evidence in the terrestrial biosphere (such as scorched ground, radioactivity, mutilated animals, etc.) that is associated with the alien craft's visit. This type of sighting has been named a close encounter of the second kind. Finally, in the last type of UFO report, a close encounter of the third kind, the observer claims to have seen and sometimes to have been contacted by the alien visitors. Extraterrestrial contact stories range from simple sightings of "ufonauts," to communication with them

Table 1   UFO Report Classifications

A. NOCTURAL (Nighttime) LIGHT

B. DIURNAL (Daytime) DISK

C. RADAR CONTACT (Radar Visual or "RV")

D. VISUAL SIGHTING OF ALIEN CRAFT AT MODEST TO CLOSE RANGE
[also called: Close Encounter of the First Kind, CE I]

E. VISUAL SIGHTING OF ALIEN CRAFT PLUS DISCOVERY OF (hard) PHYSICAL EVIDENCE OF CRAFT'S INTERACTION WITH TERRESTRIAL ENVIRONMENT
[also called: Close Encounter of Second Kind, CE II]

F. VISUAL SIGHTING OF ALIENS THEMSELVES, INCLUDING POSSIBLE PHYSICAL CONTACT
[also called: Close Encounter of Third Kind, CE III]

SOURCE: Derived from work of Dr. J. Allen Hynek and Project Blue Book.

(usually telepathic), to cases of kidnapping and then release of the terrestrial observer. There are even some reported stories in which a terran was kidnapped and then seduced by an alien visitor—a challenging task of romantic compatibility even for an advanced starfaring species!

Despite numerous stories about such UFO encounters, not a single shred of scientifically credible, indisputable evidence has yet to be acquired! If we were to judge these reports on some arbitrary proof scale, table 2 might be used as a guide for helping us determine what type of data or testimony we will need to convince ourselves that the "little green men" have arrived in their flying saucer. Unfortunately, we do not have any convincing data to support categories 1 to 3 in table 2. Instead, all we have are large quantities of eyewitness accounts of various UFO encounters (category 4 items in table 2). Even the most sincere human testimony changes in time and is often subject to wide variations and contradiction. The scientific method puts very little weight on human testimony in validating a hypothesis.

Even from a more philosophical point of view, it is very difficult to logically accept the UFO hypothesis. Although intelligent life may certainly have evolved elsewhere in the Universe, the UFO encounters reported to date hardly reflect the logical exploration patterns and encounter sequences we might anticipate from an advanced, starfaring alien civilization.

From our current understanding of the laws of physics, interstellar travel appears to be an extremely challenging, if not technically impossible, undertaking. Any alien race that developed the advanced technologies necessary to travel across vast interstellar distances would most certainly be capable of developing sophisticated remote-sensing technologies. With these remote-sensing technologies they could study the Earth essentially undetected—unless, of course, they wanted to *be* detected. And if they wanted to make contact, they could most surely observe where the Earth's population centers are and land in places where they could communicate with competent authorities. It is insulting not only to their intelligence but to our own

human intelligence as well to think that these alien visitors would repeatedly contact only people in remote, isolated areas, scare the dickens out of them and then lift off into the sky. Why not once land in the middle of the Orange Bowl during a football game or near the site of an international meeting of astronomers and astrophysicists! And why only short, momentary intrusions into the terrestrial biosphere? After all, the *Viking* Landers we sent to Mars gathered data for years. It's really hard to imagine that an advanced culture would make the tremendous resource investment to send a robot probe or even to come themselves and then only flicker through an encounter with beings on this planet. Are we that uninteresting? If that's the case, then why so many reported visits? From a simple exercise of logic, the UFO hypothesis just doesn't make sense—terrestrial or extraterrestrial!

Hundreds of UFO reports have been made since the late 1940s. Again, why are we so interesting? Are we at a galactic crossroads? Are our outer planets an "interstellar truck stop" where alien starships pull in and refuel? (Some people have already proposed this hypothesis.) Let's play a simplified interstellar traveler game to see if so many reported visits are realistic—even if we are very interesting. First, we'll assume that our Galaxy of over 100 billion stars contains about 100,000 different starfaring alien civilizations which are more or less uniformly dispersed. (This is a very *optimistic* number according to the Drake Equation and scientists who have speculated about the likelihood of Kardashev Type II civilizations.) Then each of these ET civilizations has, in principle, one million other star systems to visit without interfering with any other civilization. Yes, the Galaxy is a big place! What do you think the odds are of two of these civilizations both visiting our Solar System and only casually exploring the planet Earth during the last three decades? The only realistic conclusion that can be drawn is that the UFO reports are not credible indications of extraterrestrial visitations!

See also: **ancient astronauts; Drake Equation; extraterrestrial civilizations; interstellar contact**

Universe Everything that came into being at the moment of the Big Bang, and everything that has evolved from that initial mass or energy; everything that we can (in principle) observe. All energy (radiation), all matter, and the space that contains them.

See also: **"Big Bang" theory; closed universe; cosmology; open universe**

Uranian Of or relating to the planet Uranus; (in science fiction) a native of the planet Uranus.

Uranus Unknown to ancient astronomers, the planet Uranus was discovered by Sir William Herschel in 1781. Herschel attempted to name the new planet *Georgium Sidus* (or "George's star") after the reigning King of England, George III. Other 18-century astronomers chose to call the newly found celestial object "Herschel," in honor

Table 2 Proposed "Proof Scale" to Establish Existence of UFOs

Highest Value*

(1)	The alien visitors themselves; or the alien spaceship
(2)	Irrefutable physical evidence of a visit by aliens or the passage of their spaceship
(3)	Indisputable photograph of an alien spacecraft or one of its occupants
(4)	Human eyewitness reports

Lowest Value

*from a standpoint of the scientific method and validating the UFO hypothesis with "hard" technical data.

SOURCE: Based on work of Dr. J. Allen Hynek and Project Blue Book.

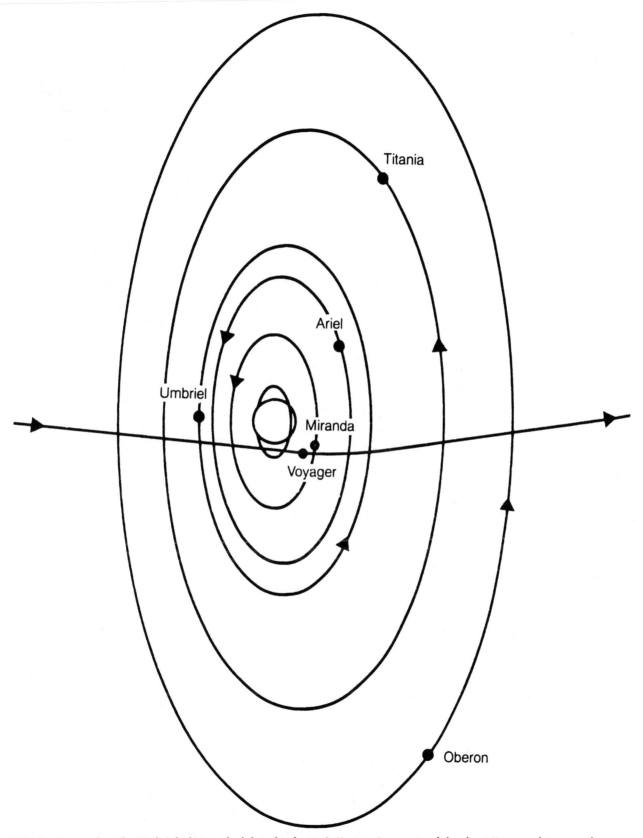

Fig. 1.   As seen from the Earth (which is on the left in this drawing), *Voyager 2* encountered the planet Uranus and its moons by passing to the right of the planet and behind it on January 24, 1986, with an encounter velocity of over 14 kilometers per second. The giant celestial bull's-eye appearance of the Uranian system occurs because the planet is actually tipped over on its side (with the south pole facing the Sun and the Earth during the encounter). (Courtesy of NASA.)

of its discoverer. In the end, however, the use of mythological names for the planets prevailed, and by the middle of the 19th century the seventh planet from the Sun was called Uranus after the ancient Greek god of the sky and father of the Titan, Cronos (Saturn in Roman mythology).

At nearly 3 billion kilometers from the Sun, Uranus is too distant from Earth to permit telescopic imaging of its features. Because of the methane in its upper atmosphere, the planet appears as only a blue-green disk or blob in the most powerful of terrestrial telescopes. On January 24, 1986, a revolution took place in our understanding and knowledge about the planet, as the *Voyager 2* spacecraft encountered the Uranian system at a relative velocity of over 14 kilometers per second. What we know about Uranus today is largely the result of that spectacular planetary encounter. (See fig. 1.)

Uranus has one particularly interesting property—its axis of rotation lies in the plane of its orbit rather than vertical to the orbital plane as occurs with the other planets. Because of this curious situation, Uranus moves around the Sun like a barrel rolling along on its side rather than like a top spinning on its end. In other words, Uranus is tipped over on its side, with its orbiting moons and rings creating the appearance of a giant celestial bull's-eye. The northern and southern polar regions are alternatively exposed to sunlight or to the darkness of space during the planet's 84-year-long orbit around the Sun.

At its closest approach, *Voyager 2* came within 81,500 kilometers (50,600 mi) of Uranus's cloudtops. The spacecraft telemetered back to Earth thousands of spectacular images and large quantities of other scientific data about the planet, its moons, rings, atmosphere and interior.

Voyager 2's images of the five largest (previously known) moons around Uranus revealed complex features indicative of varying geologic pasts. The spacecraft's cameras also detected ten previously unknown moons and two additional rings. *Voyager 2* data showed that the planet's rate of rotation is 17.23 hours. The spacecraft also found a Uranian magnetic field that is both large and unusual. Instead of following the usual north–south axis as found on other planets, the Uranian magnetic field is tilted 60 degrees and offset from the planet's center. On Earth, the equivalent phenomenon would be like having the north magnetic pole in New York and the south magnetic pole in Jakarta, Indonesia.

Uranus's upper atmosphere consists mainly of hydrogen (about 85 percent) and helium (some 12 to 15 percent), with small amounts of methane (CH_4), water vapor (H_2O) and ammonia (NH_3). It is the methane in the upper atmosphere of Uranus and its preferential absorption of red light that gives the planet its overall blue–green color.

The sunlit south pole is shrouded in some type of photochemical "smog" which scientists believe is composed of acetylene (HC_2H) and other sunlight-generated chemicals. This sunlit hemisphere also was found to radiate large amounts of ultraviolet light, a phenomenon that Voyager scientists have called "dayglow." The average temperature

on Uranus is about 60 degrees Kelvin (−350 degrees Fahrenheit).

One of the more surprising results of the *Voyager 2* encounter was the development of a new interior model for Uranus. The new proposed structure has a large core composed of a mixture of rock, ice and gas, covered by a thick atmosphere of hydrogen, helium and heavier gases. The planet's core may either possess many layers with different compositions or else is characterized by a continuous density and composition gradation within the core. There may or may not be a very small molten "rocky" core.

Voyager 2 also found radiation belts at Uranus of an intensity similar to those at Saturn, but different in composition. The Uranian radiation belts appear to be dominated by hydrogen ions, without any evidence of heavier ions. These radiation belts are sufficiently intense that exposure to them (for perhaps 100,000 years) would darken

Table 1   Selected Physical and Dynamic Properties of the Planet Uranus

Diameter (equatorial)	51,120 km
Mass (estimated)	8.7×10^{25} kg
"Surface" gravity	8.6 m/sec²
Mean density (estimated)	1.3 g/cm³
Albedo	0.4–0.5
Temperature (average)	60°K (−350°F)
Magnetic field	Yes, intermediate strength (field tilted 60° with respect to axis of rotation)
Atmosphere	Hydrogen (85%), helium (12–15%), methane (1–3%)
"Surface" features	Bland and featureless (except for some discrete methane clouds)
Escape velocity	22 km/sec
Radiation belts	Yes (intensity similar to those at Saturn)
Rotation period	17.23 hours (retrograde)
Eccentricity	0.047
Mean orbital velocity	6.8 km/sec
Sidereal year (a Uranian year)	84 years
Inclination of planet's equator to its orbit around the Sun	98°
Number of natural satellites	15
Rings	Yes (11)
Average distance from Sun	2.871×10^9 km (19.19 AU) [159.4 light-min]
Solar flux at average distance from Sun	3.7 W/m² (approx.)

SOURCE: Adapted by author from NASA data.

Table 2 Selected Physical and Dynamic Properties of the Major Uranian Moons

Name	Discoverer/Year of Discovery	Diameter (km)	Period (day)	Distance from Center of Uranus (km)	Albedo	Average Density (g/cm³)
Miranda	Kuiper (1948)	500	1.414	129,900	0.2–0.4	1.3–1.6
Ariel	Lassell (1851)	1,160	2.520	190,900	0.4	1.3–1.6
Umbriel	Lassell (1851)	1,190	4.144	266,000	0.2	1.3–1.6
Titania	W. Herschel (1787)	1,610	8.706	436,300	0.25–0.3	1.6–1.7
Oberon	W. Herschel (1787)	1,550	13.463	583,400	0.25–0.3	1.6–1.7

SOURCE: Adapted by author from NASA data.

Table 3 Selected Physical and Dynamic Properties of the Minor Uranian Moons

Name	Year of Discovery	Diameter (km)	Period (day)	Distance from Center of Uranus (km)
Cordelia (1986U7)[a]	1986	40	0.330	49,700
Ophelia (1986U8)[a]	1986	50	0.372	53,800
Bianca (1986U9)	1986	50	0.433	59,200
Juliet (1986U3)	1986	60	0.463	61,800
Desdemona (1986U6)	1986	60	0.475	62,700
Rosalind (1986U2)	1986	80	0.493	64,600
Portia (1986U1)	1986	80	0.513	66,100
Cressida (1986U4)	1986	60	0.558	69,900
Belinda (1986U5)	1986	60	0.622	75,300
Puck (1985U1)	1985	170	0.750	86,000

[a] Shepherding moon.

SOURCE: Adapted by author from NASA data.

(by radiation damage) any methane trapped in the icy surfaces of the "irradiated" Uranian moons. Table 1 presents selected physical and dynamic properties of Uranus.

Two additional rings and an extensive set of dust bands were discovered during the *Voyager 2* encounter—which brings the total number of known rings to eleven. The first nine Uranian rings were discovered in 1977 by scientists performing stellar occultation observations using a telescope mounted in NASA's Kuiper Airborne Observatory (a C-141 aircraft modified for astronomical studies). In contrast to Saturn's rings, however, which are composed mostly of bright grain-sized water ice particles, the Uranian rings are essentially colorless and extremely dark, possibly because they might consist of boulder-sized chunks of irradiated (and therefore darkened) methane ice or dark organic-rich minerals.

Voyager 2 obtained clear, high-resolution images of each of the five previously known (major) moons of Uranus: Miranda, Ariel, Umbriel, Titania and Oberon. (See table 2.) In addition, ten new moons were also discovered around Uranus, between 40 and 170 kilometers (25 to 105 mi) in diameter. (See table 3.) Two of these recently discovered moons, Cordelia (1986U7) and Ophelia (1986U8) are *shepherding satellites* that gravitationally constrain the outermost Uranian ring (called "epsilon"). The other eight new moons lie between the orbits of Ophelia and Miranda. As originally started in 1852 by John Herschel (William's son), the names for the new Uranian moons discovered during

the *Voyager 2* encounter were also taken from English literature (i.e., the works of Shakespeare and Pope).

The large Uranian moons appear to be about 50 percent water ice, 20 percent carbon- and nitrogen-based materials and 30 percent rock. Their surfaces, almost uniformly dark gray in color, display varying degrees of geologic history. Very ancient, heavily cratered surfaces are apparent on some of the moons, while others show strong evidence of internal geologic activity.

Titania, for example, is marked by huge fault systems and canyons that indicate some degree of geologic activity in its history. Ariel has the brightest and possibly the geologically youngest surface in the Uranian moon system. Umbriel is ancient and dark, apparently having undergone little geologic activity. Oberon also has an old, heavily cratered surface with little evidence of internal activity other than some unknown dark material apparently covering the floors of many craters.

Finally, Miranda, the innermost of the five large Uranian moons, is considered by scientists to be one of the strangest bodies yet observed in the Solar System. *Voyager 2* images revealed an unusual world consisting of huge fault canyons as deep as 20 kilometers (12.4 mi), terraced layers and a mixture of old and young surfaces (see fig. 2). Some scientists speculate that Miranda's younger regions may have been produced by incomplete differentiation of the moon—a process in which upwelling of lighter material to the surface occurred in limited areas. Other planetary scientists

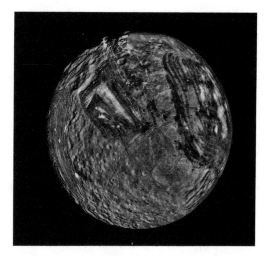

Fig. 2   Miranda, the smallest of the major moons of Uranus, is one of the strangest objects yet discovered in our Solar System. This is a mosaic image assembled from the *Voyager 2* encounter. (Courtesy of NASA/JPL.)

speculate that Miranda may be a reaggregation of material from an earlier time when the moon was fractured into pieces by a violent impact.

See also: **Voyager**

V

Venus   Venus is the second planet out from the Sun. Often called the Evening Star or the Morning Star, it is named after the Roman goddess of love and beauty. Among the planets in our Solar System, it is the only one named after a female deity. Venus is called an inferior planet because it revolves around the Sun within the orbit of the Earth. It maintains an average distance of about 0.723 astronomical unit (AU) from the Sun. The planet appears to observers on Earth as either an evening star or a morning star. In fact, ancient astronomers regarded these two bright "wandering stars" as separate objects and even gave them different names. The early Greeks, for example, called the evening star Hesperos, and the morning star Phosphoros.

In the not too distant past, it was quite popular to think of Venus as literally Earth's twin. People thought that since Venus' diameter, density and gravity were only slightly less than Earth's, it must be similar, especially since it had an obvious atmosphere and was a little nearer the Sun. Visions of a planet with oceans, tropical forests and even giant reptiles and primitive natives frequently appeared in science fiction stories. Unfortunately, spacecraft probes

soon dispelled these romantic fantasies of a prehistoric world that mirrored a younger Earth. Except for a few physical similarities of size and gravity, the Earth and Venus are very different worlds.

Why should Venus be so different from Earth? Today, the environment on Venus differs significantly from the terrestrial biosphere. Its surface is an inferno and its atmosphere is nearly 100 times as dense as that of the Earth. Also, Venus rotates much slower and in retrograde fashion. The surface of Venus is enshrouded by thick clouds. In the ultraviolet portion of the spectrum these clouds exhibit markings that appear to rotate about the planet in a period of four (Earth) days. The predominantly carbon dioxide Venusian atmosphere contains only minute amounts of water vapor. Venus does not possess a significant magnetic field, so the interaction of the planet with the solar wind is quite different from that of the Earth. Venus also does not have a natural satellite or moon. Table 1 provides

Table 1   Dynamic and Physical Properties of Venus

Diameter (equatorial)	12,100 km [≈0.95 Earth]
Mass	4.88×10^{24} kg [≈0.82 Earth]
Density (mean)	5.24 g/cm³ [Earth = 5.5]
Surface gravity	8.88 m/sec² [Earth = 9.81]
Escape velocity	10.4 km/sec [Earth = 11.2]
Albedo (over visible spectrum)	0.7–0.8 [Earth ≈ 0.4]
Surface temperature (approx.)	750°K (480°C, 890°F) [Earth ≈ 20°C (68°F)]
Surface atmospheric pressure	9616 kPa (1396 psi) [Earth = 101 kPa (14.7 psi)]
Solar flux (at top of atmosphere)	2620 W/m²
Atmosphere (main components)	CO_2 (96.4%), N_2 (3.4%)
Magnetic field	Negligible (1/25,000th that of Earth)
Radiation belts	None yet detected
Number of natural satellites	None
Average distance from the Sun	1.082×10^8 km (0.723 AU)
Rotation period (a Venusian "day")	243 days (retrograde)
Eccentricity	0.007
Mean orbital velocity	35.0 km/sec
Sidereal year (period of revolution around Sun)	224.7 days

Earth-to-Venus distances
    Maximum
    Minimum
	
	2.59×10^8 km (1.73 AU)
	0.42×10^8 km (0.28 AU)

SOURCE: NASA.

physical and dynamic data for our nearest planetary neighbor. (At closest approach, Venus is approximately 42 million kilometers from Earth).

Despite its closeness to Earth, astronomers using optical telescopes have been unable to unveil any details from the yellowish, brilliant disk of Venus. Then, on May 10, 1961, a radar signal was reflected from the planet. Analysis of the returned echo indicated that it must rotate extremely slowly. Subsequent investigations revealed that Venus rotated about its axis in 243 (Earth) days in the opposite direction (retrograde) to the way the Earth rotates. On Venus, the Sun rises in the west and sets in the east.

Why should Venus rotate so slowly? Most other planets in the Solar System rotate in periods of hours rather than days. The slow rotation of the innermost planet, Mercury, is attributed to tidal effects from the Sun; but Venus is too far from the Sun for such effects to have been significant over the lifetime of the planet. Some scientists speculate that Venus' rotation was slowed down by a grazing collision with an asteroid.

Venus is an almost perfect sphere. Planetary scientists hypothesize that its interior is similar to that of Earth—namely, a liquid core, a solid mantle and a solid crust.

Despite the dense atmosphere and clouds enveloping Venus, some sunlight does penetrate to its surface where the solar flux is estimated to be equivalent to that at the Earth's surface on an overcast day in the mid-latitudes. Table 2 describes the composition of the Venusian atmosphere at an altitude of 21.6 kilometers. Figure 1 illustrates various views of the inferno-like Venusian surface as taken by the Soviet *Venera* landers.

As the nearest planet, Venus has been the target of many probes, flybys and orbiter spacecraft from both the United States and the Soviet Union (see table 3). The American *Mariner 2* space probe, launched in 1962, was the first successful interplanetary mission to the mysterious planet. As the spacecraft zoomed by, all the romantic myths about Earth's twin were laid to rest.

Mariner 2 passed within 35,000 kilometers of Venus on December 14, 1962, and became the first spacecraft to scan another planet. Its instruments made measurements of Venus for 42 minutes. *Mariner 5*, launched by the United States in June 1967, flew much closer to the planet. Passing

within 4,000 kilometers of Venus, its instruments measured the planet's (extremely weak) magnetic field and temperatures. On its way to Mercury, *Mariner 10* flew past Venus and provided ultraviolet images that showed cloud circulation patterns in the Venusian atmosphere.

In the spring and summer of 1978, two American spacecraft, jointly called the Pioneer Venus mission, were launched to further help unravel the mystery of Venus. On December 4 the *Pioneer Venus Orbiter* was placed in orbit around the planet. Five days later the five separate components that made up the second spacecraft, called the *Pioneer Venus Multiprobe*, entered the Venusian atmosphere at different locations above the planet. These four independent probes and the main body telemetered data back to Earth on the properties of the Venusian atmosphere as they plunged to the surface.

Radar data from the *Pioneer Venus Orbiter* have helped unveil the mystery surrounding this cloud-enshrouded world. These data, as well as those from the Soviet *Venera 15* and *16* radar mappers, revealed an alien world of great mountains, expansive plateaus, enormous rift valleys and shallow basins. Three quite different regions became apparent: an ancient crust at intermediate elevations, relatively smooth lowland plains and highlands. Some 70 percent of Venus consists of upland rolling plains, on which circular features may possibly be the remains of large impact craters. The lowlands cover about 25 percent of the Venusian surface. One extensive lowland basin, called *Atalanta Planitia*, is about the size of the North Atlantic Ocean basin on Earth. It is interesting to note that except for a few surface features named after they were discovered by Earth-based radar searches, all Venusian surface features are now being given female (deity) names that follow the overall tradition of the planet's mythological name. Finally, there are only two major highland or continental masses on Venus: *Ishtar Terra* and *Aphrodite Terra*. (Ishtar was the Babylonian goddess of love, and Aphrodite the love goddess of the ancient Greeks). *Ishtar Terra* is about the size of Australia or the continental United States and possesses the highest peaks yet discovered on Venus. It consists of three geographical units: *Maxwell Montes, Lakshmi Planum* and *Freyja Montes*.

There are many intriguing questions still to be answered about Venus. What happened to the early Venusian oceans, if there ever were any? Because this planet lies on the inner edge of the Solar System's ecosphere, Venus has been considered an eventual candidate for planetary engineering. If we could reduce its currently intolerable surface temperatures and reverse the ongoing runaway greenhouse effect, Earth's twin might someday (perhaps centuries from now) become the balmy, tropical world envisioned by many science fiction writers. Until then, however, it must remain the excessively hot, desolate and waterless world revealed by our robot spacecraft in the last few decades. The latest of these spacecraft was the *Magellan*, which, in June 1991, completed a highly successful radar mapping mission of the planet's surface.

Table 2   Composition of Venusian Atmosphere at 21.6 Kilometers

Gas	Volume Mixing Ratio (Percent)
CO_2	96.4 ± 1.0
N_2	3.41 ± 0.01
H_2O	0.135
O_2	16.0 ± 7.4 × 10^{-4}
Ar	67.2 ± 2.3 × 10^{-4}
CO	19.9 ± 3.12 × 10^{-4}
Ne	4.31 ± 4 × 10^{-4}
SO_2	185 (+350 − 155) × 10^{-4}

Source: NASA.

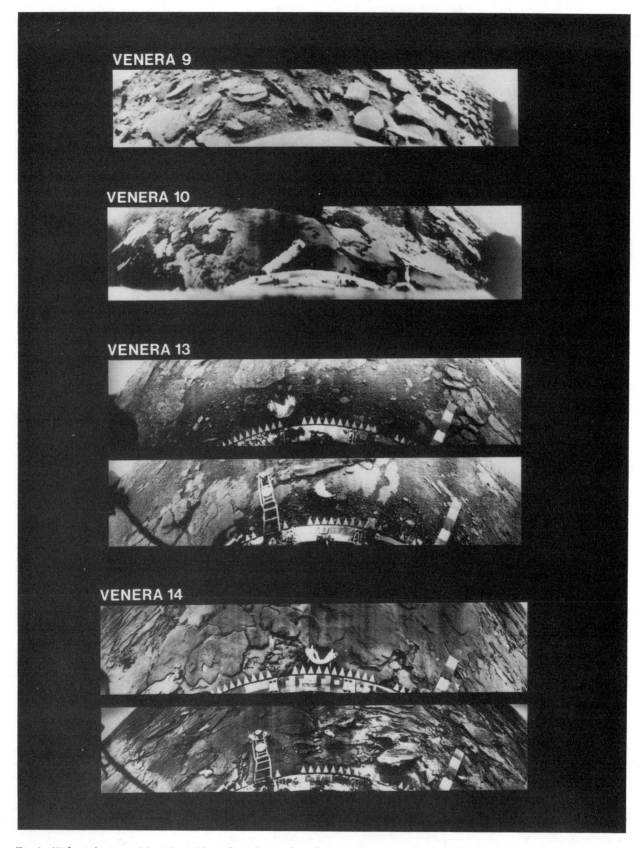

Fig. 1   Wide-angle views of the inferno-like surface of Venus from the Soviet *Venera 9, 10, 13* and *14* landers. (Courtesy of NASA.)

Table 3 Spacecraft Exploration of Venus (1961–1990)

Spacecraft	Country	Launch Date	Comments
Venera 1	USSR	12 Feb 61	Passed Venus at 100,000 km in May 61; but lost radio contact on 27 Feb 61.
Mariner 2	USA	27 Aug 62	1st successful interplanetary probe; passed Venus on 14 Dec 62 at 35,000 km.
Venera 2	USSR	12 Nov 65	Passed Venus on 27 Feb 66 at 24,000 km; but communications failed.
Venera 3	USSR	16 Nov 65	Impacted on Venus on 1 Mar 66; but communications failed earlier.
Venera 4	USSR	12 Jun 67	Probed Venusian atmosphere; fly–by spacecraft and descent module.
Mariner 5	USA	14 Jun 67	Venus fly–by on 19 Oct 67 within 3,400 km.
Venera 5	USSR	5 Jan 69	Descent probe entered atmosphere on 16 May 69.
Venera 6	USSR	10 Jan 69	Descent module entered atmosphere on 17 May 69.
Venera 7	USSR	17 Aug 70	Descent module soft landed on Venus on 15 Dec 70.
Venera 8	USSR	27 Mar 72	Descent module soft landed on Venus on 22 Jul 72.
Mariner 10	USA	3 Nov 73	Fly–by investigation of Venus on 5 Feb 74 at 5800 km; Mariner 10 continued on to Mercury.
Venera 9	USSR	8 Jun 75	Orbiter and descent module arrived at Venus on 22 Oct 75; descent module soft landed and returned picture; orbiter circled planet at 1545 km.
Venera 10	USSR	14 Jun 75	Orbiter and descent module arrived at Venus on 25 Oct 75; descent module soft landed and returned picture; orbiter circled planet at 1665 km.
Pioneer Venus			
Orbiter	USA	20 May 78	Orbited Venus on 4 Dec 78; radar mapping mission.
Multiprobe	USA	8 Aug 78	3 small probes, 1 large probe and main bus entered atmosphere on 9 Dec 78.
Venera 11	USSR	9 Sep 78	Descent module softlanded; fly–by vehicle passed planet at 35,000 km on 25 Dec 78.
Venera 12	USSR	14 Sep 78	Descent module softlanded; fly–by vehicle passed planet at 35,000 km on 21 Dec 78.
Venera 13	USSR	30 Oct 81	Orbiter and descent module; descent module softlanded on 3 Mar 82 and returned color picture.
Venera 14	USSR	4 Nov 81	Orbiter and descent module; descent module softlanded on 5 Mar 82 and returned color picture.
Venera 15	USSR	2 Jun 83	Orbiter; radar mapping mission.
Venera 16	USSR	6 Jun 83	Orbiter; radar mapping mission.
Vega 1	USSR	15 Dec 84	Venus flyby spacecraft (on way to Comet Halley encounter); Venus lander and instrumented balloon in Venusian atmosphere.
Vega 2	USSR	21 Dec 84	Venus flyby spacecraft (on way to Comet Halley encounter); Venus lander (automated soil sampling); instrumented balloon in Venusian atmosphere.
Magellan	USA	4 May 1989	Orbiter; high-resolution radar mapping mission.

SOURCE: NASA.

See also: **Magellan Mission; Pioneer Venus; planetary engineering**

Venusian　Of or relating to the planet Venus; (in science fiction) a native of the planet Venus.

Very Large Array (VLA)　The Very Large Array is a major radio astronomy facility located on the Plains of San Agustin in central New Mexico about 80 kilometers (50 mi) west of Socorro. The VLA consists of 27 large dish-shaped antennas assembled in a flexible Y-pattern configuration to form a single large radio telescope. This facility collects the faint radio waves emitted by celestial objects and produces radio images of these objects with as much clarity and resolution as the photographs from some of the world's largest optical telescopes.

The Very Large Array is one of several radio telescopes operated by the National Radio Astronomy Observatory (NRAO), which is the primary national research center in the United States for the study of radio-wave emissions from cosmic objects. In addition to the VLA, the NRAO (under funding from the National Science Foundation) operates radio telescopes at its sites in Green Bank, West Virginia, and on Kitt Peak in Arizona. The NRAO headquarters is located in Charlottesville, Virginia. In November 1988, the large 91-meter (300-ft) diameter radio telescope at Green Bank collapsed due to structural failure. This NRAO facility will be replaced by a new, 100-meter (330-ft) diameter radio telescope that is now scheduled for full operation in 1995.

The technique of focusing and combining the signals from a distributed array of smaller telescopes to simulate the resolution of a single, much larger telescope is called *aperture synthesis*. In astronomy, a radio telescope is used to measure the intensity of the weak, static-like cosmic radio waves coming from some particular direction in the Universe. *Sensitivity* is defined as the radio telescope's ability to detect these very weak radio signals; while *resolution* is defined as the telescope's ability to locate the source of these signals. The sensitivity of a distributed array of telescopes, such as the VLA, is proportional to the sum of the collecting area of all the individual elements; while the array's resolution is determined by the area over which the array elements can be spread. Each of the 25-meter (82-ft) diameter dish antennas in the VLA was specially designed with aluminum panels formed into a parabolic surface accurate to 0.5 millimeter (20 thousandths of an inch)—a design condition that enables the antennas to focus radio signals as short as one centimeter wavelength. The VLA antennas can track a radio source across the sky with an accuracy of 10 seconds of arc.

In the VLA array, each antenna collects incoming radio signals and sends them to a central location where they are combined. The sensitive VLA receivers can be tuned to wavelengths of 90, 20, 6, 3.5, 2 and 1.3 cm. These receivers are cooled to 18 degrees Kelvin (−428 degrees Fahrenheit) to reduce internally generated noise, which tends to mask the very weak radio signals from space. These incoming signals are then amplified several million times and sent to the VLA's Control Building by means of a waveguide.

The data collected by the VLA are stored on magnetic tape. An astronomer can, therefore, evaluate a radio image days or even years after the actual observation was made. Different astronomical observations require different resolution capabilities. A single highly detailed radio image might involve over 40 hours of observation, while a crude "radio signal snapshot" of a particular source may only require 10 minutes of observing time. For example, a high-resolution image would be needed to explore the inner core of a radio galaxy, while only a low-resolution image could reveal the large faint overall radio emissions features of that galaxy.

The VLA is used to produce radio images with as much detail as those made by an optical telescope. To accomplish this, the VLA's 27 dish-shaped antennas are arranged in a giant Y-pattern, with the southeast and southwest arms of the Y-pattern each 21 kilometers (13 mi) long, and the north arm 19 kilometers (11.8 mi) long. The resolution of this radio telescope array can then be varied by changing the separation and spacing of its 27 antenna elements. The VLA is generally found in one of four standard configurations. In the smallest antenna dispersion configuration (a low-resolution configuration), the 27 individual antennas are clustered together within 0.6 kilometers (2,000 ft) of the array's center. In the largest antenna dispersion configuration (the high-resolution configuration) the individual antennas stretch out in a giant Y-pattern some 21 kilometers (13 mi) from the center.

The 235-ton VLA antennas are carried along the Y-pattern array arms by a special transporter that moves on two parallel sets of railroad tracks. Generally, it takes about two hours to move an antenna from one station (pedestal) to another and about a week is needed to reconfigure the entire VLA array.

The high-resolution and high-sensitivity radio images produced by the Very Large Array make it one of the world's leading radio telescope facilities.

See also: **Arecibo Observatory; interstellar medium; radio astronomy**

Vesta Missions　Two identical space systems planned for launch in 1996 that will encounter between ten and twenty asteroids and at least two comets over a five-year period. These missions are named after Vesta, one of the largest of the asteroids, and represents a trilateral cooperative effort by the European, French and Soviet space agencies (ESA, CNES and Intercosmos, respectively). The Vesta missions will build upon and extend the small bodies of the Solar System program that began with the Comet Halley flybys in 1986 by the *Giotto* (ESA) and *Vega* (Soviet) spacecraft.

Each Vesta space system will consist of a Flyby Module and an Approach Module. The Vesta Flyby Module will contain a high-instrumented scanning platform and will encounter (fly by) several asteroids and cometary nuclei at distances of from 500 to 2,000 kilometers and at relative encounter velocities of from 2 to 15 kilometers per second. The Vesta Approach Module will be jettisoned in the vicinity of a target asteroid, reduce its relative encounter velocity to less than 0.1 kilometers per second and then release two penetrators that will anchor themselves to the target asteroid and then perform on-site measurements.

See also: **asteroid; comet; Comet Rendezvous and Asteroid Flyby Mission; European Space Agency; Rosetta Mission**

Viking Project   The Viking Project was the culmination of a series of American missions to explore the planet Mars. This series of interplanetary missions began in 1964 with *Mariner 4*, and continued with the *Mariner 6* and *7* flyby missions in 1969, and then the *Mariner 9* orbital mission in 1971 and 1972.

Viking was designed to orbit Mars and to land and operate on the surface of the Red Planet. Two identical spacecraft, each consisting of a lander and an orbiter, were built.

The Orbiters carried the following scientific instruments:

1. A pair of cameras with 1,500-millimeter focal length that performed systematic searches for landing sites; then looked at and mapped almost 100 percent of the Martian surface. Cameras onboard the *Viking 1* and *2* Orbiters took more than 51,000 photographs of Mars.

2. A Mars atmospheric water detector that mapped the Martian atmosphere for water vapor and tracked seasonal changes in the amount of water vapor.

3. An infrared thermal mapper than measured the temperatures of the surface, polar caps and clouds; it also mapped seasonal changes. In addition, although the Viking Orbiter radios were not considered scientific instruments, they were used as such. By measuring the distortion of radio signals as these signals traveled from the Orbiter spacecraft to the Earth, scientists were also able to measure the density of the Martian atmosphere.

The *Viking Landers* carried the following instruments:

1. The biology instrument, consisting of three separate experiments designed to detect evidence of microbial life in the Martian soil. There was always a chance that larger life-forms could be present on Mars. But exobiologists thought the life-forms most likely to be there (if any at all) would be microorganisms.

2. A gas chromatograph/mass spectrometer (GCMS) that searched the Martian soil for complex organic molecules. These organic molecules could be the precursors or the remains of living organisms.

3. An X-ray fluorescence spectrometer that analyzed samples of the Martian soil to determine its elemental composition.

4. A meteorology instrument that measured air temperature and wind speed and direction at the landing sites. These instruments returned the first extraterrestrial weather reports in the history of meteorology.

5. A pair of slow-scan cameras that were mounted about one meter apart on the top of each Lander. These cameras provided black-and-white, color and stereo photographs of Mars.

6. A seismometer had been designed to record any "Marsquakes" that might occur on the Red Planet. Such information would have helped planetologists determine the nature of the planet's internal structure. Unfortunately, the seismometer on Lander 1 did not function after landing and the instrument on Lander 2 observed no clear signs of internal (tectonic) activity.

7. An upper atmosphere mass spectrometer that conducted its primary measurements as each Lander plunged through the Martian atmosphere on its way to the landing site. The Lander's first important scientific discovery—the presence of nitrogen in the Martian atmosphere—was made by this instrument.

8. A retarding potential analyzer that measured the Martian ionosphere, again during entry operations.

9. Accelerometers, a stagnation pressure instrument and a recovery temperature instrument, that helped determine the structure of the lower Martian atmosphere as the Landers approached the surface.

10. A surface sampler boom that employed its collector head to scoop up small quantities of Martian soil to feed the biology, organic chemistry and inorganic chemistry instruments. It also provided clues to the soil's physical properties. Magnets attached to the sampler, for example, provided information on the soil's iron content.

11. The Lander radios were also used to conduct scientific experiments. Physicists were able to refine their estimates of Mars' orbit by measuring the round trip time for radio signals to travel between Mars and Earth. The great accuracy of these radio-wave measurements also allowed scientists to confirm portions of Einstein's General Theory of Relativity.

Both Viking spacecraft were launched from Cape Canaveral, Florida. *Viking 1* was launched on August 20, 1975, and *Viking 2* on September 9, 1975. The Landers were sterilized before launch to prevent contamination of Mars were terrestrial microorganisms. These spacecraft spent nearly a year in transit to the Red Planet. *Viking 1* achieved Mars orbit on June 19, 1976; and *Viking 2* began orbiting Mars on August 7, 1976. *Viking 1* performed the first landing on Mars on July 20, 1976, on the western slope of *Chryse Planitia* (the Plains of Gold) at 22.46 degrees north latitude, 48.01 degrees west longitude. The *Viking 2* Lander landed on September 3, 1976, at *Utopia Planitia* located at 47.96 degrees north latitude, 225.77 degrees west longitude.

The Viking mission was planned to continue for 90 days after landing. Each Orbiter and Lander, however, operated far beyond its design lifetime. *Viking Orbiter 1* exceeded four years of active flight operations in orbit around Mars.

The Viking Project's primary mission ended on November 15, 1976, just 11 days before Mars passed behind the Sun (an astronomical event called a "superior conjunction"). After conjunction, in mid-December 1976, telemetry and command operations were reestablished and extended mission operations began.

The *Viking Orbiter 2* mission ended on July 25, 1978, due to an exhaustion of attitude-control system gas. The Orbiter 1 spacecraft also began to run low on attitude-control-system gas, but through careful planning it was possible to continue collecting scientific data (at a reduced level) for another two years. Finally, which its control gas supply exhausted, the Orbiter 1's electrical power was commanded off on August 7, 1980, after 1,489 orbits of Mars.

The last data from the *Viking 2* Lander were received on April 11, 1980. The Lander 1 made its final transmission to Earth on November 11, 1982. After over six months of effort to regain contact with the Lander 1, the Viking mission came to an end on May 21, 1983.

With the single exception of the seismic instruments, the scientific instruments of the Viking Project acquired far more data than ever anticipated. The seismometer on Lander 1 did not function after touchdown, while the seismometer on Lander 2 detected only one event that might have been of seismic origin. Nevertheless, this instrument still provided data on surface wind velocity at Landing Site 2 (supplementing the meteorology experiment) and also indicated that the Red Planet has a very low level of seismicity.

The primary objective of the Lander was to determine whether life exists on Mars. The evidence provided by the Landers is still being interpreted by most scientists as strongly indicative that life does *not* now exist on Mars!

Three of the Lander instruments were capable of detecting life on Mars. The Lander cameras could have photographed living creatures large enough to be seen with the human eye. These cameras could also have observed growth in organisms such as plants and lichens. Unfortunately, the cameras at both sides observed nothing that could be interpreted as living.

The gas chromatograph/mass spectrometer (GCMS) could have found organic molecules in the soil. (Organic compounds combine carbon, nitrogen, hydrogen and oxygen.) These compounds are present in all living matter on Earth. The GCMS was programmed to search for heavy organic molecules, those large molecules that contain complex combinations of carbon and hydrogen and are either life precursors or the remains of living systems. To the surprise of exobiologists, the GCMS (which easily detects organic matter in the most barren terrestrial soils) found no trace of any organic molecules in the Martian soil samples.

Finally, the Lander Biology Instrument was the primary device used to search for extraterrestrial life. It was a one cubic foot box, loaded with the most sophisticated scientific instrumentation yet built and flown in space. The Biology Instrument actually contained three smaller instruments that examined the Martian soil for evidence of metabolic processes like those used by bacteria, green plants and animals on Earth.

The three biology experiments worked flawlessly on each Lander. All showed unusual activity in the Martian soil—activity that mimicked life—but exobiologists here on Earth needed time to understand the strange behavior of the Red Planet's soil. Today, according to most scientists who helped analyze these data, it appears that the chemical reactions were not caused by living things.

Furthermore, the immediate release of oxygen, when the Martian soil contacted water vapor in the Biology Instrument, and the lack of organic compounds in the soil, indicate that oxidants are present in both the Martian soil and atmosphere. Oxidants, such as peroxides and superoxides, are oxygen-bearing compounds that break down organic matter and living tissue. Consequently, even if organic compounds evolved on Mars, they would be quickly destroyed.

Evaluation of the Martian atmosphere and soil has revealed that all the elements essential for life (as we know it on Earth)—carbon, hydrogen, nitrogen, oxygen and phosphorus—are also present on the Red Planet. However, exobiologists also consider the presence of liquid water as an absolute requirement for the evolution and continued existence of life. The Viking Project discovered ample evidence of Martian water in two of its three phases—vapor and solid (ice), and even evidence of large quantities of permafrost. But under current environmental conditions on Mars, it is impossible for water to exist as a liquid on the planet's surface.

Therefore, the conditions now known to occur on and just below the surface of the Red Planet do not appear adequate for the existence of living (carbon-based) organisms. However, exobiologists, though disappointed in their first serious search for extraterrestrial life, add that the case for life sometime in the past history of Mars is still open.

While the gas chromatograph/mass spectrometer found no sign of organic chemistry at either landing site, it did provide a precise and definitive analysis of the composition of the Martian atmosphere. The GCMS, for example, found previously undetected trace elements. The Lander X-ray fluorescence spectrometer measured the elemental composition of the Martian soil.

The two Landers continuously monitored weather at the landing sites. The midsummer Martian weather proved repetitious but in other seasons the weather varied and became more interesting. Cyclic variations in Martian weather patterns were observed. Atmospheric temperatures at the southern (*Viking 1*) landing site were as high as -14 degrees Celsius ($+7$ degrees Fahrenheit) at midday; while the predawn summer temperature was typically -77 degrees Celsius (-107 degrees Fahrenheit). In contrast, the diurnal temperatures at the northern (*Viking 2*) landing site during the midwinter dust storm varied as little as 4 degrees Celsius (7.2 degrees Fahrenheit) on some days. The lowest observed predawn temperature was -120 degrees Celsius (-184 degrees Fahrenheit), which is about the frost point of carbon dioxide. A thin layer of water frost covered the ground near the *Viking 2* Lander each Martian winter.

The barometric pressure was observed to vary at each landing site on a semiannual basis. This occurred because carbon dioxide (the major constituent of the Martian atmosphere) freezes out to form an immense polar cap—alternately at each pole. The carbon dioxide forms a great cover of "snow" and then evaporates (or sublimes) again with the advent of Martian "spring" in each hemisphere. When the southern cap was largest, the mean daily pressure observed by Lander 1 was as low as 6.8 millibars; while at other times during the Martian year it was as high as 9.0 millibars. Similarly, the pressures at the Lander 2 site were 7.3 millibars (full northern cap) and 10.8 millibars. (For comparison, the sea-level atmospheric pressure on Earth is about 1,000 millibars or 1 bar.)

Martian surface winds were also typically slower than anticipated. Scientists had expected these winds to reach speeds of hundreds of kilometers per hour. But neither Lander recorded a wind gust in excess of 120 kilometers (74 miles) per hour; and average speeds were considerably lower.

Photographs of Mars from the *Viking* landers and Orbiters surpassed all expectations in both quantity and quality. The Landers provided over 4,500 images and the Orbiters over 52,000. The Landers provided the first close-up view of the surface of the Red Planet, while the Orbiters mapped almost 100 percent of the Martian surface, including detailed images of many intriguing surface features.

The infrared thermal mapper and the atmospheric water detector on board the Orbiters provided essentially daily data. Through these data it was determined that the residual northern polar ice cap that survives the northern summer is composed of water ice, rather than frozen carbon dioxide (dry ice), as scientists once believed.

The Orbiters also provided high-quality photographs of the two Martian satellites, Phobos and Deimos.

Today, after all the Viking robot explorers have fallen silent, we are heir to billions of bits of valuable scientific data about Mars and now possess over 50,000 outstanding photographs from this project alone.

But intriguing questions about the Red Planet still remain. Is there a remote possibility that life exists in some crevice or biological niche on this mysterious world? Did life once evolve there, only to have vanished millions of years ago? And how did climatic conditions change so radically that great floods of water, which apparently raged over the Martian plains, have now vanished, leaving behind the dry, sterile world found by the Viking Project explorers? Only further exploration, including human expeditions, will resolve these intriguing questions.

See also: **Mars**

Voyager Once every 176 years, the giant outer planets Jupiter, Saturn, Uranus and Neptune—align themselves in such a pattern that a spacecraft launched from Earth to Jupiter at just the right time might be able to visit the other three planets on the same mission, using a technique called "gravity assist." NASA space scientists named this multiple giant planet encounter mission the "Grand Tour" and took advantage of a unique celestial alignment opportunity in 1977 by launching two sophisticated spacecraft, called *Voyager 1* and *2*. (See fig. 1.)

Each Voyager spacecraft has a mass of 825 kilograms (1,815 lb-mass) and carries a complement of scientific instruments to investigate the outer planets and their many moons and intriguing ring systems. These instruments, provided electric power by a long-lived nuclear system called a radioisotope generator (RTG), recorded spectacular closeup images of the giant outer planets and their interesting moon systems, explored complex ring systems and measured properties of the interplanetary medium.

Taking advantage of the 1977 Grand Tour launch window, the *Voyager 2* spacecraft lifted off from Cape Canav-

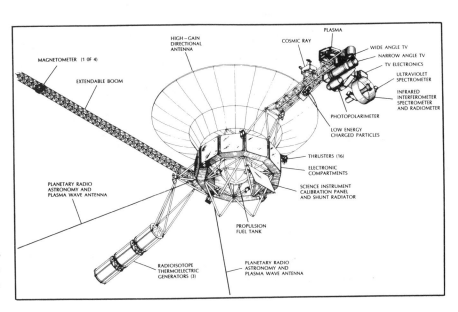

Fig. 1 The 825-kilogram *Voyager* spacecraft and its complement of sophisticated scientific instruments. (Courtesy of NASA.)

eral, Florida on August 20, 1977, on board a Titan–Centaur rocket. (NASA called the first Voyager spacecraft launched *Voyager 2*, because the second Voyager spacecraft to be launched would eventually overtake it and become *Voyager 1*). *Voyager 1* was launched on September 5, 1977. This spacecraft followed the same trajectory as its *Voyager 2* twin and overtook its sistership, just after entering the asteroid belt in mid-December 1977.

Voyager 1 made its closest approach to Jupiter on March 5, 1979, and then used Jupiter's gravity to swing itself to Saturn. On November 12, 1980, *Voyager 1* successfully encountered the Saturnian system and was then flung up out of the ecliptic plane on an interstellar trajectory. The *Voyager 2* spacecraft successfully encountered the Jovian system on July 9, 1979 (closest approach), and then used the gravity assist technique to follow *Voyager 1* to Saturn. On August 25, 1981, *Voyager 2* encountered Saturn, and then went on to successfully encounter both Uranus (January 24, 1986) and Neptune (August 25, 1989). (See fig. 2.) Space scientists consider the end of *Voyager 2*'s encounter of the Neptunian system as the end of a truly extraordinary epoch in planetary exploration. In the 12 years since they were launched from Cape Canaveral, these incredible

spacecraft contributed more to our understanding of the giant outer planets of our Solar System than was accomplished in over three millennia of Earth-based observations. Following its encounter with the Neptunian system, *Voyager 2* was also placed on an interstellar trajectory.

VOYAGER INTERSTELLAR MISSION (VIM)

Following the Neptune encounter, *Voyager 2* (like its *Voyager 1* twin) now continues to travel outward from the Sun. As the influence of the Sun's magnetic field and solar wind grow weaker, both Voyager spacecraft will eventually pass out of the heliosphere and into the interstellar medium. Through NASA's Voyager Interstellar Mission (VIM) (which officially began on January 1, 1990) the two Voyager spacecraft will continue to be tracked on their outward journey, just as the *Pioneer 10* and *11* spacecraft are being tracked.

The two major objectives of the Voyager Interstellar Mission are (1) an investigation of the interplanetary and interstellar media, and a characterization of the interaction between the two; and (2) a continuation of the successful Voyager program of ultraviolet astronomy. During the Voyager Interstellar Mission the spacecraft will search for the

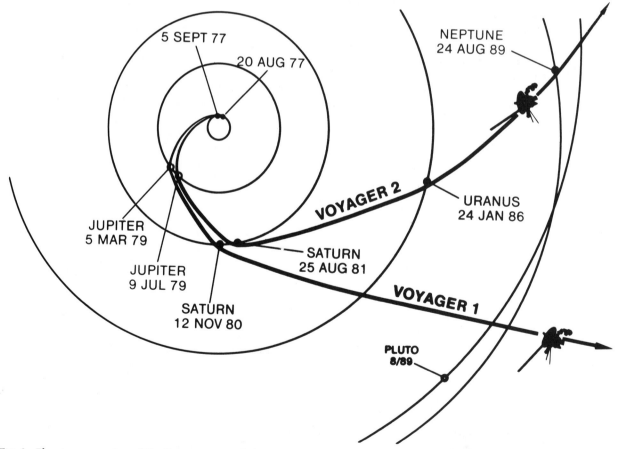

Fig. 2   Planetary encounters of the Voyager spacecraft (1977–1989)—a Grand Tour exploration epoch of the outer Solar System. (Courtesy of NASA.)

Table 1   "Nearby" Stars That Will Be Encountered by *Voyager 2* in the Next Million Years

	Year of Closest Approach	Voyager 2-to-Star Distance (Light-years)	Sun-to-Voyager 2 Distance (Light-years)	Sun-to-Star Distance (Light-years)
Barnard's Star	8,571	4.03	0.42	3.80
Proxima Centauri	20,319	3.21	1.00	3.59
Alpha Centauri	20,629	3.47	1.02	3.89
Lalande 21185	23,274	4.65	1.15	4.74
Ross 248	40,176	1.65	1.99	3.26
DM−36 13940	44,492	5.57	2.20	7.39
AC+79 3888	46,330	2.77	2.29	3.76
Ross 154	129,084	5.75	6.39	8.83
DM+15 3364	129,704	3.44	6.42	6.02
Sirius	296,036	4.32	14.64	16.58
DM−5 4426	318,543	3.92	15.76	12.66
44 Ophiuchi	442,385	6.72	21.88	21.55
DM+27 1311	957,963	6.62	47.38	47.59

SOURCE: NASA/JPL.

heliopause (the outermost extent of the solar wind, beyond which lies interstellar space). Figure 2 on page 220 presents a model diagram of what space scientists think the outermost regions of the Solar System look like. This figure also shows the trajectories of the Voyager and Pioneer spacecraft, as well as the suspected location of the heliopause.

Hopefully one of the Voyager or Pioneer spacecraft will still be functioning when it penetrates the heliopause, and provides us with our first true sampling of the interstellar environment. Barring a catastrophic failure on board either Voyager spacecraft, their individual supply of hydrazine (for attitude control, that is, keeping the spacecraft platform pointed in the proper direction) should last until about 2035 and their nuclear power systems should provide useful levels of electric power until at least 2015. (The older Pioneer spacecraft do not have as long an active lifetime expectancy.)

Since both Voyager spacecraft would eventually journey beyond the Solar System, their designers placed a special interstellar message (a record entitled: "The Sounds of Earth") on each spacecraft in the hope that perhaps millions of years from now some intelligent alien race will find either spacecraft drifting quietly through the interstellar void. If they are able to decipher the instructions for using this record, they will learn about our contemporary terrestrial civilization and the men and women who sent Voyager on its stellar journey! (See table 1.)

See also: **images from space; Jupiter; Neptune; orbits of objects in space; Saturn; Uranus; Voyager record**

Voyager record   The *Voyager 1* and *2* spacecraft, launched during the summer of 1977, will eventually leave our Solar System. As these spacecraft wander through the Milky Way Galaxy over the next million or so years, each has the potential of serving as an interstellar ambassador—since each carries a special message from Earth on the chance that an intelligent alien race might eventually find one of the spacecraft floating among the stars. The *Voyager's* interstellar message is a phonograph record called "The Sounds of Earth." Electronically imprinted on it are words, photographs, music and illustrations that will tell an extraterrestrial civilization about our planet. Included are greetings in over fifty different languages, music from various cultures and periods and a variety of natural terrestrial sounds such as the wind, the surf and different animals. The *Voyager* record also includes a special message from former president Jimmy Carter. Dr. Carl Sagan describes in detail the full content of this "phonograph message to the stars" in his delightful book, *Murmurs of Earth* (see table 1).

Table 1   A Partial List of the Contents of the Voyager Record, *The Sounds of Earth*

A. SOUNDS OF EARTH

Whale, volcanoes, rain, surf, cricket frogs, birds, hyena, elephant, chimpanzee, wild dog, laughter, fire, tools, Morse code, train whistle, Saturn V rocket liftoff, kiss, baby.

B. MUSIC

Bach: *Brandenberg Concerto #2*, 1st movement; Zaire: "Pygmy Girls" initiation song; Mexico: mariachi band playing: "El Cascabel"; Chuck Berry: "Johnny B. Goode"; Navajo: night chant; Louis Armstrong: "Melancholy Blues"; China (zither) "Flowering Streams"; Mozart: Queen of the Night aria (*Magic Flute*); Beethoven: Symphony #5, 1st movement.

C. GREETINGS (in 55 different languages)

(example) English: "Hello from the children of planet Earth"

D. PICTURES (digital data to be reconstructed into images)

Calibration circle; solar location map; the Sun; Mercury; Mars; Jupiter; Earth; fetus; birth; nursing mother; group of children; sequoia (giant tree); snowflake; seashell; dolphins; eagle; Great Wall of China; Taj Mahal; U.N. Building; Golden Gate Bridge; radio telescope (Arecibo); Titan Centaur launch; astronaut in space.

Each record is made of copper with gold plating and is encased in an aluminum shield that also carries instructions on how to play it (see fig. 1). Look at figure 1, without reading beyond this paragraph. Can you decipher the instructions we've given to alien civilizations? No, it does not mean "Batteries not included!" If you can decipher these instructions, congratulations—you qualify as an extraterrestrial interpreter. If not, please do not feel disappointed; we shall now explore them together.

In the upper left is a drawing of the phonograph record and the stylus carried with it. Written around it in binary notation is the correct time for one rotation of the record, 3.6 seconds. Here, the time units are 0.70 billionths of a second, the time period associated with a fundamental transition of the hydrogen atom. The drawing further indicates that the record should be played from the outside in. Below this drawing is a sideview of the record and stylus, with a binary number giving the time needed to play one side of the record (approximately one hour).

The information provided in the upper-right portion of the instructions is intended to show how pictures (images) are to be constructed from the recorded signals. The upper-right drawing illustrates the typical wave form that occurs at the start of a picture. Picture lines 1, 2 and 3 are given in binary numbers and the duration of one of the picture "lines" is also noted (about 8 milliseconds). The drawing immediately below shows how these lines are to be drawn vertically, with a staggered interlace to give the correct picture rendition. Immediately below this is a drawing of an entire picture raster, showing that there are 512 vertical

lines in a complete picture. Then, immediately below this is a replica of the first picture on the record. This should allow extraterrestrial recipients to verify that they have properly decoded the terrestrial pictures. A circle was selected for this first picture to guarantee that any aliens who find the message use the correct aspect ratio in picture reconstruction.

Finally, the drawing at the bottom of the protective aluminum shield is that of the same pulsar map drawn on the *Pioneer 10* and *11* plaques. (These spacecraft are also headed on interstellar trajectories.) The map shows the location of our Solar System with respect to 14 pulsars, whose precise periods are also given. The small drawing with two circles in the lower-right-hand corner is a representation of the hydrogen atom in its two lowest states, with a connecting line and digit 1. This indicates that the time interval associated with the transition from one state to the other is to be used as the fundamental time scale, both for the times given on the protective aluminum shield and in the decoded pictures.

If you were making up a message to be included on an interstellar phonograph record, what would you like to say to some distant alien civilization?

See also: **interstellar communication; interstellar contact; Pioneer plaque; Voyager**

Vulcan   A planet that some 19th-century astronomers believed existed in an extremely hot orbit between Mercury and the Sun. Named after the Roman god of fire and metalworking craftsmanship, Vulcan's existence was postulated to account for gravitational perturbations observed in the orbit of Mercury. Modern astronomical observations have failed to reveal this celestial object and Einstein's theory of relativity has enabled 20th-century astronomers to account for the observed irregularities in Mercury's orbit. As a result, the planet Vulcan, created out of theoretical necessity by the 19th-century astronomers, has now quietly disappeared in contemporary discussions about our Solar System.

W

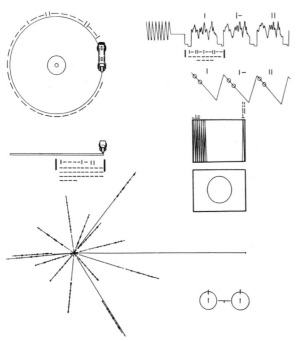

Fig. 1   The set of instructions to any alien civilization that might find the *Voyager 1* or *2* spacecraft, explaining how to operate the *Voyager* record and where the spacecraft and message came from. (Courtesy of NASA.)

water hole   A term used in the search for extraterrestrial intelligence (SETI) to describe a narrow portion of the electromagnetic spectrum that appears especially appropriate for interstellar communications between emerging and advanced civilizations. This band lies in the radio frequency (RF) part of the spectrum between 1420 megahertz frequency (21.1 cm wavelength) and 1660 megahertz frequency (18 cm wavelength).

Hydrogen (H) is abundant throughout interstellar space. When hydrogen experiences a "spin-flip" transition (due to an atomic collision), it emits a characteristic 1420 megahertz frequency (or 21.1 cm wavelength) radio wave. Any intelligent race throughout the Galaxy that has risen to the technological level or radio astronomy will eventually detect these missions. Similarly, there is another grouping of characteristic spectral emissions centered near the 1660 megahertz frequency (18 cm wavelength) that are associated with hydroxyl (OH) radicals.

As we all know from elementary chemistry: H + OH = H_2O. So we have, as suggested by Dr. Bernard Oliver and other SETI investigators, two radio-wave emission signposts associated with the dissociation products of water that "beckon all water-based life to search for its kind at the age-old meeting place of all species: the water hole."

Is this high regard for the 1420 to 1660 megahertz frequency band reasonable, or simply a case of terrestrial chauvinism? Well, many exobiologists currently feel that if other life exists in the Universe, it will most likely be carbon-based life and water is essential for carbon-based life as we know it. In addition, for purely technical reasons, if we scan all the decades of the electromagnetic spectrum in search for suitable frequency at which to send or receive interstellar messages, we would arrive at the narrow microwave region between 1 and 10 gigahertz as the most suitable candidate for conducting interstellar communication. The two characteristic emissions of dissociated water, namely 1420 megahertz for H and 1660 megahertz for OH, are situated right in the middle of this optimum communication band.

Based on this type of reasoning, the water hole has been strongly recommended by scientists currently engaged in SETI projects. They generally feel that this portion of the electromagnetic spectrum represents a logical starting place for us to listen for interstellar signals from other intelligent civilizations.

See also: **search for extraterrestrial intelligence**

wormhole Some scientists speculate that matter falling into a black hole may actually survive! They suggest that under very special circumstances such matter might be conducted by means of passageways, called "wormholes," to emerge in another place or time in this Universe or in another universe. In forming this concept, scientists are theorizing that black holes can play "relativistic tricks" with space and time.

See also: **black holes**

X-ray A penetrating form of electromagnetic radiation of very short wavelength (approximately 0.01 to 10 nanome-

ters or 0.1 to 100 angstroms) and high photon energy (approximately 100 electronvolts to some 100 kiloelectronvolts). X-rays are emitted when either the inner orbital electrons of an excited atom return to their normal energy states (these photons are called characteristic X-rays) or when a fast-moving charged particle (generally an electron) loses energy in the form of photons upon being accelerated and deflected by the electric field surrounding the nucleus of a high atomic number element (this process is called "bremsstrahlung" or braking radiation). Unlike gramma rays, X-rays are non-nuclear in origin.

See also: **Advanced X-Ray Astrophysics Facility; astrophysics; X-ray astronomy**

X-ray astronomy Since the Earth's atmosphere absorbs most of the X-rays coming from celestial phenomena, astronomers must use orbiting space platforms to study these interesting emissions, which are usually associated with very energetic, violent processes occurring somewhere in our Universe. In fact, X-ray emissions carry detailed information about the temperature, density, age and other physical conditions of the interesting and dynamic celestial objects that have produced them. X-ray astronomy is the most advanced of the three general disciplines associated with high-energy astrophysics: X-ray, gamma ray and cosmic ray astronomy. The observation of X-ray emissions has been very valuable in the study of high-energy events, such as mass transfer in binary star systems, the interaction of supernovae remnants with interstellar gas and the functioning of quasars.

Space-based X-ray observatories provide data allowing scientists to study and understand (1) stellar structure and evolution, including binary star systems, supernovae remnants, pulsars and black hole candidates; (2) large-scale galactic phenomena, including the interstellar medium itself and the soft X-ray emissions of local galaxies; (3) the nature of active galaxies, including the spectral characteristics and time variation of X-ray emissions from the central (or nuclear) regions of such galaxies; and (4) rich clusters of galaxies, including their associated X-ray emissions.

See also: **Advanced X-ray Astrophysics Facility; astrophysics; X-Ray Timing Explorer**

X-Ray Timing Explorer (XTE) A planned NASA explorer-class payload for the study of the temporal variability of X-ray-emitting objects. The main instrument on the XTE, called a large area proportional counter, will be applied to the detailed observation of individual celestial objects over long exposure times, while an all-sky monitor will record all the remaining strong X-ray emitters simultaneously over longer time scales with decreased sensitivity. In this way, the XTE can sample all time scales from microseconds to years for the entire X-ray source catalog. The XTE will be launched in 1994 and will be placed into a 400 kilometer circular orbit at a 28.5 degree inclination. A three- to four-year mission lifetime is anticipated.

See also: **astrophysics; X-ray astronomy**

Z

zero-gravity aircraft   An aircraft that flies a special parabolic trajectory to create low-gravity conditions (typically 0.01 g) for short periods of time (10 to 30 seconds), where one g here represents the acceleration due to gravity at the Earth's surface (9.8 m/sec^2).

See also: **microgravity**

zodiac   The word zodiac comes from the ancient Greek language and means circle of figures or circle of animals. In astronomy, zodiac refers to a band in the sky, extending about nine degrees to each side of the ecliptic. Since the earliest times, the zodiac has been divided into intervals of 30 degrees along the ecliptic, with each of these sections being designated by a "sign of the zodiac." The annual revolution of the Earth around the Sun causes the Sun to appear to enter a different constellation of the zodiac each month. These twelve constellations (or signs) are Aries (Ram), Taurus (Bull), Gemini (Twins), Cancer (Crab), Leo (Lion), Virgo (Maiden), Libra (Scales), Scorpius (Scorpion), Sagittarius (Archer), Capricornus (Sea-goat), Aquarius (Water-bearer) and Pisces (Fish). Although the signs of the zodiac originally corresponded in position to the twelve constellations just named, because of the phenomenon of precession, the zodiacal signs do not presently coincide with these constellations. For example, when people today say the Sun enters Aries at the vernal equinox, it has actually shifted forward (from ancient times) and is now actually in the constellation Pisces. Finally, looking ahead into the distant future, the vernal equinox will move into Aquarius sometime near 3000 A.D.

zodiacal light   A faint cone of light extending upward from the horizon in the direction of the ecliptic. Zodiacal light is seen from the tropical latitudes for a few hours after sunset or before sunrise. It is due to sunlight being reflected by tiny pieces of interplanetary dust in orbit around the Sun.

Zoo hypothesis   One response to the Fermi paradox. It assumes that intelligent, technically very advanced species do exist in the Galaxy, but that we cannot detect or interact with them because they have set the Solar System aside as a perfect zoo or wildlife preserve.

The reasoning followed in establishing this hypothesis goes something like this. Technically advanced beings exert a great deal of control over their environment. For example, human beings have a far greater influence on the biosphere than all the other creatures that co-inhabit the planet with us. Occasionally, we have decided not to exert this influence, but rather to set aside certain regions of the Earth as zoos, wildlife sanctuaries or wilderness areas, where other species can develop naturally with little or no human interaction. In fact, the perfect wildlife sanctuary or zoo is one set up so that the species within are not even aware of the presence or existence of their zookeepers. Thus, in response to the question: "Where are they?," the Zoo hypothesis suggests that we cannot and will never be able to detect "them," because our "extraterrestrial zookeepers" have set aside the Solar System as a perfect zoo or wildlife sanctuary and want us to develop "naturally" without awareness of or interaction with them.

See also: **extraterrestrial civilizations; Fermi paradox; Laboratory hypothesis**

Appendix A:
Special Reference List

Table A-1　Special units for astronomical investigations

astronomical unit (AU): The mean distance from the Earth to the Sun—approximately $1.495\ 979 \times 10^{11}$ meters.

light year (ly): The distance light travels in one year's time—approximately $9.460\ 55 \times 10^{15}$ meters.

parsec (pa): The parallax shift of one second of arc (3.26 light years)—approximately $3.085\ 768 \times 10^{16}$ meters.

speed of light (c): $2.997\ 9 \times 10^8$ meters per second.

SOURCE: NASA.

Table A-2　International System (SI) Units and their conversion factors

Quantity	Name of unit	Symbol	Conversion factor
Distance	meter	m	1 km = 0.621 mile 1 m = 3.28 ft 1 cm = 0.394 in. 1 mm = 0.039 in. $1\ \mu m = 3.9 \times 10^{-5}$ in. $= 10^4$ Å 1 nm = 10 Å
Mass	kilogram	kg	1 tonne = 1.102 tons 1 kg = 2.20 lb 1 gm = 0.0022 lb = 0.035 oz $1\ mg = 2.20 \times 10^{-6}$ lb $= 3.5 \times 10^{-5}$ oz
Time	second	sec	$1\ yr = 3.156 \times 10^7$ sec 1 day = $8.64\ 10^4$ sec 1 hr = 3600 sec
Temperature	kelvin	K	273 K = 0° C = 32° F 373 K = 100° C = 212° F
Area	square meter	m²	$1\ m^2 = 10^4\ cm^2 = 10.8\ ft^2$
Volume	cubic meter	m³	$1\ m^3 = 10^6\ cm^3 = 35\ ft^3$
Frequency	hertz	Hz	1 Hz = 1 cycle/sec 1 kHz = 1000 cycles/sec $1\ MHz = 10^6$ cycles/sec
Density	kilogram per cubic meter	kg/m³	$1\ kg/m^3 = 0.001\ gm/cm^3$ $1\ gm/cm^3$ = density of water
Speed, velocity	meter per second	m/sec	1 m/sec = 3.28 ft/sec 1 km/sec = 2240 mi/hr
Force	newton	N	$1\ N = 10^5$ dynes = 0.224 lbf
Pressure	newton per square meter	N/m²	$1\ N/m^2 = 1.45 \times 10^{-4}$ lb/in²
Energy	joule	J	1 J = 0.239 calorie
Photon energy	electronvolt	eV	$1\ eV = 1.60 \times 10^{-19}$ J; $1\ J = 10^7$ erg
Power	watt	W	1 W = 1 J/sec
Atomic mass	atomic mass unit	amu	$1\ amu = 1.66 \times 10^{-27}$ kg
Customary Units Used With the SI Units			
Quantity	Name of unit	Symbol	Conversion factor
Wavelength of light	angstrom	Å	1 Å $= 0.1$ nm $= 10^{-10}$ m
Acceleration of gravity	g	g	$1\ g = 9.8\ m/sec^2$

SOURCE: NASA.

Table A-3   Recommended SI Unit prefixes

Prefix	Abbreviation	Factor by which unit is multiplied
tera	T	10^{12}
giga	G	10^{9}
mega	M	10^{6}
kilo	k	10^{3}
hecto	h	10^{2}
centi	c	10^{-2}
milli	m	10^{-3}
micro	μ	10^{-6}
nano	n	10^{-9}
pico	p	10^{-12}

SOURCE: NASA.

Table A-4   Common Metric/English conversion factors (for space technology activities)

	Multiply	By	To Obtain
Length:	inches	2.54	centimeters
	centimeters	0.393 7	inches
	feet	0.304 8	meters
	meters	3.281	feet
	statute miles	1.609 3	kilometers
	kilometers	0.621 4	statute miles
	kilometers	0.54	nautical miles
	nautical miles	1.852	kilometers
	kilometers	3 281.	feet
	feet	0.000 304 8	kilometers
Weight and Mass	ounces	28.350	grams
	grams	0.035 3	ounces
	pounds	0.453 6	kilograms
	kilograms	2.205	pounds
	tons	0.907 2	metric tons
	metric tons	1.102	tons
Liquid Measure	fluid ounces	0.029 6	liters
	gallons	3.785 4	liters
	liters	0.264 2	gallons
	liters	33.814 0	fluid ounces
Temperature	degrees Fahenheit plus 459.67	0.555 5	kelvins
	degrees Celsius plus 273.16	1.0	kelvins
	kelvins	1.80	degrees Fahrenheit minus 459.67
	kelvins	1.0	degrees Celsius minus 273.16
	degrees Fahrenheit minus 32	0.555 5	degrees Celsius
	degrees Celsius	1.80	degrees Fahrenheit plus 32
Thrust (Force)	pounds force	4.448	newtons
Pressure	newtons	0.225	pounds
	millimeters mercury	133.32	pascals (newtons per square meter)
	pounds per square inch	6.895	kilopascals (1000 pascals)
	pascals	0.007 5	millimeters mercury at 0° C
	kilopascals	0.145 0	pounds per square inch

SOURCE: NASA.

Appendix B:
Selected Reading List

General Subject Coding: [1] Space science and space technology; [2] people in space; [3] our Solar System; [4] our extraterrestrial civilization; [5] search for extraterrestrial intelligence (SETI); life in the Universe; [6] unidentified flying objects (UFOs); and [7] global change, Mission to Planet Earth.

Adelman, Saul J., and Adelman, Benjamin. *Bound for the Stars.* Englewood Cliffs, NJ: Prentice Hall, 1981. [4]

Akins, Faren, Conners, Mary, and Harrison, Albert. *Living Aloft: Human Requirements for Extended Spaceflight.* NASA SP-483. Washington, DC: NASA, 1985. [2]

Allen, Joseph. *Entering Space; An Astronaut's Odyssey.* New York: Stewart, Tabori and Chang, 1984. [2]

American Institute of Aeronautics and Astronautics (AIAA.) *U.S. Civil Space Program: An AIAA Assessment.* Washington, DC: AIAA, March 1987. [1]

Angelo, Joseph A., Jr. *The Dictionary of Space Technology.* New York: Facts On File, 1982. [1]

Angelo, Joseph A., Jr., and Buden, David. *Space Nuclear Power* (2nd ed). Melbourne, FL: Orbit Book Co., 1991. [1]

Angelo, Joseph A., Jr., Buden, David, and Lee, James. *Aerospace Nuclear Safety.* Melbourne, FL: Orbit Book Co., 1991. [1]

Asimov, Isaac. *Extraterrestrial Civilizations.* New York: Crown, 1979. [5]

Barney, Gerald O. *The Global 2000 Report to the President: Entering the 21st Century.* Volumes 1 and 2. Washington, DC: Council on Environmental Quality and U.S. State Department, 1977. [7]

Batson, R. M., et al. *Atlas of Mars.* NASA SP-438. Washington, DC: NASA, 1979. [3]

Baugher, Joseph F. *On Civilized Stars.* Englewood Cliffs, NJ: Prentice-Hall, 1985. [5]

Beatty, J. Kelly, O'Leary, Brian, and Chaikin, Andrew (eds). *The New Solar System* (3rd ed). New York: Cambridge University Press, 1990. [3]

Belwe, L. F. (ed). *Skylab, Our First Space Station.* NASA SP-400. Washington, DC: NASA, 1977. [2]

Bernal, John. *The World, the Flesh and the Devil.* London: K. Paul, Trench, Trubner, 1929. [4]

Billingham, John (ed). *Life In the Universe.* NASA CP-2156. Washington, DC: NASA, 1981. [5]

Billingham, John, Gilbreath, William, and O'Leary, Brian (eds). *Space Resources and Settlements.* NASA SP-428. Washington, DC: NASA, 1979. [4]

Black, David C. (ed). *Project Orion: A Design Study of a System for Detecting Extrasolar Planets.* NASA SP-436. Washington, DC: NASA, 1980. [5]

Brooks, C. G., et al. *Chariots for Apollo: A History of Manned Lunar Spaceflight.* NASA SP-4205. Washington, DC: NASA, 1979. [2]

Calder, Nigel. *Spaceships of the Mind.* New York: Viking, 1978. [4]

Carr, M. H., and Evans, N. *Images of Mars: The Viking Extended Mission.* NASA SP-444. Washington, DC: NASA, 1980. [3]

Cocconi, G., and Morrison, P. "Searching for Interstellar Communications." *Nature* **184** (1959): 844. [5]

Collins, Michael. *Liftoff.* New York: Grove Press, 1988. [2]

Committee on Earth Sciences of the Federal Coordinating Committee for Science, Engineering, and Technology. *Our Changing Planet: The FY 1990 Research Plan.* Washington, DC: Executive Office of the President (Office of Science and Technology Policy), 1989. [7]

Committee on Earth Sciences of the Federal Coordinating Committee for Science, Engineering, and Technology. *Our Changing Planet: A. U.S. Strategy for Global Change Research.* Washington, DC: Executive Office of the President (Office of Science and Technology Policy), 1989. [7]

Condon, Edward U. *Scientific Study of Unidentified Flying Objects.* New York: Bantam, 1969. [6]

Cooper, Henry S. F., Jr. *The Search for Life on Mars: The Evolution of an Idea.* New York: Holt, Rinehart, Winston, 1980. [3,5]

Cortright, Edgar (ed). *Apollo Expeditions to the Moon.* NASA SP-350. Washington, DC: NASA, 1975. [2]

Crick, Francis. *Life Itself.* New York: Simon & Schuster, 1981. [5]

Davies, M. E., et al. *Atlas of Mercury.* NASA SP-423. Washington, DC: NASA, 1978.   [3]

DeFrees, D., et al. (eds). *Exobiology in Earth Orbit.* NASA, SP-500. Washington, DC: NASA, 1989. [4,5]

Disney, Michael. *The Hidden Universe.* New York: Macmillan, 1984.   [1]

Dole, Stephen H. *Habitable Planets for Man* (2nd ed). New York: American Elsevier, 1970.   [5]

Dunne, J. A., and Burgess, E. *The Voyager of Mariner 10.* NASA SP-424, 1978.   [3]

Dyson, Freeman. *Infinite in All Directions.* New York: Harper & Row, 1988.   [4]

Ezell, Edward C., and Ezell, Linda Neuman. *The Partnership: A History of the Apollo-Soyuz Test Project.* Washington, DC: NASA, 1978.   [2]

Ezell, Edward C., and Ezell, Linda Neuman. *On Mars: Exploration of the Red Planet 1958–1978.* NASA SP-4212. Washington, DC: NASA, 1984.   [3]

Feinberg, G., and Shapiro, R. *Life Beyond Earth: The Intelligent Earthling's Guide to Life in the Universe.* New York: Morrow, 1980.   [5]

Fimmel, R. O., *Pioneer Odyssey.* NASA SP-349. Washington, DC: NASA, 1977.   [3]

Fimmel, R. O., et al. *Pioneer: First to Jupiter, Saturn, and Beyond.* NASA SP-446, 1980.   [3]

Fimmel, R. O., et al. *Pioneer Venus.* NASA SP-461. Washington, DC: NASA, 1983.   [3]

Finney, Ben R., and Jones, Eric M. (eds). *Interstellar Migration and the Human Experience.* Berkeley, California: University of California Press, 1985.   [4,5]

Forward, Robert L., and Davis, Joel. *Mirror Matter.* New York: Wiley, 1988.   [1]

French, B. M., and Maran, S. P. (eds). *A Meeting with the Universe.* NASA EP-177. Washington, DC: NASA, 1982.   [1,2,3]

Friedman, Louis. *Starsailing.* New York: Wiley, 1988.   [1,4]

Froehlich, Walter. *Space Station: The Next Logical Step.* NASA EP-213. Washington, DC: NASA, 1985. [2,4]

Gale, William (ed). *Life in the Universe: The Ultimate Limits to Growth.* Boulder, CO: Westview Press, 1979.   [5]

Gibson, E. G. *The Quite Sun.* NASA SP-303. Washington, DC: NASA, 1973.   [3]

Ginsberg, Irving W., and Angelo, Joseph A., Jr. (eds). *Earth Observations and Global Change Decisionmaking: A National Partnership.* Melbourne, FL: Krieger, 1990.   [7]

Glenn, J. C., and Robinson, G. S. *Space Trek: The Endless Migration.* New York: Warner Books, 1980.   [4]

Goldsmith, Donald. *The Quest for Extraterrestrial Life.* Mill Valley, CA: University Science Books, 1980. [5]

Goldsmith, Donald. *Nemesis.* New York: Walker, 1985.   [3,5]

Goldsmith, Donald, and Owen, Tobias. *The Search for Life in the Universe.* Menlo Park, CA: Benjamin/Cummings, 1980.   [5]

Greenstein, George. *The Symbiotic Universe.* New York: Morrow, 1988.   [5]

Grosser, M. *The Discovery of Neptune.* New York: Dover, 1979.   [3]

Hartmann, William K., Miller, Ron, and Lee, Pamela. *Out of the Cradle (Exploring the Frontiers Beyond Earth).* New York: Workman, 1984.   [4]

Heppenheimer, Thomas A. *Colonies in Space.* New York: Warner Books, 1978.   [4]

Heppenheimer, Thomas A. *Toward Distant Suns.* New York: Fawcett Columbine, 1979.   [4]

Horowitz, Norman. *To Utopia and Back.* San Francisco: W.H. Freeman, 1986.   [3]

Hoyle, Fred. *The Intelligent Universe.* New York: Holt, Rinehart, & Winston, 1983.   [5]

Hoyle, Fred, and Wickramasinghe, N.C. *Lifecloud.* New York: Harper & Row, 1978.   [5]

Hoyle, Fred, and Wickramasinghe, N.C. *Diseases from Space.* New York: Harper & Row, 1979.   [5]

Hoyle, Fred, and Wickramasinghe, N.C. *Evolution from Space.* New York: Simon & Schuster, 1981.   [5]

Hynek, J. A. *The Hynek UFO Report.* New York: Dell, 1977.   [6]

Hynek, J. A. *The UFO Experience: A Scientific Inquiry.* New York: Ballantine, 1975.   [6]

International Council of Scientific Unions (ICSU). *The International Geosphere-Biosphere Program: A Study in Global Change—A Plan for Action.* Washington, DC: National Academy Press, 1988.   [7]

Jastrow, Robert. *Until the Sun Dies.* New York: Norton, 1977.   [4,5]

Jastrow, Robert. *Journey To The Stars.* New York: Bantam, 1989.   [4,5]

Johnson, Richard D., and Holbrow, Charles (eds). *Space Settlements: A Design Study.* NASA SP-413. Washington, DC: NASA, 1977.   [4]

Kaplan, David (compiler). *Environment of Mars, 1988.* NASA TM-100470. Washington, DC: NASA, October 1988.   [3]

Kohlhase, Charles (ed). *The Voyager Neptune Travel Guide.* NASA/JPl 89-24. Pasadena, CA: Jet Propulsion Laboratory, June 1, 1989.   [3]

Kosofsky, L. J., and El-Baz, F. *The Moon as Viewed by Lunar Orbiter.* NASA SP-200. Washington, DC: NASA, 1970.   [3]

Lewis, John S., and Lewis, Ruth A. *Space Resources: Breaking the Bonds of Earth.* New York: Columbia University Press, 1987.   [4]

Lovelock, James. *GAIA—A New look at Life On Earth.* Oxford: Oxford University Press, 1987.   [5]

Lovelock, James, and Allaby, Michael. *The Greening of Mars.* New York: St. Martin's, 1984.   [3,4]

Mallove, E. *The Quickening Universe.* New York: St. Martin's, 1988.   [5]

Mallove, E., and Forward, R. L. "Bibliography of Interstellar Travel and Communication." *Journal of British Interplanetary Society* 33 (1980): 201–48.   [5]

Margulis, Lynn, and Sagan, Dorion. *Microcosmos.* New York: Summit Books, 1986.   [5]

Masursky, H., et al (eds). *Apollo Over the Moon: A View from Orbit.* NASA SP-362. Washington, DC: NASA, 1978.   [3]

McDonough, Thomas R. *The Search for Extraterrestrial Intelligence (Listening for Life in the Cosmos).* New York: Wiley, 1987.   [5]

McDonough, Thomas R. *SPACE—The Next Twenty-Five Years* (revised and updated). New York: Wiley, 1989.   [1,4]

Mendell, W. W. *Lunar Bases and Space Activities of the 21st Century.* Houston: Lunar and Planetary Institute, 1985.   [3,4]

Moore, P. *Mission to the Planets: The Illustrated Story of Man's Exploration of the Solar System.* New York: Norton, 1990.   [3]

Moore, P., and Tombaugh, C. *Out of the Darkness: The Planet Pluto.* Harrisburg, PA: Stackpole Books, 1980.   [3]

Morrison, D. *Voyages to Saturn.* NASA SP-451. Washington, DC: NASA, 1982.   [3]

Morrison, D., and Samz, J. *Voyager to Jupiter.* NASA SP-439. Washington, DC: NASA, 1980.   [3]

Morrison, Philip, Billingham, John, and Wolfe, John (eds). *The Search for Extraterrestrial Intelligence.* NASA SP-419. Washington, DC: NASA, 1977.   [5]

NASA. *Life Beyond Earth and the Mind of Man.* NASA SP-328. Washington, DC: NASA, 1973.   [5]

NASA. *Project Cyclops: A Design Study of a System for Detecting Extraterrestrial Intelligent Life.* NASA CR-114445 (rev ed). Moffett Field, CA: NASA Ames Research Center, July 1973.   [5]

NASA. *A Forecast of Space Technology 1980–2000.* NASA SP-387. Washington, DC: NASA, January 1976.   [1]

NASA. *Why Man Explores.* NASA EP-123. Washington, DC: NASA, 1976.   [1,4]

NASA. *Skylab Explores the Earth.* NASA SP-380. Washington, DC: NASA, 1977.   [2,7]

NASA. *The Martian Landscape—Viking Mars Mission and Photographs.* NASA SP-425. Washington, DC: NASA, 1978.   [3]

NASA. *Voyager 1 Encounters Saturn.* NASA JPL 400-100. Pasadena, CA: NASA/JPL, 1980.   [3]

NASA. *Advanced Automation for Space Missions.* NASA CP-2255. Washington, DC: NASA, 1982.   [1,4]

NASA. *Beyond Earth's Boundaries (Human Exploration of the Solar System in the 21st Century).* Washington, DC: NASA (Office of Exploration), 1988.   [2,3,4]

NASA. *Exploring the Living Universe.* Washington, DC: NASA, 1988.   [2,5]

NASA. *1989 Long-Range Program Plan.* Washington, DC: NASA, 1989.   [1,2,3,4,5]

NASA. *Civil Space Technology Initiative.* Washington, DC: NASA (The Office of Aeronautics, Exploration & Technology), August 1990.   [1]

NASA. *The Gamma-Ray Observatory.* NASA NP-124. Washington, DC; NASA, 1990.   [1]

NASA. *Mission to Planet Earth.* Washington, DC: NASA, 1990.   [7]

NASA Advisory Council. *Planetary Exploration Through Year 2000: Part One: A Core Program.* Washington, DC: NASA (Advisory Council–Solar System Exploration Committee), 1983.   [3]

NASA Advisory Council. *Earth System Science Overview: A Program for Global Change.* Washington, DC: NASA (Advisory Council–Earth Systems Sciences Committee), 1986.   [7]

NASA Advisory Council. *Planetary Exploration Through Year 2000: Part Two: An Augmented Program.* Washington, DC: NASA (Advisory Council–Solar System Exploration Committee), 1986.   [3]

NASA Life Sciences Working Group. *Advanced Missions with Humans in Space.* Washington, DC: NASA, January 1987.   [2]

National Academy of Sciences. *Life Beyond the Earth's Environment–The Biology of Living Organisms In Space.* Washington, DC: National Academy Press, 1979.   [5]

National Academy of Sciences. *Scientific Plan for the World Climate Research Program.* Washington, DC: National Academy Press, 1988.   [7]

National Academy of Sciences. *Space Science in the Twenty-First Century.* Washington, DC: National Academy Press, 1988.   [1]

National Commission On Space. *Pioneering the Space Frontier.* New York: Bantam, 1986.   [4]

Nicogossian, Arnauld, and Parker, James. *Space Physiology and Medicine.* NASA SP-447. Washington, DC: NASA, 1982.   [2]

Oberg, James E. *New Earths.* Harrisburg, PA: Stackpole, 1981.   [4]

Oberg, James E. *Mission to Mars: Plans and Concepts for the First Manned Landing.* Harrisburg, PA: Stackpole, 1982.   [3,4]

Oberg, James E., and Oberg, Alcestis R. *Pioneering Space.* New York: McGraw-Hill, 1986.   [4]

O'Neill, Gerald K. *The High Frontier: Human Colonies in Space.* New York: Morrow, 1977.   [4]

Nicks, O. W. (ed). *This Island Earth.* NASA SP-250. Washington, DC: NASA, 1970.   [3,7]

Papagiannis, Michael D. (ed). *The Search for Extraterrestrial Life: Recent Developments.* Boston: D. Reidel, 1985.   [5]

Ponnamperuma, Cyril, *The Origins of Life.* New York: Dutton, 1972.   [5]

Ponnamperuma, Cyril, and Cameron, A. G. W. (eds). *Interstellar Communication: Scientific Perspectives.* Boston: Houghton Mifflin, 1974.   [5]

Powers, Robert M. *Mars.* Boston: Houghton Mifflin, 1986.   [3,4]

Presidential Commission on the Space Shuttle Challenger Accident. *Report,* 5 volumes. Washington, DC: The White House, 1986.   [2]

Ride, Sally K. *Leadership and America's Future in Space.* Washington, DC: NASA, August 1987.   [4]

Ridpath, Ian. *Messages from the Stars.* New York: Harper & Row, 1978.   [5]

Rood, Robert T., and Trefil, James S. *Are We Alone?* New York: Scribners, 1981.   [5]

Sagan, Carl. *The Cosmic Connection.* New York: Doubleday, 1973.   [5]

Sagan, Carl. *Cosmos.* New York: Random House, 1980.   [1,3,5]

Sagan, Carl, et al. *Murmurs of Earth.* New York: Ballantine, 1978.   [5]

Shklovskii, I. S., and Sagan, Carl. *Intelligent Life in the Universe.* New York: Dell, 1966.   [5]

Short, N. M., et al. *Mission to Earth: Landsat Views the World.* NASA SP-360. Washington, DC: NASA, 1976.   [3,7]

Spitzer, C. R. *Viking Orbiter Views of Mars.* NASA SP-441. Washington, DC: NASA, 1980.   [3,7]

Stine, G. Harry. *The Third Industrial Revolution.* New York: Ace Publishing, 1979.   [1,4]

Sullivan, Walter. *We Are Not Alone.* New York: McGraw-Hill, 1966.   [5]

Tucker, Wallace, and Tucker, Karen. *The Dark Matter.* New York: Morrow, 1988.   [1]

U.S. Department of Commerce. *Space Commerce—An Industry Assessment.* Washington, DC: Department of Commerce, May 1988.   [1,4]

U.S. Office of Technology Assessment. *Civilian Space Stations and the U.S. Future in Space.* Washington, DC: Office of Technology Assessment, November 1984.   [2,4]

U.S. Office of Technology Assessment. *U.S./Soviet Cooperation in Space: A Technical Memorandum.* Washington, DC: Office of Technology Assessment, July 1985.   [1]

Vajk, J. Peter. *Doomsday Has Been Cancelled.* Culver City, CA: Peace Press, 1978.   [4]

Von Braun, Wernher. *Space Frontier.* New York: Holt, Rinehart & Winston, 1963.   [1]

Von Braun, Wernher, Ordway, Frederick, and Dooling, Dave. *Space Travel: A History.* New York: Harper & Row, 1985.   [1]

Wilford, John Noble. *Mars Beckons.* New York: Knopf, 1990.   [3,4]

Wood, John A., and Chang, Sherwood (eds). *The Cosmic History of the Biogenic Elements and Compounds.* NASA SP-476. Washington, DC: NASA, 1985.   [5]

Index